T0363124

MARINE PLANTS OF AUSTRALIA

First published in 2019 and this
updated edition in 2023 by
UWA Publishing
Crawley, Western Australia 6009
www.uwap.uwa.edu.au

UWAP is an imprint of UWA Publishing,
a division of The University of Western Australia.

ISBN: 978-1-76080-260-8

A catalogue record for this
book is available from the
National Library of Australia

Design by Upside Creative
Cover image and inside front and back:
Detail from the painting *Elizabeth Reef* by Angela Rossen.
Acrylic on Canvas 1500mm x 8000mm
Printed in China by Imago

John M. Huisman

MARINE PLANTS OF AUSTRALIA

revised and updated edition

CONTENTS

Shallow reef at Neds Beach, Lord Howe Island, with a variety of green, brown and red algae (photo A.J. Kendrick).

(left) Intertidal reef flat at Point Lonsdale, Victoria. The rocky shores of southern Australia support a variety of marine plants. Here a luxuriant growth of *Hormosira banksii* can be seen on the exposed rock and several species of other brown algae, including *Sargassum*, are visible in the rock pool.

(far left) Cape Peron, south of Perth, Western Australia, at low tide. The area has a diversity of habitats that supports a vast array of marine plants.

Preface and Acknowledgements

The first edition of *Marine Plants of Australia* was very much a labour of love, my attempt to present the marine plants the way I saw them – not as smelly beach drift but as attractive and unusual plants, certainly worthy of greater attention and appreciation than had been previously bestowed on them. That book included 300 species, a target not chosen randomly but intended to emulate the first richly illustrated treatment of Australian seaweeds, the five-volume *Phycologia Australica* by the Irish botanist William Henry Harvey, published in 1858–63. Since Harvey's time, our knowledge of Australian seaweeds has advanced considerably, but their appreciation remains limited in comparison to the more iconic marine organisms such as fish and corals. I hope that the first edition went some way to redress that imbalance.

In 2019 the revised edition of *Marine Plants of Australia* was published, which greatly expanded the coverage to include over 600 species, illustrated in the most part by a set of newly acquired images but also incorporating the line drawings and updated text from the original book. The popularity of that book meant that the print run was quickly exhausted, and the need for a reprint presented the opportunity for further additions and updates. This 'revised and updated' edition now includes some 645 species, with many of the new species being added to existing genera, but others representing 12 genera included for the first time. The updates also include taxonomic revisions, which saw the species of some genera (e.g. *Epiphloea, Gelinaria, Hydropuntia*) being moved to other genera. Marine plant taxonomy is an active field of study and AlgaeBase should be consulted for recent updates.

The contributions of the many people acknowledged in the previous editions are of no less significance here, and I thank them once again. More recently, I have been fortunate to have been given numerous opportunities to photograph marine plants in remote locations. In the first instance, this has been through the Western Australian Museum's Marine Biodiversity Surveys to the Kimberley region of north-west Australia, generously supported by Woodside Energy. The second major opportunity was with the CReefs program, which allowed me to visit Lizard Island and Heron Island in the Great Barrier Reef. Despite the opportunities afforded by these expeditions, some of my favourite locations are considerably closer to home. Cape Peron, south of Perth, has a diversity of habitats in easy reach and mostly at snorkel depths, which has permitted a frequency of visits that would not be feasible for the more remote locations.

Several people deserve special thanks: Dr Roberta Cowan (Murdoch University and Western Australian Herbarium) continued to provide a sympathetic ear as well as advice on coralline algae;
Ian McKernan (Murdoch University) co-ordinated much of the field work and always lent patient diving support; Alan Kendrick (Department of Biodiversity, Conservation and Attractions) was also a supportive dive partner and advisor; Rainbo Dixon (formerly Murdoch University)

was an enthusiastic student and guided several of the taxonomic decisions presented here; and Mike van Keulen (Murdoch University) was always supportive and contributed several images of seagrasses. The scans from Harvey's *Phycologia Australica* were kindly provided by Olivier De Clerck and Frederik Leliaert (Ghent University), and Graham Edgar (University of Tasmania) and Mike Guiry (AlgaeBase) contributed several images. Angela Rossen kindly allowed me to reproduce her wonderful painting *Elizabeth Reef* for the cover. Thanks also to the staff of the Western Australian Herbarium (Department of Biodiversity, Conservation and Attractions), in particular Kevin Thiele, Cheryl Parker, Karina Knight, Julia Percy-Bower and Sean Moylan, who all kept me on track during the book's gestation. I also acknowledge the support of Mike Guiry and AlgaeBase, the authoritative web-based resource that I rely on heavily and could not function without.

Further financial support during the preparation of the manuscript was provided by the Australian Biological Resources Study, which also contributed towards publication costs. All photographs and drawings are by the author, except where indicated.

Lastly, I dedicate this book to my wife, Leanne, and daughter, Sophie, as thanks for their ongoing love and support.

John Huisman

The subtidal at Cape Peron with a range of brown, red and green seaweeds.

Sirophysalis trinodis, as depicted by William Dampier in his *A Voyage to New Holland, &c., in the Year 1699* (1703, tab. 2, fig. 2, top right).

Introduction

Few people could deny the simple pleasure of a visit to the seashore. It is certainly very much a part of the Australian way of life – the vast majority of Australians live within reach of the coast and we quite happily promote an image overseas of ourselves as sun-bronzed beach-lovers, even if for most of us it is a long way from the truth. But what is it about the seashore that is so attractive? For some people the enjoyment lies in relaxation and recreation. The combination of fresh sea air, warming sunshine, and the occasional dip in cooling water provides an experience that is hard to surpass. For others, the seashore caters to the urge to collect, to search for that 'hidden treasure' that the sea so often provides. Fossicking around in the cast-up remains of marine organisms in the hope of finding that one perfect shell or an unusually shaped sponge is an activity whose popularity never seems to wane. Inevitably the seashore also becomes a learning experience, for that perfect shell is never entirely satisfying until it has a name, and in the process of consulting textbooks and field guides it is easy to become familiar with a vast number of different organisms.

But the rewards of beachcombing only provide a glimpse of the world hidden beneath the waves, and many people feel the urge to explore the marine environment more closely. Viewed in their natural habitat, marine creatures present a myriad of life forms, from colourful and fast-moving fish to sedentary sponges and other invertebrates. Furnishing a constant backdrop to this busy environment are the unsung providers, the marine plants. Marine plants flourish in shallow coastal regions and, in doing so, they also provide the basic energy source for almost all other organisms. The process of growth involves converting the sun's energy into something more tangible, the plant body, which is then an attractive food for all manner of animals. Coastal regions are particularly suitable for the growth of plants. The depth of water is insufficient to stop the penetration of light and the seawater itself bathes the area with a constant supply of nutrients. These features combine to ensure maximum productivity and a particularly rich environment.

Marine plants fall into two broad ecological categories: those that are attached, either to the sea bottom or to other plants and animals (these are known as the 'benthic' plants); and those that are unattached, their travels being largely at the whim of currents and waves (these are known as the 'phytoplankton'). It is the former category that is the subject of this book. While the phytoplankton have great ecological importance, they are mostly small, single-celled organisms that are generally invisible to the naked eye. As this book is intended as a field guide, the phytoplankton cannot be adequately treated.

The benthic plants can be further divided into two broad groups: the algae (singular 'alga') and the seagrasses. Although they share similar habitats, the algae and the seagrasses are only remotely related. Seagrasses are true flowering plants, with roots, stems and vascular tissue; their closest relations are found among the land plants. The algae, on the other hand, lack these features. Although they can reproduce sexually, algae do not have prominent flowers, and it is mostly impossible to tell whether they are reproductively mature without resorting to the use of a microscope. The term 'algae' also encompasses several groups that are only distantly related to one another; those included here are the red algae (Division Rhodophyta), green algae (Division Chlorophyta), blue-green algae (Division Cyanobacteria) and brown algae (Division Ochrophyta).

This book is devoted to the benthic marine plants of Australia. Some 645 species are illustrated, mostly photographed in their natural habitat. Although this represents only a small portion of the entire marine plant flora, the majority of the species included are those that are commonly encountered. Many of the photographs were taken in Western Australia, but the genera and species they depict are, nevertheless, mostly widespread in Australia.

History of Marine Botany in Australia

The recognition and documentation of Australia's marine plants are processes that began well before the European occupation of the country and continue to this day. Aboriginal use and cognisance of marine plants has been documented by Thurstan *et al.* (2018), who recorded instances of seaweeds being used as food, medicines, clothing, and as part of ceremonial activities. One example relates to the 'bull kelp' *Durvillaea potatorum* (p. 315), from which Tasmanian Aborigines fashioned containers to carry food and water (Thurstan *et al.*, 2018). The name *potatorum* (= 'potable') was apparently coined in recognition of this behaviour. Numerous Aboriginal words describing marine plants are known, but these have rarely been adopted in modern taxonomic usage. Many have, however, been utilised in other ways. For example, the name of the Sydney coastal suburb of Coogee is supposedly derived from an Aboriginal word for 'smelly', apparently referring to the rotting seaweed that occurs there.

Cryptonemia kallymenioides – as illustrated by the Irish botanist William Henry Harvey (1859a, pl. 103, as *Halymenia cliftonii*).

The first of Australia's marine plants to be recorded in print was collected from Shark Bay on the Western Australian coast. William Dampier, the famous buccaneer turned explorer, described and drew many species of Australian wildlife during his visits to Australia in the seventeenth century (Dampier, 1703; George, 1999). One of his specimens, now housed in the Oxford herbarium, was illustrated by Dampier as '*Fucus foliis capillaceis brevissimis, vesiculis minimis donatis*' and is the brown alga now known as *Sirophysalis trinodis* (see p. 331).

Subsequent voyages of discovery by English expeditions generally included trained botanists. The first of these to collect marine plants was Archibald Menzies, as part of the expedition led by Captain George Vancouver in the ships *Discovery* and *Chatham* (1791–95). Menzies collected specimens at Point D'Entrecasteaux, in south-western Australia. He did not describe the plants himself but entrusted them to Dawson Turner, who published them as part of his *Fuci*. Turner also received plants from Robert Brown, the botanist who accompanied Captain Matthew Flinders on board the *Investigator* (1801–05). This last collection included many plants from King George Sound, Western Australia, and from Tasmania and Victoria. Turner's *Fuci* is the first publication to include descriptions of new species of plants collected in Australian waters, for example *Hormosira banksii* (p. 316), *Platythalia quercifolia* (p. 325) and *Scytothalia dorycarpa* (p. 334), although at the time they were all included in the genus *Fucus*. In total, some 55 species of Australian algae were described from collections made during these early English expeditions.

During this period, Australia was also visited by several French scientific expeditions. Their contribution to the recognition of the area's marine plants (and the wildlife in general) is immense. The first French expedition to collect marine plants was that led by Admiral Bruny D'Entrecasteaux in the ships *Recherche* and *Espérance* (1791–94). The botanist Labillardière was on board and, following his return to France, he published his *Novae Hollandiae Plantarum Specimen*, an account of the plants collected during this and later French expeditions. Only nine algae were included, all collected from the Tasmanian coast. The seagrass *Amphibolis antarctica* (see p. 420) was first described by Labillardière from material collected from Esperance Bay, Western Australia, but it is not known whether he himself collected the material or whether it came from a later expedition. Several other algal species collected by Labillardière were later described by other European botanists. Following this, an expedition led by Captain Nicolas Baudin in the ships *Géographe* and *Naturaliste* visited Western Australia, in particular Rottnest Island and Shark Bay, before moving on to Tasmania and eastern Australia. Shark Bay was also visited in 1817 by an expedition led by Louis de Freycinet, in the ship *Uranie*.

Although a complete account of the marine plants collected during the French expeditions has never been published, the material was subsequently examined by several European botanists and it formed the basis of numerous publications. Many new species were described by a variety of authors, most notable of whom are Jean Vincent Lamouroux (e.g. *Acetabularia caliculus*, p. 405), Jean Baptiste Lamarck (e.g. *Amphiroa anceps*, p. 53; *Metagoniolithon stelliferum*, p. 59) and Carl Adolph Agardh (e.g. *Caulocystis uvifera*, p. 318) and his son Jacob Georg Agardh. The material was instrumental in forming many of the concepts of algal classification, structure and function that survive to this day.

Following this period of exploration, most subsequent collections of Australian marine plants were made by residents of Australia. The first of these was Charles Fraser (colonial botanist in Sydney), who collected material from near the Swan River mouth. A substantial collection was made by Dr Johann Ludwig Preiss, who lived in Western Australia from 1839 to 1842. He collected numerous botanical specimens of both higher plants and algae, with the latter represented by some 200 species that were subsequently described by Otto Wilhelm Sonder (1845, 1846–1848). Of these, 84 were believed new to science.

Sonder had earlier worked with the Irish botanist William Henry Harvey on the South African flora. Harvey's influence on the Australian algal flora was already being felt as he had published several accounts of Tasmanian algae, but his impact was yet to be fully realised. Harvey visited Western Australia during the mid-nineteenth century, arriving at Albany on 7th January 1854. By this stage, he had already published his *Nereis Australis Or*

Algae of the Southern Ocean (1847–49) and numerous other books dealing with British and southern hemisphere plants. His reputation was already well established. Harvey's first days in Albany were disappointing. The calm summer seas cast up very little in the way of drift plants and his dredging yielded little other than the seagrasses *Amphibolis* and *Zostera*. Towards the end of his stay, however, a storm cast up numerous plants and Harvey busily collected and pressed some 700 specimens in a single day. After a relatively fruitless trip to Cape Riche, Harvey set out for the Swan River Colony, arriving on 13th April 1854. He stayed in Perth for a few days before moving to Fremantle, where he was able to dredge offshore and make three excursions to Garden Island. He then visited Rottnest Island for six weeks, collecting numerous specimens. Harvey returned to Albany at the end of July before setting out for Melbourne late in August 1854. While at sea he wrote his paper 'Some account of the marine botany of the colony of Western Australia', in which 352 species were catalogued, including 132 that were new to science (Harvey, 1855).

Harvey's visit to Western Australia not only resulted in one of the more extensive collections of marine plants, but also inspired others to collect algae for him. William Ashford Sanford was the colonial secretary of Western Australia at the time of Harvey's visit and he accompanied Harvey on several trips to Fremantle. Sanford subsequently sent specimens to Harvey in Dublin and Harvey named the red alga *Asparagopsis sanfordiana* (now known as *Asparagopsis taxiformis*, see p. 44) in his honour. Sanford also arranged for Harvey to be assisted by George Clifton, then superintendent of water police, during his stay in Fremantle. Clifton supplied boats for Harvey's trips to Garden and Rottnest Islands and became so interested in Harvey's work that for the next nine years he sent thousands of specimens to Harvey in Dublin. Clifton's collections included many plants new to science, and the genera *Cliftonaea* (named by Harvey in Clifton's honour, see p. 223), *Bindera* (now known as *Webervanbossea*, see p. 149) and *Encyothalia* (see p. 305) are based on collections made by Clifton. In addition, 10 species were named for Clifton (e.g. *Coelarthrum cliftonii*, p. 161).

Upon his return to Trinity College, Dublin, Harvey embarked on what was to become his most significant contribution to the study of Australian marine plants: the five-volume *Phycologia Australica* (1858–63). These lavish volumes include colour plates of 300 species of Australian algae, and to this day they remain important references for students of Australian marine algae.

Harvey, while well versed in the results of the early English expeditions and in possession of a 'tolerably perfect set' (Harvey, 1855, p. 525) of the plants collected by Preiss and described by Sonder, was unfortunately unaware of the results of the earlier French expeditions. As a result, many of his names for Australian algae have been reduced to synonymy. This by no means devalues his contribution to Australian phycology and Harvey well deserves the title 'father of Australian phycology', which was bestowed upon him by botanist Sophie Ducker (1990).

During the second half of the nineteenth century, progress in Australian phycology was greatly

William Henry Harvey c.1850 by Thomas Herbert Macguire (1821–1895), lithograph. Collection: National Portrait Gallery, Canberra; purchased 2012.

promoted by Ferdinand von Mueller, who emigrated to Australia from Germany to both pursue his botanical career and to enjoy the health benefits of the climate. Mueller became a naturalised British subject and, in 1853, was appointed government botanist of Victoria. While Mueller did not publish extensively on algae himself, he did send numerous specimens to phycologists such as Sonder and Harvey. His willingness to exchange material, to be studied by those perhaps more fully versed in phycology, meant that Mueller's contribution to Australian phycology was greater than his publication record in the algae would indicate.

This same period and the early- to mid-twentieth century saw significant contributions from resident schoolteachers. This group includes John Bracebridge Wilson, who made substantial collections from the Port Phillip Heads region of Victoria, and Arthur Henry Shakespeare Lucas, who published the only floristic accounts of Australian algae to appear in the first half of the twentieth century [*Seaweeds of South Australia*, Part I (Green and Brown Algae) (Lucas 1936) and Part II (Red Algae) (Lucas & Perrin, 1947)]. Until the floras of H.B.S. ('Bryan') Womersley (see following), these two volumes were the only readily accessible guides to the marine flora.

For further reading on the history of phycology in Australia, see the excellent account by Cowan and Ducker (2007).

Recent Studies

The most significant of recent floristic studies of marine plants are those of Professor Bryan Womersley (University of Adelaide and the State Herbarium of South Australia), with his series of books *The Marine Benthic Flora of Southern Australia* (Womersley, 1984, 1987, 1994, 1996, 1998, 2003). These excellent volumes are the summation of a lifetime's dedication to the study of the algae and will be the benchmark for many years to come. They should be the first references to be consulted for detailed descriptions of southern Australian marine plants. Professor Womersley passed away in 2011, but his influence continues through the work of his students, many of whom have continued floristic and evolutionary studies of algae.

The marine flora of Australia's tropics has recently received considerable attention, with the publication of several volumes in the *Algae of Australia* series. These have documented the green and brown algae of the southern Great Barrier Reef and Lord Howe Island (Kraft, 2007, 2009), the green, brown and red algae of north-western Australia (Huisman, 2015, 2018), and the Nemaliales of Australia (Huisman, 2006). The cold-water flora has not been ignored, with the publication of the excellent colour guide *Marine Plants of Tasmania* by Fiona Scott (2017).

Other contributions to documenting Australian marine flora include studies by Alan Millar (previously of the National Herbarium of New South Wales) and Gerald Kraft (of the University of Melbourne, now retired), who have greatly enhanced our knowledge of the marine algae of New South Wales and Lord Howe Island (Millar, 1990; Millar & Kraft, 1993, 1994a, 1994b). Millar and Kraft

Caulerpa cactoides – as illustrated by Harvey (1858, pl. 26).

(1993, 1994b) have also given an historical account of phycological studies in New South Wales. In addition, both authors have made forays into other geographical areas and have contributed Australia-wide monographs of numerous taxa. Alan Cribb (of the University of Queensland, now retired) published several studies of the algae of Queensland as well as accounts describing the red algal flora of the southern Great Barrier Reef (Cribb, 1983) and the marine algae of Queensland (Cribb, 1996). The brown algae of Queensland have been catalogued by Julie Phillips (University of Queensland) and Ian Price (James Cook University, now semi-retired) (Phillips & Price, 1997), and Price and Fiona Scott (formerly of James Cook University) have provided an account of the turf Rhodophyta of the Great Barrier Reef (Price & Scott, 1992). Lewis (1984, 1985, 1987) has catalogued the macroalgae recorded from northern Australia. Studies of Western Australian algae have been catalogued by Huisman and Walker (1990) and Huisman (1993, 1997), and the algal species recorded from the Indian Ocean by Silva, Basson and Moe (1996). In addition to these floristic accounts and catalogues, many monographs by other authors have been published. Most of these are catalogued in the above publications, but special mention should be made of the works by Bill Woelkerling (Latrobe University; Acrochaetiales and Corallinales), Gerald Kraft (all groups), Gary Saunders (University of New Brunswick; all groups), Alan Millar (Rhodophyta), Bryan Womersley (all groups), Elise Wollaston (University of Adelaide; Ceramiales), Elizabeth Gordon-Mills (University of Adelaide; Ceramiales), Margaret Clayton (Monash University; Phaeophyceae), Julie Phillips (Monash University, University of Queensland; Chlorophyta and Dictyotales), Murray Parsons (formerly Landcare Research, New Zealand; Dasyaceae), Sophie Ducker (University of Melbourne; *Chlorodesmis* and *Metagoniolithon*), Robert King (University of New South Wales; Rhodomelaceae), John West (University of Melbourne), and Valerie May (National Herbarium of New South Wales).

The advent and now common use of DNA sequence analyses in marine plant studies has had a spectacular impact on our understanding of evolutionary relationships, as well as facilitating species identification through the establishment of 'barcodes'. Recent studies on Australian algae by post-graduate students Kyatt Dixon (University of New Brunswick; Peyssonneliales), Rainbo Dixon (Murdoch University; Fucales), Gareth Belton (Adelaide University; *Caulerpa),* and Yola Metti (National Herbarium of New South Wales; *Laurencia* group) have utilised molecular methods to guide revisions of difficult taxa, and this approach is now firmly established. Ideally, each species will ultimately be characterised by a unique genetic sequence, but completion of this goal is still some years away. Molecular methods have uncovered previously obscured diversity and often indicate the presence of cryptic species, those that are not separable based on morphology. One example is this specimen of the red algal genus *Gibsmithia*, photographed in deep water at the base of Wistari Channel adjacent to Heron Island. This plant would previously have been identified as *Gibsmithia hawaiiensis*, a species thought to be widespread in the tropical Indo-Pacific. DNA sequence analyses have shown

H.B.S. 'Bryan' Womersley at work in his laboratory (photo: Board of the Botanic Gardens and State Herbarium, Adelaide, South Australia).

G. hawaiiensis to be restricted to the Hawaiian Islands, and this plant represents an undescribed species (Gabriel *et al.*, 2017). Observations such as this are a regular occurrence, indicating that there is still much to be done to document the Australian marine flora.

How to Use This Book

The process of accurately identifying many of the marine plants included in this book is a complicated and arduous task that can frustrate even the trained scientist. Over the years, an extensive body of literature that can be consulted has accumulated, but by and large this is the preserve of the specialist as the majority is written using esoteric terms describing structures and processes alien to the casual observer. While this literature is essential for a meaningful understanding of the relationships within the algae and seagrasses, it is also daunting for those merely wishing to give an accurate name to a specimen at hand.

The most common tool used in the process of identification is the taxonomic key, which generally comprises a series of linked questions by which the reader can be eventually led to the name of the plant. Unfortunately, most available keys have been written by specialists and require a degree of familiarity with the subject that can be difficult to acquire. Thus, in practice, the most common approach is to avoid using the keys and to identify specimens simply by matching them with available pictures. However, this process can also lead to the adoption of incorrect names if the supporting literature is not consulted.

This book has been designed for those who wish to identify marine plants accurately, but who are unfamiliar with the specialist literature and are therefore wary of delving into it. The photographs clearly depict the species in question and, in most cases, will lead to the correct name. Where there is likely to be some confusion, a line drawing is given of an important feature, either a section of the plant or detail of branches. Where appropriate, illustrations from Harvey's *Phycologia Australica* (1858–1863) have been included. For those who wish to go further (or who have a plant that appears similar but not identical to the one pictured), a reference is given to a treatment in the scientific literature.

Arrangement and Use of the 'Systematic Section'

Three divisions of algae (Chlorophyta, the green algae; Ochrophyta, herein only the brown algae; and Rhodophyta, the red algae), one of photosynthetic bacteria (Cyanobacteria, the blue-green algae) and one of seagrasses (Magnoliophyta) are included. Each division is divided into several classes (ending -phyceae), orders (ending -ales) and then families (ending -aceae). The taxonomic arrangement follows Huisman (2015, 2018), wherein most recent revisions were adopted. The major variation from most published taxonomic hierarchies is the inclusion of the brown algae as the class Phaeophyceae in the order Ochrophyta. Earlier texts recognise the brown algae as their own division, the

Gibsmithia sp.: A new species from deep water in Wistari Channel, Queensland.

Phaeophyta, then subsequently as a class in the division Heterokontophyta. Usage of the latter name has been limited, and the more popular Ochrophyta is adopted here. Within each family, the genera are arranged in alphabetical order. Generally, one or two species are included in each genus, although there are some exceptions where a large number of common species occur (e.g. *Caulerpa*). The higher classification of each species is given in the banner heading; for example, the genus *Trichogloea* (p. 25) belongs to the class Florideophyceae, order Nemaliales, family Liagoraceae.

Information given for each species includes the type locality, distribution, and a reference for further reading. The type locality is the place from which the type specimen was collected, the type specimen being the single plant that the person who named the species designated as representative of the species. If a specimen was not designated at the time the species was named, it can be done at a later date. This locality is especially useful to know if you wish to collect fresh material of a particular species and to be relatively confident of its identity. Information on distribution varies between species. For those with an essentially southern distribution, the limits are quite precise, attributable largely to the work of H.B.S. Womersley, with most details taken from his 'Flora' series (Womersley, 1984, 1987, 1994, 1996, 1998, 2003). For northern species, the limits are vaguer, in most part because of the lack of information regarding the tropical algal flora. Distribution data has been taken primarily from Kraft (2007, 2009) and Huisman (2015, 2018). Many tropical species are probably widely distributed, but their precise geographical limits are as yet unknown. In many cases, the distribution of tropical species is given herein as 'from...across northern Australia to...'. Where this type of distribution is given, it is likely that the species is actually known from northern Western Australia and northern Queensland, and its presence in the intervening region is assumed. Following the distribution data for each species, the 'Further Reading' section gives a reference to a (hopefully) recent text that includes the species; this should be consulted if verification of identifications is necessary.

To use the 'Systematic Section' (pp. 1–431), you must first decide whether the plant at hand is a seagrass, an alga or a photosynthetic bacterium. The seagrasses are all green in colour and generally form dense beds in shallow sandy areas. They often have thick stems or long, strap-like leaves. Internally, seagrasses have vascular tissue (veins), which will show up in a section as several bundles of thick-walled cells. The algal divisions are separated by their colour, with the Chlorophyta including the green algae, the Ochrophyta the brown algae and the Rhodophyta the red algae. While this visual clue is generally sufficient to place the alga in a division, some care must be taken. The green algae are generally straightforward as they are usually grass-green in colour, with any variations still retaining a greenish hue. The brown colour in the brown algae also shows little variation, although some dried specimens can be nearly black. The red algae are generally light pink to deep red or purple. They are the most likely to confuse, as many species can appear green or even

The type specimen of *Halimeda versatilis*, collected by W.H. Harvey from Cape Riche, Western Australia, and distributed by him (incorrectly) as *Halimeda macroloba*. The uniqueness of this species was recognised by J. Agardh, who named it in 1887 (see Cremen et al., 2016).

Author John Huisman collecting seaweeds at the Rowley Shoals, WA (photo E. Matson).

black, especially those growing in the intertidal region. Special care must be taken with drift specimens, as these will often have started to decompose and will have faded. In the red algae, it is usually the red pigments that fade first, so that the underlying greenish hue can be seen. Likewise with specimens viewed underwater – red light is the first to disappear as the depth of water increases, so that red algae might not appear red at all. All of the photographs included have been taken with an electronic flash, which restores the red wavelengths, and the plants are close to their natural colour, if not to their appearance at depth. The photosynthetic bacteria (the blue-green algae) are all small and either unicellular, colonial or filamentous. The only time they are conspicuous is when they form large colonies. Cells of the Cyanobacteria lack the structural complexity found in the other groups included here.

Once the division is chosen, it is a matter of matching the specimen to the picture. Features such as branching pattern are generally consistent and can be used with confidence. A certain amount of variation in colour and general appearance can be expected, although in extreme cases the line drawing and the references listed under 'Further Reading' should be consulted to confirm the identification. The line drawings are of 'sections' – these are thin slivers of the plant, mounted on a glass slide such that the internal construction can be viewed with a microscope. Those included here are generally cut in a transverse direction (i.e. at right angles to the direction of growth) and taken from a mature portion of the plant (at least several centimetres from the growing tip). Occasionally, longitudinal (i.e. in the direction of growth) sections are shown. Unless otherwise stated, sections are transverse. Other line drawings included are of the plant apex (the growing tip) or the surface of the thallus – in such cases an unsectioned portion of the plant is mounted in its entirety. All drawings are of features observed with a microscope. For each drawing a scale bar is included and the length of the bar is indicated in the caption, usually in micrometres (μm), where one micrometre equals one thousandth of a millimetre.

What's in a Name?

The term 'plants' in the title of this book is defined in a very general sense as applying to photosynthetic organisms, and herein includes the seagrasses, algae, and photosynthetic bacteria. As such, it is used as a term of convenience. Many taxonomists who adhere to strict evolutionary principles would argue that such a usage is misleading, that the term 'plants' should only include the more advanced green photosynthetic organisms. This argument is based on the principle that collective names (such as 'plants') should only be given to groups of closely related organisms. This is an entirely valid argument and it is true that many of the organisms included in this book are only remotely related to one another, and only the seagrasses would qualify as 'plants' as strictly defined. However, adopting a very strict naming scheme would be very unwieldy, and, for the purposes of this book, the broad, convenient definition given above will suffice. Benthic marine algae are also known as 'seaweeds', an unfortunate term because 'weeds' suggests that they are undesirable invaders. In the scientific community, the term 'seaweeds' is rarely used, although, curiously, it remains popular with those involved in the more applied sciences, such as algal farming, where the term would seem to be particularly inappropriate.

In general, the marine plants are poorly known and, as a consequence, very few common names have been applied to them. Some exceptions include the seagrasses *Halophila* (paddleweed), *Amphibolis* (wireweed) and *Posidonia* (strapweed), and the green algae *Ulva* (sea lettuce) and *Codium* (dead man's fingers). The use of scientific names throughout this book may appear daunting, but this system is the only one that can guarantee the accurate and consistent application of names. For example, 'sea lettuce' is generally applied to all members of the genus *Ulva*, but there is no indication of the species involved. Thus, 'sea lettuce' in some parts of the world might mean *Ulva lactuca*, whereas in Australia it might refer to several species of *Ulva*, only one of which is *Ulva lactuca*. The choice of names and the process of naming plants follow a very precisely defined set of rules that are set out in the *International Code of Nomenclature* (Turland *et al*., 2018). This ensures that each species has only a single valid name that can be applied to it. Scientific names are composed of two parts – the genus to which the plant belongs followed by the specific epithet. In *Trichogloea requienii*, for example, the name *Trichogloea* refers to the genus, whereas *requienii* is the specific epithet. To be accurate, the two have to be used in sequence. *Trichogloea* on its own might refer to a number of different species that belong to the genus, while *requienii* on its own is meaningless. The name of each genus is unique – there can only be one genus *Trichogloea*, although it might include several different species. Within a genus, each species is also unique, although the same specific epithet can be used for a different species if it is also in a different genus. For example, *Rhodopeltis australis* and *Dictyopteris australis* are two unrelated species, despite sharing the epithet '*australis*'.

Scientific names are generally derived from Latin or Greek, and often describe some feature of the plant. To use the previous example, the name of the genus *Rhodopeltis* is derived from the Greek '*rhodo*', meaning 'red', and '*peltis*', meaning 'shield'. The name was chosen to allude to the colour and form of the frond. The specific epithet '*australis*' means 'southern', thus *Rhodopeltis australis* is the red shield from the south. Many other names commemorate the achievements of distinguished botanists or the person who collected the plant. For example, *Webervanbossea* is named for the Dutch botanist Anna Weber-van Bosse, who published numerous papers dealing with Indo-Pacific algae in the early part of the twentieth century. An understanding of the meanings of names can be useful when recognising the features of a particular species, but on occasions care must be taken. The epithets '*australis*' and '*borealis*' mean, respectively, 'southern' and 'northern', and are usually used to indicate the hemisphere in which the plants occur. This was probably true when the species *Rhodopeltis australis* and *Rhodopeltis borealis* were described, but they can be found growing side by side at Rottnest Island in Western Australia.

Hypoglossum dendroides – as illustrated by Harvey (1860, pl. 137, as *Delesseria dendroides*).

SYSTEMATIC SECTION

Division Rhodophyta

THE RED ALGAE

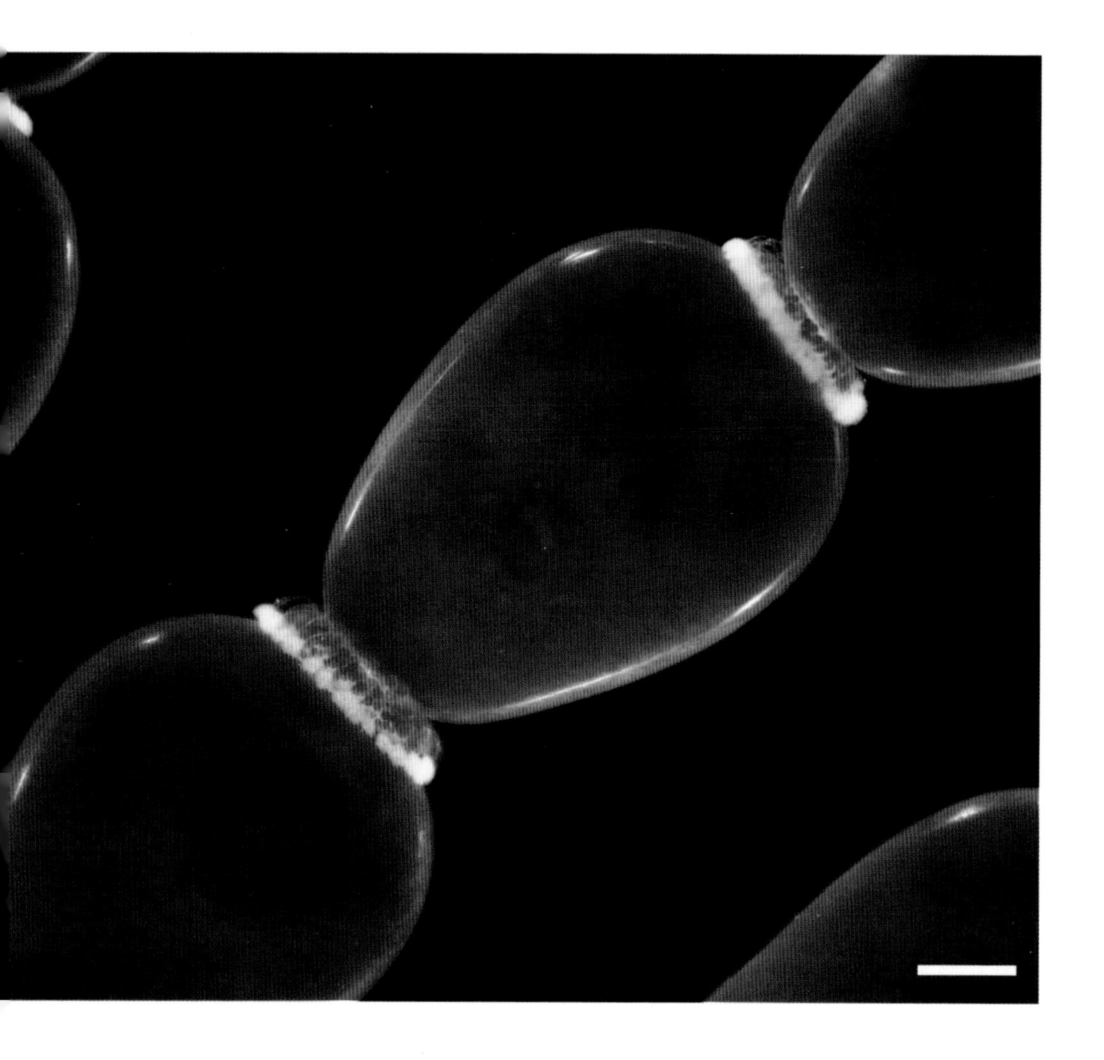

Griffithsia monilis – detail of filaments with tetrasporangia. Scale = 500 µm.

The Rhodophyta (from the Greek *rhodon* meaning 'red rose' and *phyton* meaning 'plant') includes around 5,000–5,500 species worldwide, of which some 1,300 occur in Australian seas. Like all divisions of the algae, the Rhodophyta is defined by a combination of biochemical and structural features. Its members contain the green pigment chlorophyll *a*, but this is masked by the red accessory pigment phycoerythrin. Unlike most other algae, the red algae never have motile gametes or spores. In most Australian seas, the number of species of Rhodophyta outweighs the Chlorophyta (green algae) and Ochrophyta (brown algae) by about 3:1. This might suggest a particularly conspicuous element of the flora, but, in reality, the red algae are mostly of a small to medium size and never form extensive beds. A large number are filamentous epiphytes or thin, encrusting calcified algae.

Despite their small stature, the red algae display an enormous variety of thallus forms and include many spectacular and attractive plants.

Recognition of the Rhodophyta can be difficult, with some thalli appearing anything but red, and experience no doubt being the best guide. Some hints include any trace of red colouration, and, on a microscopic level, the presence of pit-connections between vegetative cells. This latter feature is only found in the more advanced Rhodophyta, but that group includes the vastly greater proportion of the species likely to be encountered. Browsing through the following pages will give a good indication of what to expect.

Stylonema

The inconspicuous thalli of *Stylonema* have a simple filamentous construction in which the cells are regularly arranged in a mucilaginous sheath. They are invariably minute epiphytes, less than a millimetre tall, and are often overlooked. *Stylonema alsidii* is widespread in Australian waters and is distinguished from other species of the genus by its filaments remaining only one-cell thick.

Stylonema alsidii

TYPE LOCALITY: Adriatic Sea.
DISTRIBUTION: Cosmopolitan.
FURTHER READING: Womersley (1994); Huisman et al. (2007); Zuccarello et al. (2008); Huisman (2018).

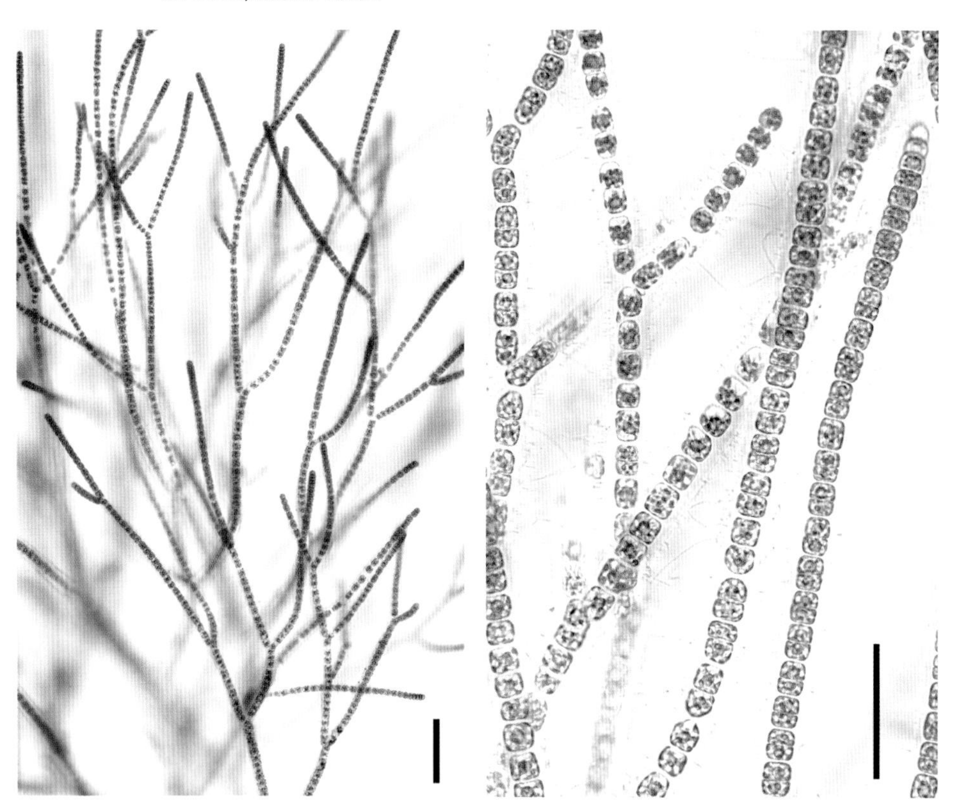

Stylonema alsidii (Cape Peron, WA). Scale = 100 µm.

(above) *Stylonema alsidii* – detail of cells (Cape Peron, WA). Scale = 50 µm.

Erythrotrichia

A genus of small filamentous plants often found growing epiphytically on a variety of algae and seagrasses. The filaments are upright and unbranched, attached by a lobed basal cell, and are usually one cell broad but can become up to four cells thick as they mature. The cells have a thick gelatinous wall and a stellate plastid with a central pyrenoid. Vegetative reproduction is by monosporangia that are cut off by a curved wall, but sexual reproduction can also occur.

Erythrotrichia carnea

TYPE LOCALITY: Llwchwr, Glamorgan, Wales.
DISTRIBUTION: Cosmopolitan as an epiphyte on larger algae.
FURTHER READING: Womersley (1994).

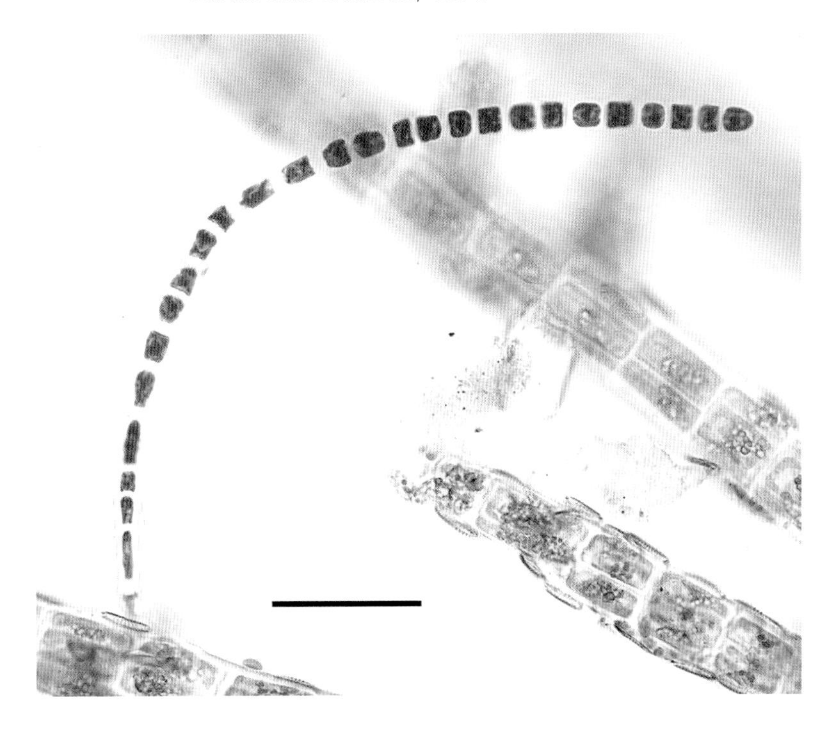

Erythrotrichia carnea – epiphytic on a species of *Polysiphonia* (Coral Bay, WA). Scale = 50 µm.

Sahlingia

Plants of *Sahlingia* form small encrusting discs on a variety of larger algae (in the photo on the green alga *Lychaete herpestica*). They can be circular or irregular in outline and are mostly less than half a millimetre in diameter, but often several plants will grow together and coalesce at the margins where they abut one another. The thallus of the common *Sahlingia subintegra* is mostly only one-cell-layer thick, but the central cells can divide further and form a thickened region. The oval shapes in the photo are diatoms.

Sahlingia subintegra

TYPE LOCALITY: Møllegrund, Skagerrak, off Hirshals, Denmark.
DISTRIBUTION: Widespread in most seas.
FURTHER READING: Huisman et al. (2007); Huisman (2018).

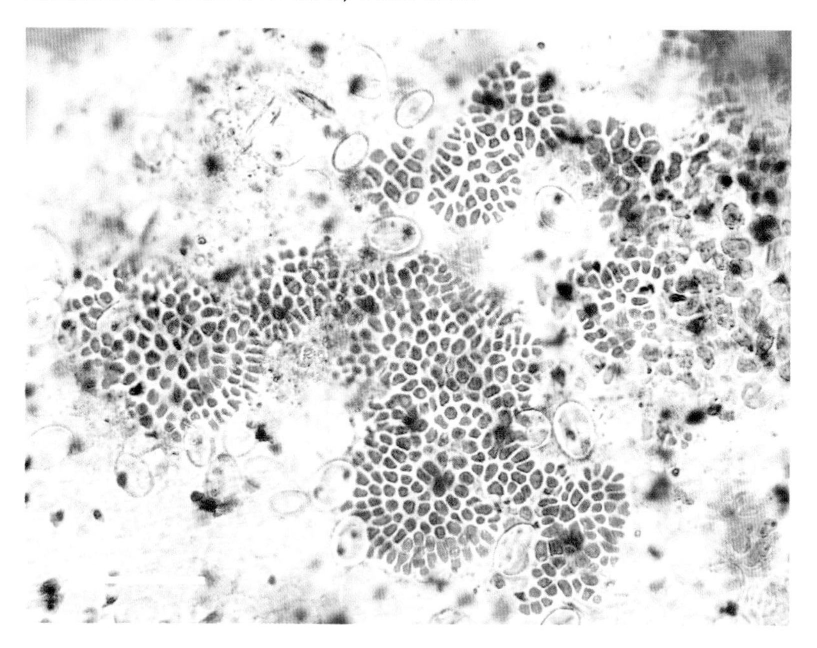

Sahlingia subintegra (Dampier Archipelago, WA). Scale = 50 µm.

Porphyropsis

In Australia, *Porphyropsis* is represented by the single species *Porphyropsis minuta*, a small, greenish brown to purple epiphyte that grows on other algae on rough water coasts. Young plants start out as tiny bladder-like thalli that, as they mature, rupture to form a filmy membrane that is only one-cell-layer thick and has a convoluted margin. The cells in surface view have a distinctive pattern and are often arranged in regular rows at right angles, although this is not consistent throughout the thallus. Reproduction in *Porphyropsis minuta* is by vegetative monospores that are released from whole cells at the margins of the membrane. Sexual reproduction is thought to occur in other species in the genus, but has not been observed in *Porphyropsis minuta*.

Porphyropsis minuta

TYPE LOCALITY: Pearson Island, South Australia.
DISTRIBUTION: Garden Island, Western Australia, around southern Australia to Collaroy, New South Wales.
FURTHER READING: Womersley & Conway (1975); Womersley (1994).

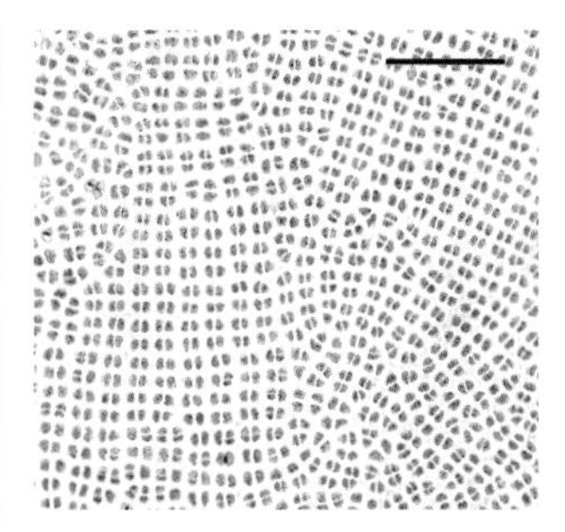

Porphyropsis minuta – epiphytic on *Laurencia* (Cape Peron, WA). Scale = 500 µm.

(above) *Porphyropsis minuta* – surface view of thallus (Cape Peron, WA). Scale = 50 µm.

Bangia

Bangia includes small filamentous plants that are commonly found on intertidal rocks along southern Australian shores. Two species (*Bangia fuscopurpurea* and *Bangia atropurpurea*) are virtually indistinguishable in appearance, but are separated based on chromosome numbers and DNA sequence analyses. The former is a marine species, the latter freshwater.

Bangia fuscopurpurea

TYPE LOCALITY: Glamorganshire, Wales.
DISTRIBUTION: Cosmopolitan in temperate marine and freshwater habitats on shoreline rock or timber.
FURTHER READING: Womersley (1994, as *Bangia atropurpurea*).

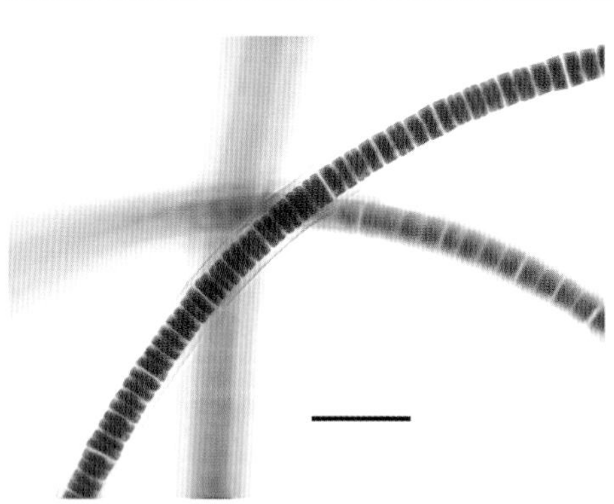

Bangia fuscopurpurea – growing on intertidal rock (Safety Cove, Tas.; Photo: M.D.Guiry)

(left) *Bangia fuscopurpurea* – detail of filaments (Canning River, WA). Scale = 50 µm.

Porphyra

Thalli of *Porphyra* are flattened and membranous, mostly unbranched and generally only one- or two-cells thick. Cells of the frond lack the intercellular connections found between cells in the 'higher' red algae. This feature, and the relatively simple thallus construction, ensure the genus is readily recognised. Three species of *Porphyra* are known to occur in Australia. *Porphyra lucasii* is a common inhabitant of the upper intertidal zone in areas of the south coast that experience calm to moderately rough seas. Species of *Porphyra* and the closely related *Pyropia* are a popular food in a number of countries – perhaps most familiar is the *nori* of Japanese cuisine, but they are also eaten as *laver* in Great Britain and Ireland and *karengo* by the Maoris of New Zealand.

Porphyra lucasii

TYPE LOCALITY: Bunbury, Western Australia.
DISTRIBUTION: Marmion, Western Australia, around southern Australia to Western Port, Victoria, and the north and east coasts of Tasmania.
FURTHER READING: Womersley (1994); Farr et al. (2003).

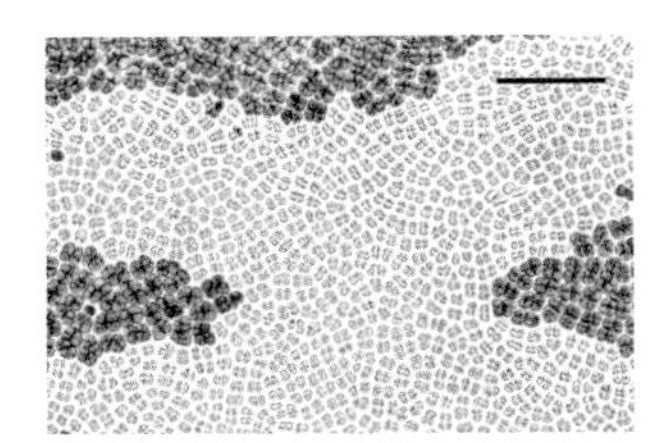

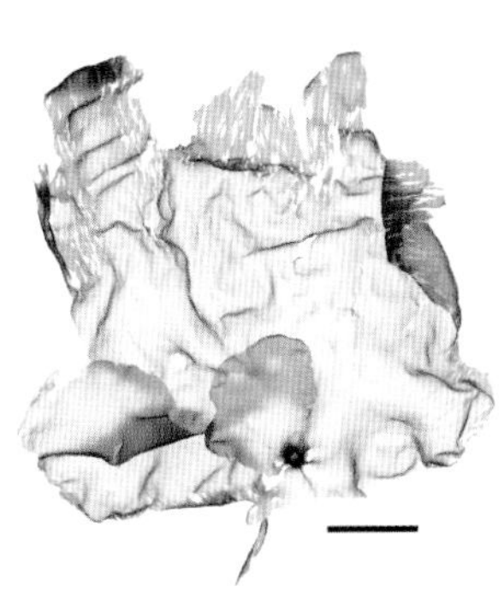

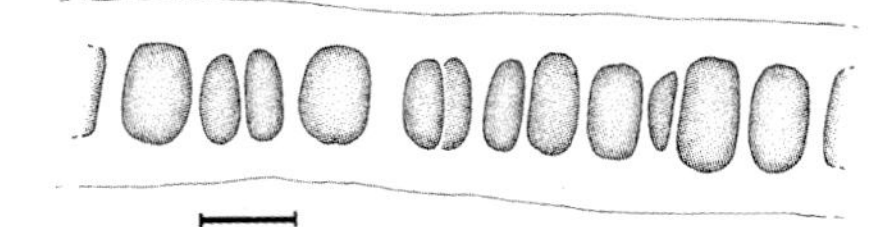

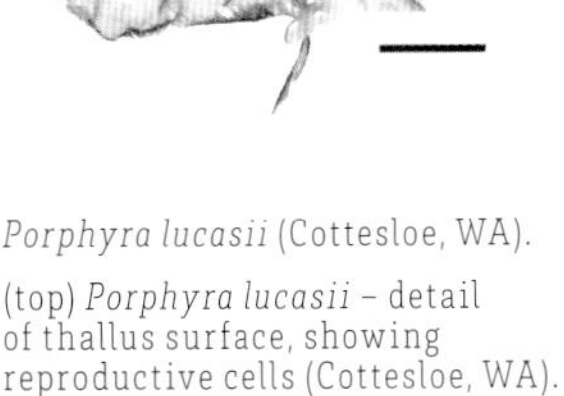

Porphyra lucasii (Cottesloe, WA).

(top) *Porphyra lucasii* – detail of thallus surface, showing reproductive cells (Cottesloe, WA). Scale = 100 µm.

(above) *Porphyra lucasii* – detail of plant (Cottesloe, WA). Scale = 1 cm.

(right) *Porphyra lucasii* – section of thallus (Marmion, WA). Scale = 20 µm.

Acrochaetium

Species of the genus *Acrochaetium* form minute tufts on rock and a wide variety of algae and seagrasses. Thalli are filamentous and uniaxial, with branching either profuse or absent, depending on the species. Elongate hairs are often present at the apices of filaments. Asexual reproduction is via monospores that serve to vegetatively propagate the plant, while sexual reproduction is known in some species but rarely observed. One common species is included here: *Acrochaetium microscopicum* can be recognised by its arcuate filaments with barrel-shaped cells. Species of *Acrochaetium* (and the similar looking *Colaconema*, see opposite) were previously included in a more broadly defined genus *Audouinella*, but DNA sequence analyses led them to being split into several smaller genera.

Acrochaetium microscopicum

TYPE LOCALITY: Torquay, England.
DISTRIBUTION: Throughout Australia. Cosmopolitan.
FURTHER READING: Woelkerling & Womersley (1994), as *Audouinella*; Harper & Saunders (2002); Huisman et al. (2007); Huisman & Woelkerling (2018a).

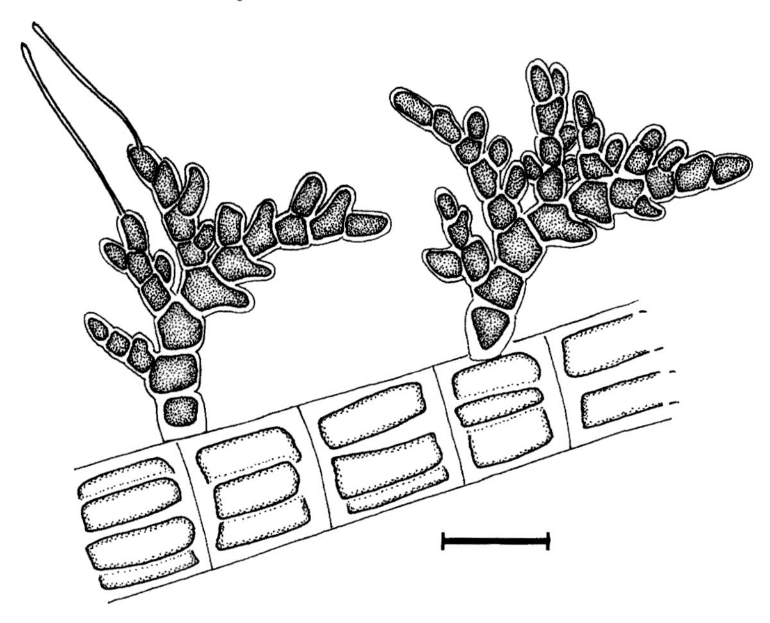

Acrochaetium microscopicum – two plants growing on *Sphacelaria* (Houtman Abrolhos, WA). Scale = 20 µm.

Colaconema

Colaconema is another small, filamentous genus that grows epiphytically on other algae and seagrasses or epizoically on invertebrates. Thalli are generally less than a centimetre tall and can reproduce vegetatively by monospores. *Colaconema daviesii* typically has straight, branched filaments that can give plants a bristly appearance. As with *Acrochaetium*, species of *Colaconema* were previously included in *Audouinella*, a broadly defined genus, until DNA sequence studies indicated that they belonged to separate genera and orders.

Colaconema daviesii

TYPE LOCALITY: Wales.
DISTRIBUTION: Throughout Australia. Cosmopolitan.

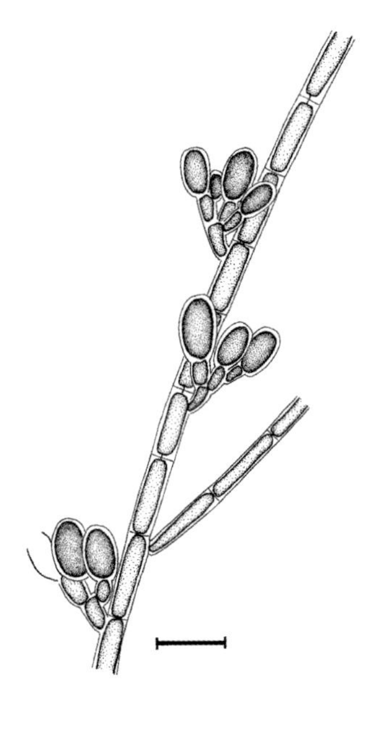

Colaconema savianum

TYPE LOCALITY: Genoa, Italy.
DISTRIBUTION: Throughout Australia. Cosmopolitan.
FURTHER READING: Woelkerling & Womersley (1994), as *Audouinella*; Harper & Saunders (2002); Huisman et al. (2007); Huisman & Woelkerling (2018b).

Colaconema daviesii – detail of thallus with clustered monosporangia (Rottnest Island, WA). Scale = 20 µm.

Colaconema savianum – plant growing on *Halimeda* (James Price Point, WA). Scale = 500 µm.

Akalaphycus

In Australia, *Akalaphycus* is currently known only from warmer waters of the east coast. It is one of several calcified genera in the Liagoraceae, most of which are similar in external appearance and can only be reliably identified by an examination of their internal structures. In *Akalaphycus* the outermost tissue has several layers of cells that are considerably larger than those above and below them, the largest about four cells from the surface. This pattern is not found in other genera. Plants of *Akalaphycus liagoroides* occur in shallow water on reefs and are often a brick-red colour, with branches that are slightly compressed.

Akalaphycus liagoroides

TYPE LOCALITY: Imuta, Koshiki Islands (Koshiki-jima), Kagoshima Prefecture, (Satsuma Province), Japan.

DISTRIBUTION: Southern Great Barrier Reef, Queensland, and Lord Howe Island, New South Wales. Japan. Mauritius.

FURTHER READING: Huisman et al. (2004), (2007).

Akalaphycus liagoroides (Heron Island, Qld).

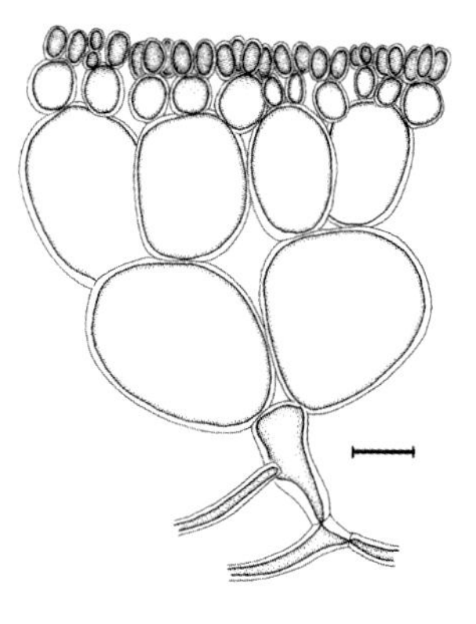

(above) *Akalaphycus liagoroides* – partial section of cortical filaments (One Tree Island, Qld). Scale = 30 µm.

Dotyophycus

Dotyophycus is one of the many genera of the Liagoraceae with a terete, dichotomously divided, calcified thallus. Structurally the axes have a core of longitudinal filaments that give rise to anticlinal cortical fascicles. These fascicles are generally dichotomously divided and form a loose outer cortex. Recognition of *Dotyophycus* is based on reproductive features, in particular the diffuse gonimoblast that arises following fertilisation and lack of sterile cells associated with the mature cystocarp. The genus is represented in Australia by two species: *Dotyophycus abbottiae*, thus far known only from a few localities in south-western Australia, and *Dotyophycus damaru* from the southern Kimberley (not included). In the field, *Dotyophycus abbottiae* can be recognised by its almost creamy-white appearance, which is due to the clear outer mucilage and smooth calcification.

Dotyophycus abbottiae

TYPE LOCALITY: Point Clune, Rottnest Island, Western Australia.
DISTRIBUTION: Rottnest Island and the Houtman Abrolhos, Western Australia.
FURTHER READING: Kraft (1988a); Huisman (2018).

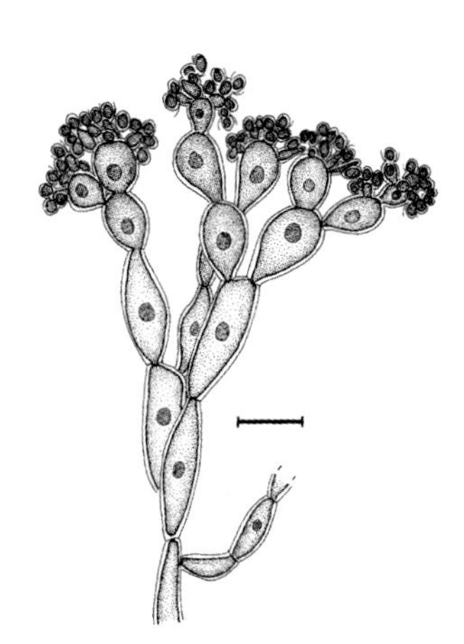

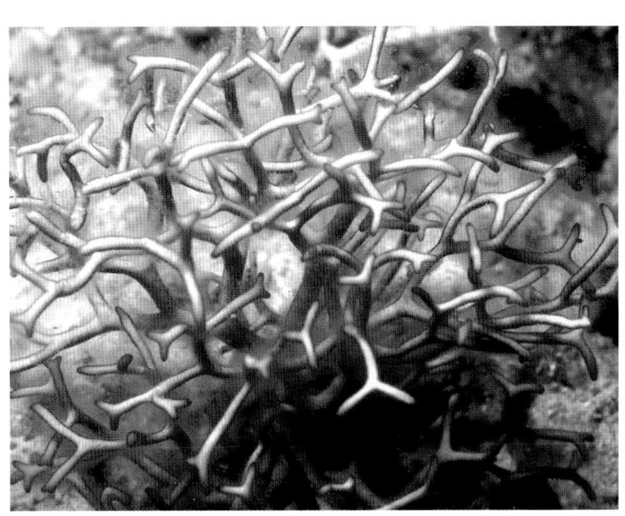

Dotyophycus abbottiae (Houtman Abrolhos, WA).

(above) *Dotyophycus abbottiae* – cortical filament with terminal spermatangia (Houtman Abrolhos, WA). Scale = 20 µm.

Ganonema

Thalli of *Ganonema* are calcified, generally dichotomously branched, with terete axes. Structurally the thallus is multiaxial, with a filamentous medulla and radiating cortical fascicles. *Ganonema* is closely related to *Liagora*, but differs in details of its vegetative structure and reproductive development. In *Ganonema,* the basal cells of cortical fascicles are isodiametric, as compared to the elongate cells in *Liagora*. Reproductive differences include the occasional production of carpogonial branches in clusters and the production of spermatangia in dense heads. Recognition of *Ganonema farinosum* is generally straightforward, as the species has distinctive cortical filaments that are composed of cylindrical cells. Cortical cells in most other genera of the Liagoraceae are ovoid or spherical. Two species are included here: the widespread *Ganonema farinosum* and *Ganonema robustum*. *Ganonema robustum* is a species from offshore reefs in the Indian Ocean that is considerably larger than *Ganonema farinosum* and has broader branches, although structurally the two species are very similar.

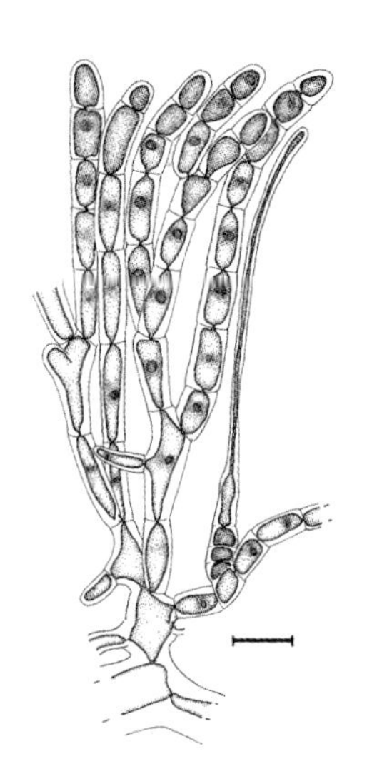

Ganonema farinosum

TYPE LOCALITY: Red Sea, near Suez.
DISTRIBUTION: Tropical to warm temperate seas worldwide.

Ganonema robustum

TYPE LOCALITY: Ashmore Reef, Australia.
DISTRIBUTION: Known from Ashmore Reef, the Rowley Shoals, and Scott Reef.
FURTHER READING: Huisman & Kraft (1994); Huisman (2002), (2006); Huisman & Lin (2018a); Lin et al. (2014).

Ganonema farinosum (Heron Island, Qld).

(top) *Ganonema farinosum* – detail of cortical fascicle with carpogonial branch (Barrow Island, WA). Scale = 30 µm.

Ganonema robustum (Ashmore Reef).

Gloiotrichus

Thalli of *Gloiotrichus* are extremely mucilaginous with a calcified inner region. They generally have several terete main axes that bear lateral branches at regular intervals – these lateral branches develop in a manner similar to the main axes and give the plant a pyramidal outline. Structurally the axes are composed of an inner medulla of longitudinal filaments that bear cortical fascicles. Filaments of the fascicles are dichotomously branched, more frequently in the mid cortex, with the outer regions unbranched for two to five cells. The genus is characterised by the development of sterile filaments on the elongate carpogonial branches that have the appearance of normal cortical filaments. Only the single species, *Gloiotrichus fractalis*, is known in Australia, where it grows associated with coral reefs at the Houtman Abrolhos, Western Australia. The species has also been reported from the Hawaiian Islands.

Gloiotrichus fractalis

TYPE LOCALITY: Suomi Island, Houtman Abrolhos, Western Australia.
DISTRIBUTION: From the Pelsaert and Easter Groups, Houtman Abrolhos, Western Australia. Hawaiian Islands.
FURTHER READING: Huisman & Kraft (1994); Huisman & Abbott (2003); Huisman (2006).

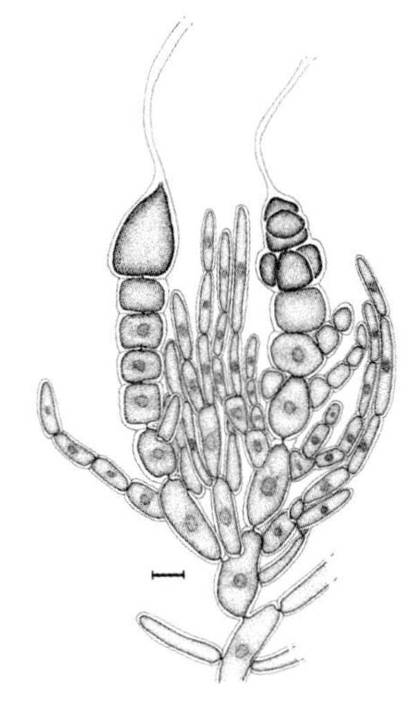

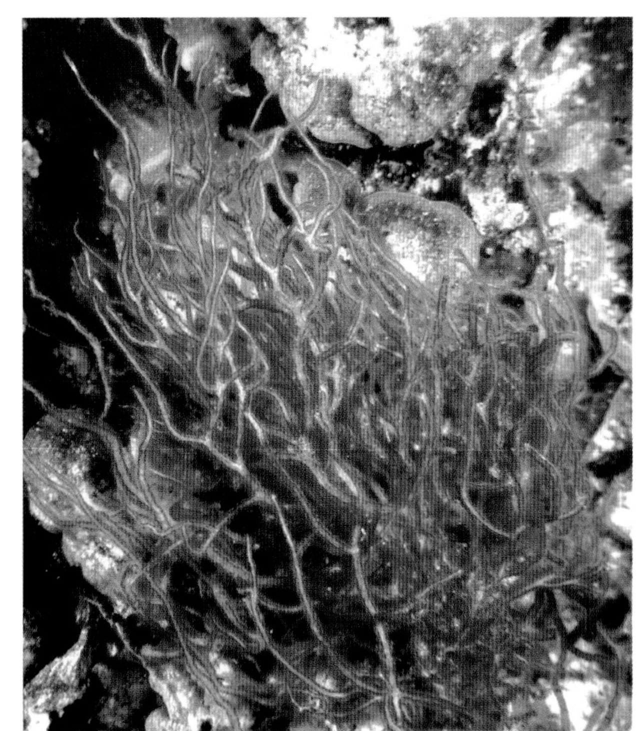

Gloiotrichus fractalis (Houtman Abrolhos, WA).

(above) *Gloiotrichus fractalis* – pre- (left) and post- (right) fertilisation carpogonial branches (Houtman Abrolhos, WA). Scale = 10 µm.

Helminthocladia

Helminthocladia is one of a number of mucilaginous genera of similar form, making precise identification difficult. An examination of internal structures is usually necessary. External features that are of some use in recognising *Helminthocladia* include its irregular branching pattern and lack of calcification. Several species occur in Australia, of which only *Helminthocladia australis* is included here. Thalli can be sporadically common in areas with some wave action, usually growing on rocks that are covered in sand. Structurally the thallus has a filamentous medulla bearing cortical fascicles, in which the terminal cells are generally the largest. The plant pictured is the gametophyte phase of the life history – the alternate phase (known as the tetrasporophyte) is a small and inconspicuous filamentous plant.

Helminthocladia australis

TYPE LOCALITY: Fremantle, Western Australia.
DISTRIBUTION: Exmouth, Western Australia, around southern mainland Australia and Tasmania to Coffs Harbour and Lord Howe Island, New South Wales. California, USA. Japan. South Africa. New Zealand.
FURTHER READING: Womersley (1965), (1994); Huisman & Womersley (2006a).

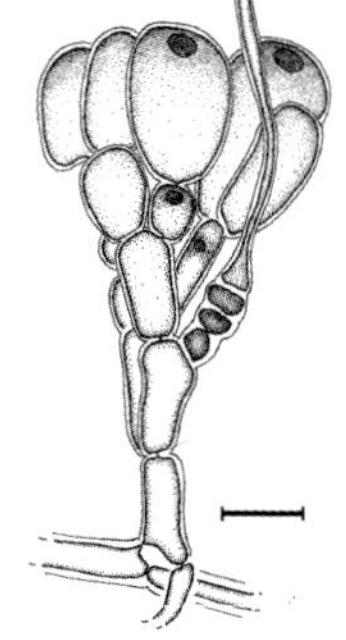

Helminthocladia australis (Cape Peron, WA).

(top) *Helminthocladia australis* – as illustrated by Harvey (1863, pl. 272).

(above) *Helminthocladia australis* – closer view of branches (Cape Peron, WA). Scale = 1 mm.

(right) *Helminthocladia australis* – cortical filaments and carpogonial branch (Rottnest Island, WA). Scale = 20 µm.

Helminthora

Helminthora is similar in appearance to *Helminthocladia*, but often has a more pyramidal shape because of the prominence of the central axis, whereas thalli of *Helminthocladia* are more irregularly branched. Both genera are mucilaginous and uncalcified. Structurally *Helminthora* differs in the form of the terminal cells of the cortical filaments, which are smaller than those below them. In *Helminthocladia*, the terminal cells are considerably larger. Two species occur in Australia, of which only *Helminthora australis* is included here. Thalli often occur epiphytically on seagrasses in shallow water, but can also grow on rock. As with all Liagoraceae, the conspicuous thalli are the gametophytes, and those of *Helminthora australis* can be common in areas with some wave action during spring and early summer.

Helminthora australis

TYPE LOCALITY: Western Port, Victoria.
DISTRIBUTION: Cottesloe, Western Australia, to Walkerville, Victoria, the north coast of Tasmania, and Kangaroo Point, Derwent River, Tasmania. New Zealand.
FURTHER READING: Womersley (1965), (1994); Huisman & Womersley (2006b).

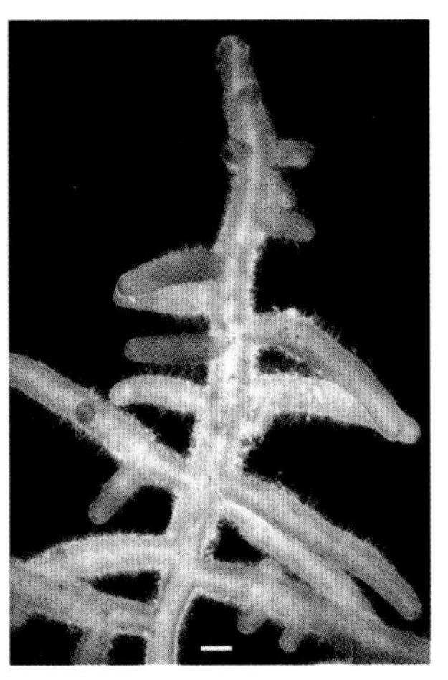

Helminthora australis (Cape Peron, WA)

(above) *Helminthora australis* – detail of upper thallus (Cape Peron, WA). Scale = 1 mm.

Hommersandiophycus

Five species of *Hommersandiophycus* are known, all from tropical waters, with three species recorded from Australia. Plants often occur in intertidal pools or shallow subtidal lagoonal habitats and can be recognised by their calcified and mucilaginous thallus, often with several percurrent primary axes. Structurally *Hommersandiophycus* is similar to *Ganonema*, with a central core of longitudinal filaments giving rise to lateral cortical fascicles. Cells of the cortical fascicles in both genera are barrel-shaped or tubular, which can assist in distinguishing them from other genera in the Liagoraceae, which mostly have ellipsoidal cortical cells. *Hommersandiophycus* is one of several recently described genera whose recognition was prompted by DNA sequence analyses.

Hommersandiophycus kraftii

TYPE LOCALITY: Ashmore Reef, Australia.

DISTRIBUTION: Known from the Dampier Archipelago north to Ashmore Reef; generally found in shallow lagoonal habitats.

Hommersandiophycus samaensis

TYPE LOCALITY: East Island, Sama, Hainan, China.

DISTRIBUTION: Known from Turtle Island, Western Australia and from Bowen and north of Mackay, Queensland. Hainan, China. Hawaiian Islands. Indian Ocean.

FURTHER READING: Huisman (2002), (2006), as *Ganonema*; Lin et al. (2014).

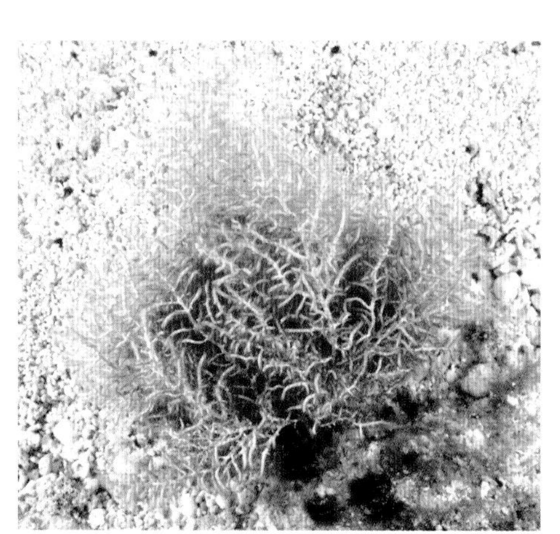

Hommersandiophycus kraftii (Ashmore Reef).

Hommersandiophycus samaensis (Turtle Island, WA).

Izziella

Izziella was named in honour of Isabella Aiona Abbott, a Hawaiian seaweed biologist who had a lifelong interest in the taxonomy of the Liagoraceae. The genus was unknown in Australia until the discovery of *Izziella vulcanensis*, growing in deep water at two remote offshore shoals in the Indian Ocean. Superficially, *Izziella* appears similar to several other calcified genera in the Liagoraceae, and an examination of reproductive structures is required for a positive identification. After fertilisation in *Izziella*, the cells of the filament that bears the carpogonial branch become enlarged and prominent, such that the carposporophyte appears to occupy a terminal position on a large filament arising from the medulla.

Izziella vulcanensis

TYPE LOCALITY: Vulcan Shoal, Western Australia.
DISTRIBUTION: Presently known only from two collections from submerged offshore shoals 110-150 km east-southeast from Ashmore and Cartier Islands.
FURTHER READING: Doty (1978); Huisman & Schils (2002); Lin et al. (2011); Huisman & Lin (2018a).

Izziella vulcanensis (Vulcan Shoal, WA).

(above) *Izziella vulcanensis* (Heywood Shoal, WA).

Liagora

Species of *Liagora* occur in warmer waters worldwide and several can be found in Australian seas. The genus can usually be recognised by its terete, dichotomously divided branches and white, often chalky appearance due to the deposition of calcium carbonate. Several species, however, produce smaller side-branches that may obscure the normal branching pattern. As with most members of the Liagoraceae, an internal examination is required for a positive identification, as several other genera display a similar form. Structurally the thallus has an inner medulla of longitudinal filaments bearing anticlinal cortical fascicles. Distinctive features of *Liagora* include the production of short carpogonial branches (two to six cells long, depending on the species) from lateral positions on cells of the cortical fascicles and a relatively compact gonimoblast. In most species, sterile filaments arise following fertilisation and surround the developing carposporophyte. *Liagora wilsoniana* and *Liagora harveyana* are essentially southern Australian species, whereas *Liagora ceranoides* is widespread in tropical seas. *Liagora izziae* is presently known only from Rottnest Island and Lord Howe Island.

Liagora ceranoides

Liagora ceranoides is a common species, growing on rock in shallow water.

TYPE LOCALITY: St Thomas, Virgin Islands.

DISTRIBUTION: From Shark Bay, Western Australia, around northern Australia to Lord Howe Island, New South Wales. Widespread in tropical seas.

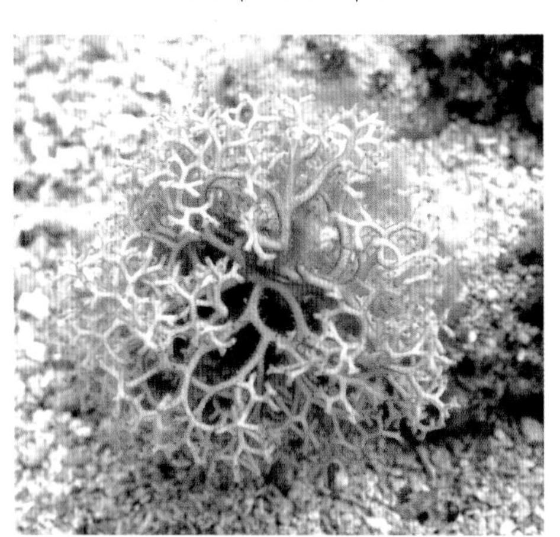

Liagora izziae

TYPE LOCALITY: Roe Reef, Rottnest Island, Western Australia.

DISTRIBUTION: Rottnest Island, Western Australia, and Lord Howe Island, New South Wales.

Liagora wilsoniana

TYPE LOCALITY: Western Port, Victoria.
DISTRIBUTION: From Rottnest Island and the Perth region, Western Australia, around southern Australia to Walkerville, Victoria.

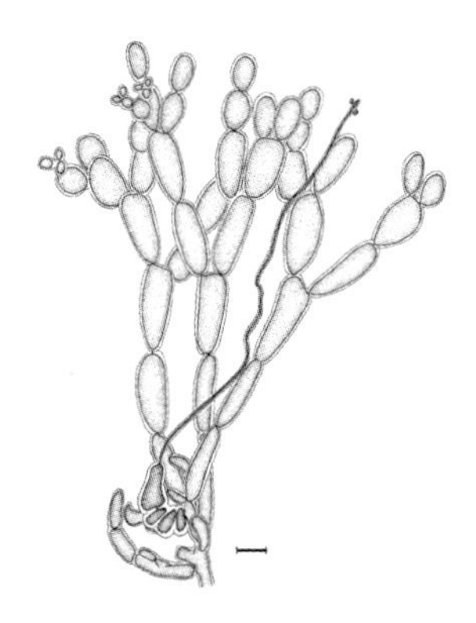

Liagora harveyana

TYPE LOCALITY: King George Sound, Western Australia.
DISTRIBUTION: Albany, Western Australia, to Walkerville, Victoria, and the north coast of Tasmania.
FURTHER READING: Womersley (1965); Huisman (2002), (2006); Lin et al. (2013).

Liagora ceranoides (Cassini Island, WA).

Liagora izziae (Rottnest Island, WA).

Liagora wilsoniana (Rottnest Island, WA).

(above) *Liagora wilsoniana* – cortex and carpogonial branch (Cosy Corner, WA). Scale = 10 µm.

Liagora harveyana (Albany, WA).

Macrocarpus

Macrocarpus is one of several genera that have been recently described to include species previously regarded as members of *Liagora*, the change driven primarily by DNA sequence analyses. The genus is also unique in having considerably larger carposporangia that are divided into four spores, a condition known as 'quadripartite'. When first described, *Macrocarpus* included only the single species, *Macrocarpus perennis* from the Hawaiian Islands, but that species, plus two others, are now known from Australia. *Macrocarpus kraftii* is included here.

Macrocarpus kraftii

TYPE LOCALITY: Barrow Island, Western Australia.
DISTRIBUTION: Known from Barrow Island and the Maret Islands, northern Western Australia, and the southern Great Barrier Reef, Queensland.
FURTHER READING: Huisman (2002), (2006), as *Liagora kraftii*; Lin et al. (2011); Huisman & Lin (2018a).

Macrocarpus kraftii (Lamont Reef, Qld).

Patenocarpus

The name *Patenocarpus* translates to 'spreading fruit' and is in reference to the unusual structure of the gonimoblast in *Patenocarpus paraphysiferus*, the single species attributed to this genus. Instead of the more common compact cluster of tissue and carpospores, the gonimoblast filaments in *Patenocarpus* spread laterally and mingle with cortical filaments. Short filaments then grow towards the surface of the thallus and produce single carpospores at their tips. In Australia, *Patenocarpus* is known from only a few locations in the Dampier Archipelago, Western Australia, where it grows attached to rock in intertidal pools and the shallow subtidal.

Patenocarpus paraphysiferus

TYPE LOCALITY: Toyohara, Iriomote Island, Ryukyu Islands, Japan.
DISTRIBUTION: Known from the Dampier Archipelago, Western Australia. Southern Japan.
FURTHER READING: Yoshizaki (1987); Huisman (2006).

Patenocarpus paraphysiferus (Dampier Archipelago, WA).

Titanophycus

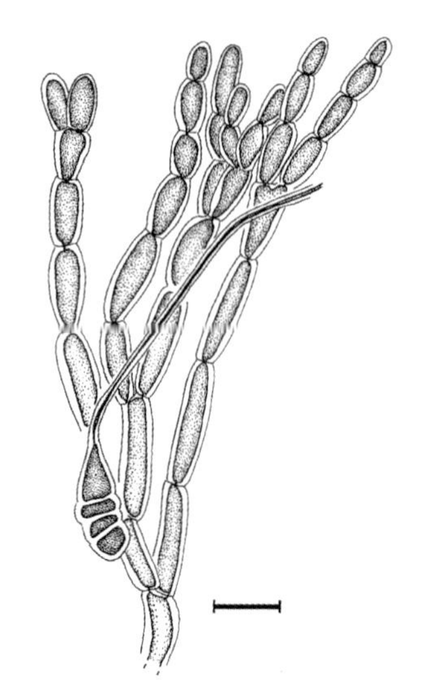

Plants of *Titanophycus* are generally heavily calcified, with dichotomously branched, terete axes. In appearance they are similar to several genera in the Liagoraceae, such as *Patenocarpus* (p. 23) and *Liagora* (p. 20), but in *Titanophycus*, the cortical filaments are only sparingly branched, with distal portions unbranched for several cells. In addition, gonimoblasts in *Titanophycus* have the appearance of an upside-down cone, subtended by a loose, cup-like basket of sterile filaments. Previous Australian records have been referred to the southern U.S. species *Titanophycus validus* in the mistaken belief that the genus included only one widespread species. Two species occur in Australia and can be separated by the shapes of their upper cortical cells, those of *Titanophycus saundersii* being more rounded than the cylindrical cells of *Titanophycus setchellii*.

Titanophycus saundersii

TYPE LOCALITY: Ashmore Reef, Australia.
DISTRIBUTION: Known from Scott Reef and Ashmore Reef.

Titanophycus saundersii (Scott Reef, WA).

Titanophycus setchellii

TYPE LOCALITY: Naha, Okinawa Island, Japan.
DISTRIBUTION: Occurs from the Kimberley coast of northern Western Australia south to Albany and east to Queensland. Japan. Hawaiian Islands.
FURTHER READING: Huisman et al. (2006); Huisman & Lin (2018a).

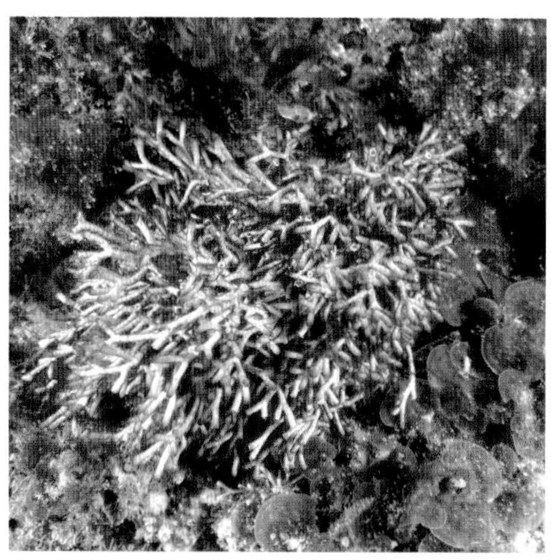

Titanophycus setchellii (Rottnest Island, WA).

(top) *Titanophycus setchellii* – cortical filaments and carpogonial branch (Houtman Abrolhos, WA). Scale = 20 µm.

Trichogloea

Species of *Trichogloea* are instantly recognisable because of their extremely mucilaginous branches and calcified central region. Only one species, *Trichogloea requienii*, occurs in Australian waters, where it is usually found associated with coral reefs. Thalli grow to approximately 20 cm in height and are irregularly branched. Structurally the thallus is multiaxial, with a central medulla of longitudinally aligned filaments that give rise to cortical fascicles. The calcified inner region can be seen as vivid white in the photograph. A distinctive feature of the genus is the presence of specialised lateral filaments on the lower cells of the carpogonial branch.

Trichogloea requienii

TYPE LOCALITY: Red Sea.
DISTRIBUTION: From the Houtman Abrolhos, Western Australia, around northern Australia to Lord Howe Island, New South Wales. Widespread in tropical seas.
FURTHER READING: Abbott & Huisman (2005); Huisman (2006).

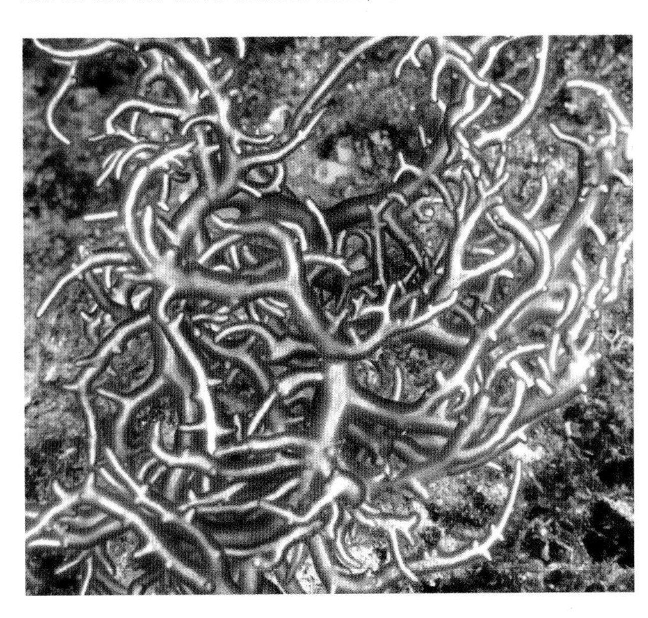

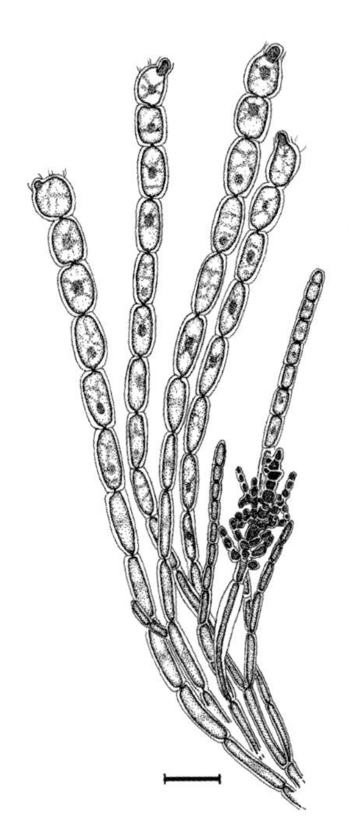

Trichogloea requienii (Heron Island, Qld).

(above) *Trichogloea requienii* – cortical filaments with developing carposporophyte (Houtman Abrolhos, WA). Scale = 30 µm.

Yamadaella

Yamadaella is known from three species, of which the widespread *Yamadaella caenomyce* is included here. Plants are commonly encountered in lower intertidal zones on reefs in tropical areas and as such are regularly exposed during low tide. Thalli are calcified and have dichotomous to irregularly divided, terete axes that are distinctly annulate. Structurally the axes have a central filamentous medulla bearing dichotomously divided cortical fascicles in which the outermost cortical cells are conspicuously larger than the lower cells. The genus is characterised by reproductive features such as a diffuse gonimoblast and the production of quadripartite carposporangia, but reproductive specimens are rarely encountered. Fortunately, the distinctive intertidal habitat and inflated outer cortical cells allow ready identification of sterile specimens. A second species, *Yamadaella australis*, was described recently from northern Western Australia.

Yamadaella caenomyce

TYPE LOCALITY: Manila, Philippines.
DISTRIBUTION: From Rottnest Island, Western Australia, around northern Australia to Lord Howe Island and Port Stephens, New South Wales. Widespread in tropical seas.
FURTHER READING: Abbott (1970); Wynne & Huisman (1998); Huisman (2006); Popolizio et al. (2015); Lin et al. (2015); Huisman & Lin (2018b).

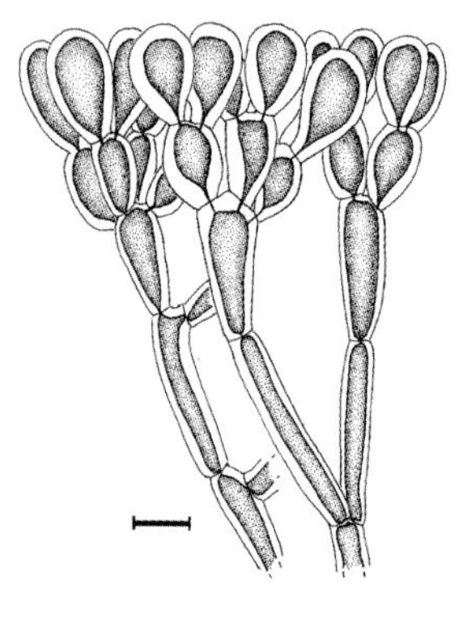

Yamadaella caenomyce (Heron Island, Qld).

(above) *Yamadaella caenomyce* – cortical filaments (Barrow Island, WA). Scale = 10 μm.

Actinotrichia

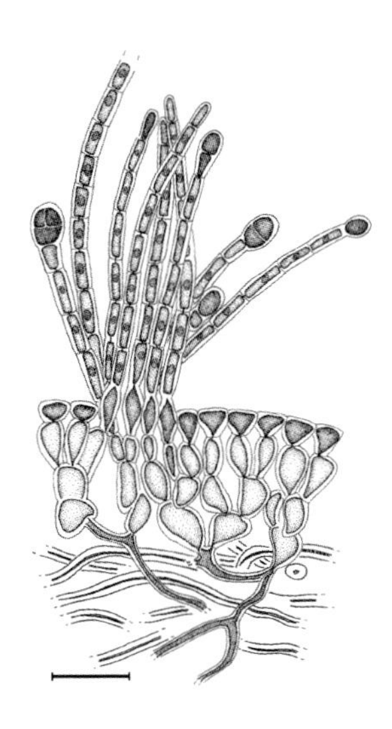

Actinotrichia is one of several dichotomously branched, calcified genera belonging to the family Galaxauraceae. It is closely related to *Galaxaura* and is superficially very similar in appearance to the sporophyte of *Galaxaura glabriuscula* (see p. 30). *Actinotrichia* can be recognised by its thinner branches (to 2 mm diameter), regular whorls of dark-red filaments, and cortex in which the outer cells have the smallest diameter. Structurally the branches have a central, multiaxial core with radiating filaments that eventually form a coherent cortex. Some cortical cells continue growth and produce the characteristic whorls of filaments. At present, five species are recognised in the genus, with two found in tropical regions of Australia, including the widespread *Actinotrichia fragilis* and the recently discovered *Actinotrichia coccinea*, which differs in its more vivid red colour and broader branching.

Actinotrichia coccinea

TYPE LOCALITY: Ashmore Reef, Australia.
DISTRIBUTION: Known from several offshore atolls off northern Western Australia (Rowley Shoals, Seringapatam Reef, Ashmore Reef).

Actinotrichia fragilis

TYPE LOCALITY: Mokha, Yemen.
DISTRIBUTION: From Ningaloo Reef, Western Australia, around northern Australia to at least the southern Great Barrier Reef, Queensland. Widespread in the tropical Indo-Pacific.
FURTHER READING: Huisman (2006); Liu & Wang (2009); Wiriyadamrikul et al. (2013b); Huisman et al. (2018c).

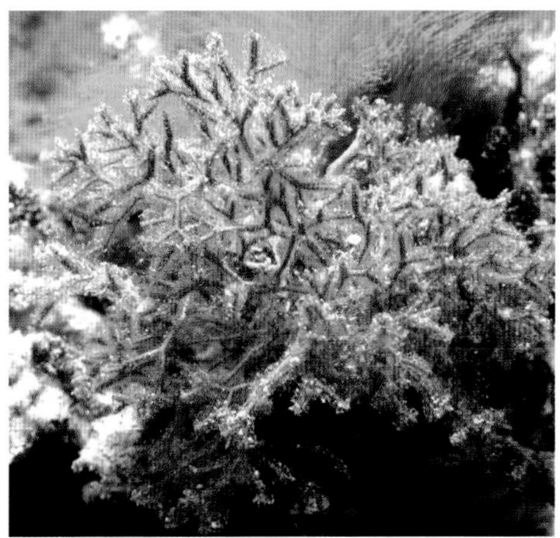

Actinotrichia coccinea (Ashmore Reef).

Actinotrichia fragilis (Heron Island, Qld).

(top) *Actinotrichia fragilis* – partial section of thallus with filaments bearing terminal tetrasporangia (Ningaloo Reef, WA). Scale = 50 µm.

Dichotomaria

The genus *Dichotomaria* was originally described in 1816, but until recently its species were included in a broadly defined *Galaxaura*. However, DNA sequence studies have shown it to be an independent genus and it was resurrected in 2004. Similar to *Galaxaura*, species of *Dichotomaria* are calcified with a chalky texture, and all are dichotomously divided, but branches can be distinctly flattened (as in *Dichotomaria sibogae*) or terete and segmented (as in *Dichotomaria obtusata*). Differences in the cortical structure of gametophytes and sporophytes, as found in *Galaxaura*, also occur in *Dichotomaria*. The genus includes 17 species, most of which are found in the tropics, but at least two flattened species, *Dichotomaria spathulata* and *Dichotomaria australis* (not illustrated), occur in southern Australia.

Dichotomaria obtusata

TYPE LOCALITY: Bahama Islands, West Indies.
DISTRIBUTION: From Augusta, Western Australia, around northern Australia to Lake Macquarie, New South Wales. Widespread in warmer seas.

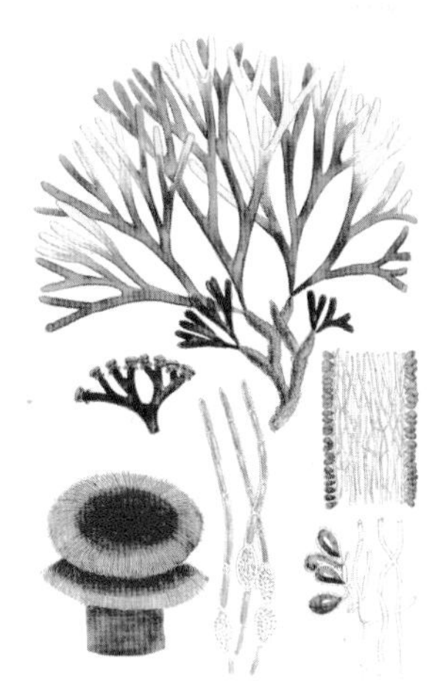

Dichotomaria spathulata

TYPE LOCALITY: Fremantle.
DISTRIBUTION: Houtman Abrolhos, Western Australia, around south-western Australia to Yorke Peninsula, South Australia.

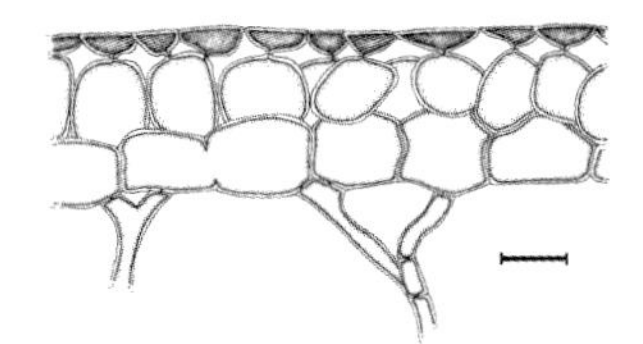

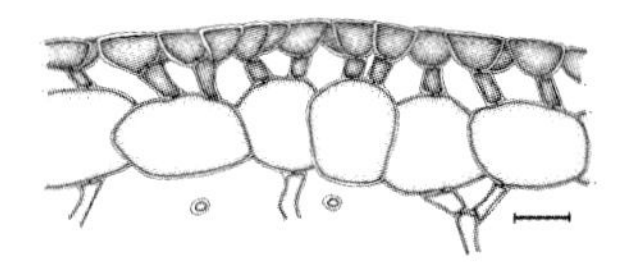

Dichotomaria sibogae

TYPE LOCALITY: Karkaralong Islands, Indonesia.
DISTRIBUTION: Tropical regions of Australia and the Indo-Pacific.
FURTHER READING: Huisman et al. (2004); Kurihara & Huisman (2006); Wiriyadamrikul et al. (2014); Huisman et al. (2018c).

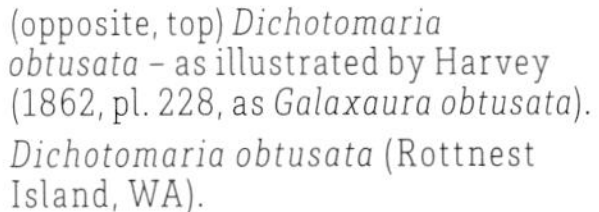

(opposite, top) *Dichotomaria obtusata* – as illustrated by Harvey (1862, pl. 228, as *Galaxaura obtusata*).

Dichotomaria obtusata (Rottnest Island, WA).

(above) *Dichotomaria obtusata* – sections of gametophyte (upper) (Cairns, Qld) and sporophyte (lower) (Rottnest Island, WA).Scales = 30 µm).

(top) *Dichotomaria spathulata* – as illustrated by Harvey (1860, pl. 136, as *Galaxaura marginata*).

Dichotomaria spathulata (Cosy Corner, WA; photo G.J. Edgar).

Dichotomaria sibogae (Ashmore Reef).

Galaxaura

Species of *Galaxaura* have been recorded in the literature as far back as the early eighteenth century (although not under their present names), and the genus has had a long and often confused history. Part of the confusion stems from the fact that in any one species, the tetrasporophyte phase can be superficially different from the gametophyte phase. In the past, this has resulted in the phases being described as separate species. *Galaxaura* is typically found in tropical regions, although one species, *Galaxaura elongata* (not illustrated), can be found in south-western Australia. Species of the genus are calcified, with a chalky texture, and are always dichotomously branched with terete branches. Structurally the thallus is multiaxial with a central filamentous medulla that grades into a pseudoparenchymatous cortex. The composition of the cortex varies, depending on the species and reproductive phase. In species such as *Galaxaura glabriuscula* and *Galaxaura elongata,* the gametophyte phase has smooth branches while those of the tetrasporophyte are covered with fine filaments, sometimes arranged in distinct whorls. The recently described *Galaxaura indica* is similar in appearance to *Galaxaura pacifica* and its recognition was based primarily on DNA sequence differences.

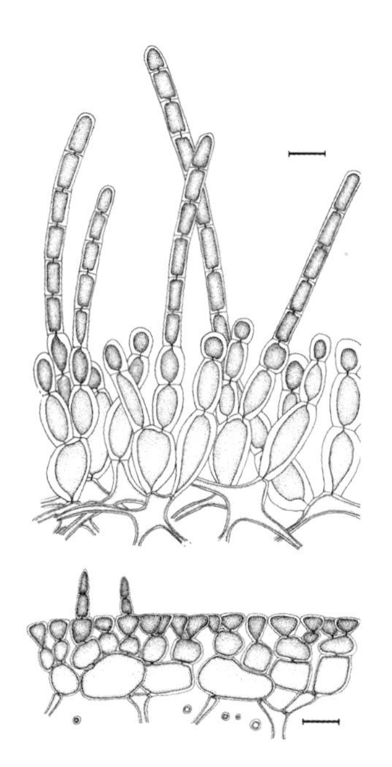

Galaxaura glabriuscula

TYPE LOCALITY: Tahiti.
DISTRIBUTION: From the Perth region, Western Australia, around northern Australia to Lord Howe Island in the south-western Pacific Ocean. Widespread in the tropical Indo-Pacific.

(left) *Galaxaura glabriuscula* – gametophyte (Scott Reef, WA).

(right) *Galaxaura glabriuscula* – sporophyte (Broomfield Reef, Qld).

(top) *Galaxaura glabriuscula* – sections of sporophyte (upper) (Lord Howe Island, NSW) and gametophytes (lower) (Rottnest Island, WA). Scales = 30 µm.

Galaxaura indica

TYPE LOCALITY: Ashmore Reef, Australia.
DISTRIBUTION: Currently only known from the type locality.

Galaxaura pacifica

TYPE LOCALITY: Haha-jima, Ogasawara Islands, Japan.
DISTRIBUTION: Widespread in the tropical Indo-west Pacific.
FURTHER READING: Huisman & Borowitzka (1990); Huisman et al. (2004); Huisman (2006); Liu et al. (2013); Huisman et al. (2018c).

Galaxaura indica (Ashmore Reef).

Galaxaura pacifica (left) and *Tricleocarpa fastigiata* (right) (Hibernia Reef, WA).

Tricleocarpa

Tricleocarpa includes seven species, of which *Tricleocarpa australiensis* and *Tricleocarpa fastigiata* are included here. The genus is closely related to *Galaxaura* and several of its species were originally placed in that genus, but were segregated into *Tricleocarpa* because of reproductive differences, a move subsequently supported by DNA analyses. As with many seaweeds, earlier concepts of fewer, widely distributed species have proven incorrect, and studies are now recognising many more species. *Tricleocarpa australiensis*, for example, was previously thought to be *Tricleocarpa cylindrica*, a West Indian species. Plants of *Tricleocarpa* are calcified with terete axes that are dichotomously branched. Structurally the thallus is multiaxial with a central core of longitudinally aligned medullary filaments. Cortical filaments arise from the medullary cells and form a filamentous outer cortex that remains coherent as a result of the calcified layer. Both cystocarps and spermatangia occur in cavities embedded in the thallus. The specimen pictured here is the gametophyte – the alternate stage, the tetrasporophyte, is a small filamentous plant. *Tricleocarpa australiensis* could be confused with species of *Amphiroa* (p. 53) or *Metagoniolithon* (p. 58), but an examination of the internal structure should allow ready identification.

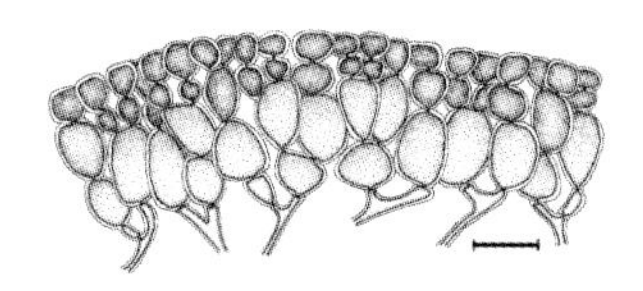

Tricleocarpa australiensis

TYPE LOCALITY: Roe Reef, Rottnest Island, Western Australia.

DISTRIBUTION: Occurs from Augusta, Western Australia, around northern Australia and down the east coast to Dee Why, New South Wales.

Tricleocarpa fastigiata

TYPE LOCALITY: Probably Timor.

DISTRIBUTION: Exact distribution unknown, possibly widespread in the Indo-Pacific.

FURTHER READING: Huisman & Borowitzka (1990); Huisman (2006); Wiriyadamrikul et al. (2013a); Liu et al. (2015); Huisman et al. (2018c).

Tricleocarpa australiensis (Rottnest Island, WA).

Tricleocarpa fastigiata (Ashmore Reef).

(top) *Tricleocarpa australiensis* – section of cortex (Rottnest Island, WA). Scale = 30 µm.

Nothogenia

Nothogenia is known in Australia by the single species *Nothogenia lingula*, which is restricted to the colder Tasmanian coasts where it grows on intertidal rock. Plants grow to about 8 cm tall and are erect and sparingly subdichotomously branched, with terete or compressed axes. Structurally the medulla is filamentous and grades into a cortex of small-celled anticlinal filaments. Cystocarps are immersed in the thallus. *Nothogenia* has a heteromorphic life history with a crustose tetrasporophyte, although this has not been demonstrated in the Australian *Nothogenia lingula*. Previous Australian records of this taxon were as the seemingly widespread *Nothogenia fastigiata*, but molecular studies have shown that species to have a more restricted distribution and supported the recognition of *Nothogenia lingula* as an independent species

Nothogenia lingula

TYPE LOCALITY: Brown's River, Tasmania.
DISTRIBUTION: Tasmania.
FURTHER READING: Huisman & Womersley (2006c, as *Nothogenia fastigata*); Parnell & Huisman (2006); Nelson (2013); Lindstrom et al. (2015).

Nothogenia lingula (Fortescue Bay, Tas; photo M.D. Guiry).

Scinaia

Species of *Scinaia* have soft, dichotomously divided thalli with (in Australian species) terete branches. Branches can be unconstricted or have regular or irregular constrictions, depending on the species. Structurally the thallus is multiaxial with a central file of entwining filaments from which radiating medullary filaments arise. At the periphery, the filaments form a cortex with an inner layer of small, pigmented cells subtending an outer layer of large, colourless cells (known as utricles). Thalli of *Scinaia* are generally found attached to rock, often with their bases covered in sand. Seven species are found in Australia, of which *Scinaia aborealis* and *Scinaia tsinglanensis* are included here. *Scinaia tsinglanensis* can be distinguished by its small size (to 10cm) and narrow branches that are generally unconstricted. *Scinaia aborealis* is a more robust plant (to 40 cm in height) and branches generally have some constrictions.

Scinaia aborealis

TYPE LOCALITY. Sorrento, Western Australia.
DISTRIBUTION: Greenough, Western Australia around southern Australia (not Tasmania) to Coffs Harbour, New South Wales. Norfolk Island.

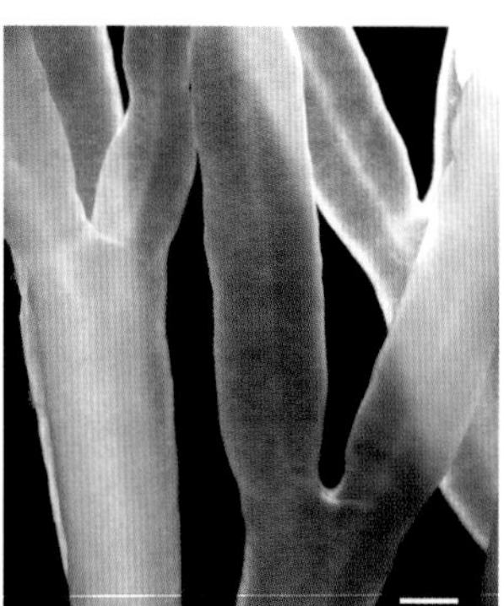

Scinaia tsinglanensis

TYPE LOCALITY: Tsinglan-Kang, Wenchang, Hainan, China.
DISTRIBUTION: Throughout Australia. China.
FURTHER READING: Huisman (1986), (2006).

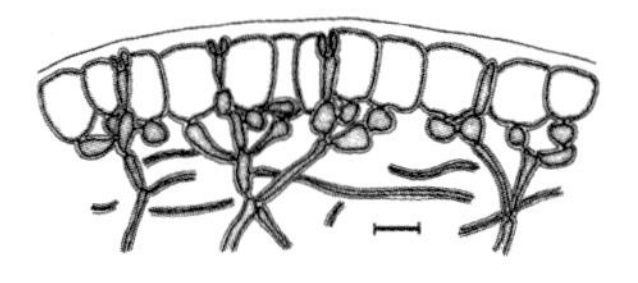

Scinaia aborealis (Rottnest Island, WA).

(left) *Scinaia aborealis* – thallus detail (Rottnest Island, WA). Scale = 1 mm.

Scinaia tsinglanensis (Heron Island, Qld).

(above) *Scinaia tsinglanensis* – section of cortex (Port Phillip, Vic). Scale = 20 µm.

Capreolia

Capreolia is a genus with only one species, *Capreolia implexa*, which forms dense mats on intertidal rock and molluscs on exposed and sheltered southern coasts. Plants are typically only a few millimetres in height but have prostrate stolons and can spread laterally for several centimetres. The thallus is structurally uniaxial and has a conspicuous apical cell. As in all Gelidiaceae, thick-walled rhizines are present in the medulla. Only tetrasporophytes of *Capreolia* have been found in the wild, but laboratory culture studies have demonstrated the existence of male and female gametophytes. Fertilisation results in a small thallus growing directly on the female gametophyte, but the carposporophyte generation, which occurs in most red algae after fertilisation, is absent from the life cycle. The small thallus attaches to the substratum and develops into a tetrasporophyte.

Capreolia implexa

TYPE LOCALITY: Sandringham, Port Phillip, Victoria.
DISTRIBUTION: Wittelbee Point, South Australia, to Broken Bay, New South Wales and around Tasmania. New Zealand.
FURTHER READING: Guiry & Womersley (1993); Womersley & Guiry (1994); Scott (2017).

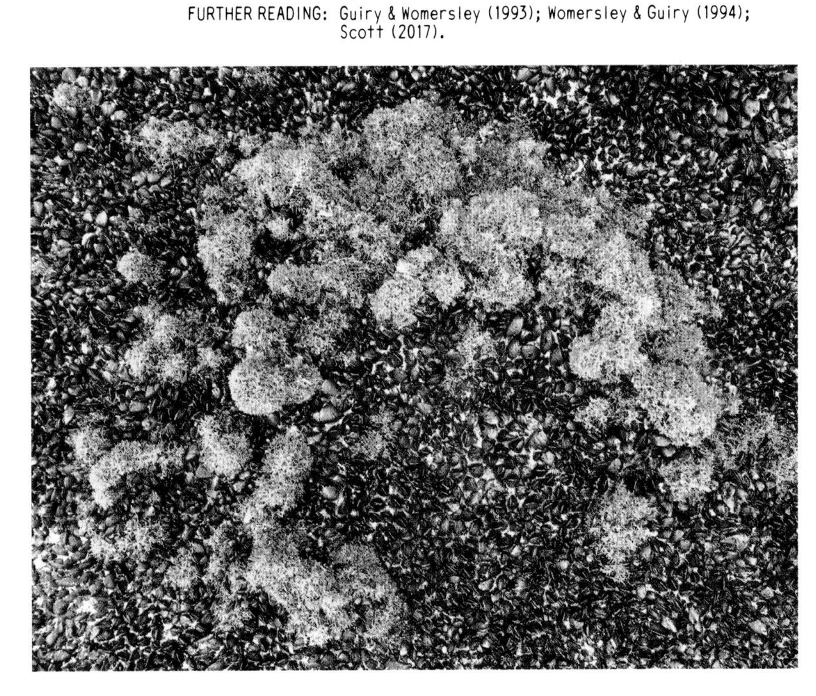

Capreolia implexa
(Point Roadknight, Vic).

Gelidium

Thalli of *Gelidium* are cartilaginous, with flattened or terete axes that are irregularly or pinnately branched. The genus is structurally similar to *Pterocladia* and *Pterocladiella*, and the genera can only be separated by an examination of cystocarpic material (see under *Pterocladia*, p. 42). Structurally the thallus is uniaxial and has a medulla of mixed longitudinal filaments with rhizines and a pseudoparenchymatous cortex. *Gelidium australe* is similar in appearance to *Pterocladiella capillacea* but is often more yellow in colour. *Gelidium crinale* forms wiry tufts and can be common on rock in the shallow subtidal. *Gelidium millarianum* occurs as low turfs on intertidal rock, growing in the same habitat that *Capreolia implexa* occupies on the colder coasts of southern Australia (see p. 35).

Gelidium australe

TYPE LOCALITY: Australia.
DISTRIBUTION: Perth region, Western Australia, around southern mainland Australia to Walkerville, Victoria, and around Tasmania.

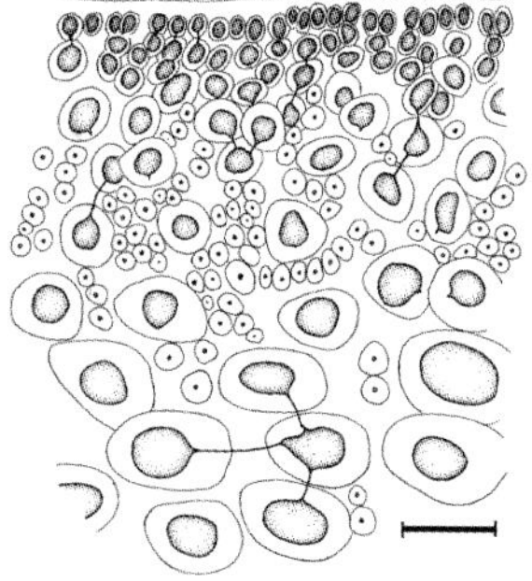

Gelidium australe (Thorny Passage, SA; photo G.J. Edgar).

(above) *Gelidium australe* – partial section of thallus (Yanchep, WA). Scale = 20 µm.

Gelidium crinale

TYPE LOCALITY: England.
DISTRIBUTION: Rottnest Island, Western Australia, around southern Australia to Point Lonsdale, Victoria.

Gelidium millarianum

TYPE LOCALITY: Neds Beach, Lord Howe Island.
DISTRIBUTION: Known from the Perth region, Western Australia, and various locations in New South Wales, including Lord Howe Island.
FURTHER READING: Womersley & Guiry (1994); Millar & Freshwater (2005); Boo et al. (2016a, 2016b); Scott (2017).

Gelidium crinale (Cape Peron, WA). Scale = 1 mm.

Gelidium crinale (Cape Peron, WA).

Gelidium millarianum (Cape Peron, WA).

Ptilophora

The genus *Ptilophora* includes three species in Australia. Plants are robust, with flattened axes that in *Ptilophora prolifera*, the species included here, can be mostly covered with a sponge. The sponge coating occurs mainly in older portions of the plant – in these regions the thallus is irregularly branched and produces numerous small surface proliferations. Younger axes near the apices are sponge-free and alternately pinnately branched. In the two other Australian species, *Ptilophora pectinata* and *Ptilophora wilsonii* (not pictured), the sponge coating is not present. Structurally the thallus is uniaxial. Mature axes have a densely filamentous medulla and a cortex composed of three distinct layers: an inner cortex of large, colourless cells, a mid cortex of filaments with interspersed bundles of rhizines, and an outer cortex of anticlinal filaments.

Ptilophora prolifera

TYPE LOCALITY: Fremantle, Western Australia.
DISTRIBUTION: From Geraldton south to Albany, Western Australia.
FURTHER READING: Womersley & Guiry (1994); Millar & Freshwater (2005).

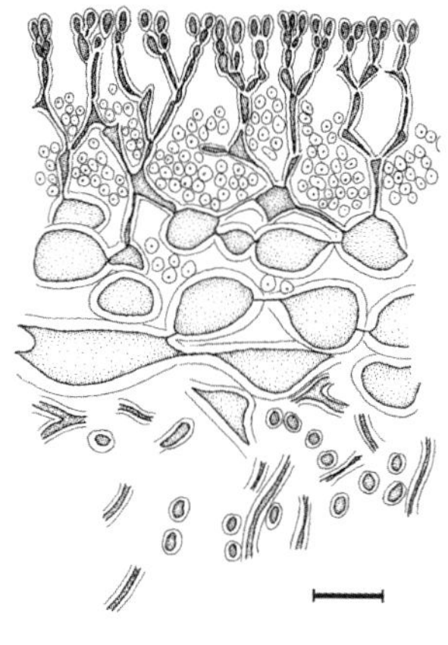

Ptilophora prolifera (Rottnest Island, WA).

(top) *Ptilophora prolifera* – as illustrated by Harvey (1862, pl. 204, as *Gelidium proliferum*).

(above) *Ptilophora prolifera* – partial section of thallus (Jurien Bay, WA). Scale = 30 µm.

Gelidiella

Gelidiella includes cartilaginous, turf-like algae that form a limited prostrate axis and upright axes that are terete or slightly flattened. Structurally the thallus is uniaxial, but the central axis is only visible near the apices. The medulla is pseudoparenchymatous and grades into a smaller celled cortex. Tetrasporangia are cruciately or tetrahedrally divided and arise in nemathecia in upper branches. Cystocarps have been reported for the genus but are not well known. *Gelidiella* is closely related to *Gelidium*, but differs in the absence of thick-walled medullary rhizoids (rhizines). Several species have been recorded from tropical Australia. *Gelidiella acerosa* is a common species that is readily recognised by its pinnate lateral branches.

Gelidiella acerosa

TYPE LOCALITY: Mokha, Yemen.
DISTRIBUTION: Coral Bay, Western Australia, around northern Australia to Lord Howe Island, New South Wales. Widely distributed in tropical seas.
FURTHER READING: Millar & Freshwater (2005); Huisman et al. (2018a).

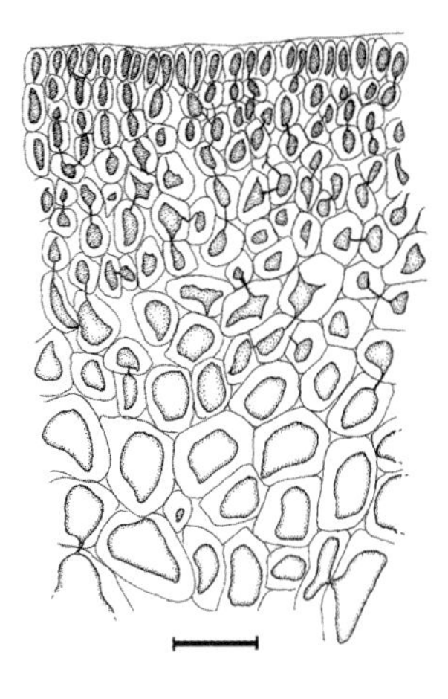

Gelidiella acerosa (Long Reef, WA).

(above) *Gelidiella acerosa* – partial section of thallus (Ningaloo Reef, WA). Scale = 20 µm.

Aphanta

Aphanta includes two species with a turf-like habit and cylindrical stolons bearing tongue-like upright branches. The genus is recognised primarily on the basis of molecular analyses, as reproductive structures are unknown in both of the included species. Only *Aphanta ligulata* occurs in Australia and it is presently known only from a single collection from Coral Bay in Western Australia. Unfortunately, no morphological features are unique to *Aphanta*, which makes separating this species from the similar looking *Pterocladiella caerulescens* (p. 43) very difficult. As a result, *Aphanta ligulata* is probably often overlooked.

Aphanta ligulata

TYPE LOCALITY: Coral Bay.
DISTRIBUTION: Only known from the type locality.
FURTHER READING: Tronchin & Freshwater (2007); Huisman et al. (2018a).

Aphanta ligulata (Coral Bay, WA).

Orthogonacladia

Orthogonacladia includes two species, both restricted to the Indian Ocean: *Orthogonacladia madagascariense* from Madagascar and *Orthogonacladia rectangularis* from Western Australia. Plants have a distinctive branching pattern in which the lateral branches arise at almost right angles to the main axes. The Australian species was previously regarded as a species of *Pterocladia*, but DNA analyses led to its placement in a segregate genus.

Orthogonacladia rectangularis

TYPE LOCALITY: Flinders Bay, Western Australia.
DISTRIBUTION: From south of Perth, Western Australia, around south-western Australia to the Isles of St Francis, South Australia.
FURTHER READING: Womersley & Guiry (1994, as *Pterocladia rectangularis*), Tronchin & Freshwater (2007); Boo et al. (2016a).

Orthogonacladia rectangularis (south of Perth, WA).

Pterocladia

Thalli of *Pterocladia* are cartilaginous, with flattened and mostly pinnately branched axes. Structurally the thallus is uniaxial and mature branches have a medulla of longitudinal filaments and a pseudoparenchymatous cortex. Thick-walled filaments, known as rhizines, are generally present in the medulla. *Pterocladia* is closely related to *Gelidium* and some species can be difficult to identify accurately if cystocarpic material is not available. In *Pterocladia*, the cystocarps are protuberant on only one side of the bearing branch, with a single ostiole (opening). In contrast, cystocarps of *Gelidium* are protuberant on both sides and have two ostioles. Unfortunately, reproductive material is generally rare and most identifications have to be based on vegetative features. Only a single species of *Pterocladia* is known in Australia, *Pterocladia lucida*, which is commonly found in shallow waters on rough coasts. In many countries, *Pterocladia* is harvested commercially as a source of agar.

Pterocladia lucida

TYPE LOCALITY: South coast of Australia.
DISTRIBUTION: Kalbarri, Western Australia, around southern mainland Australia and Tasmania to Coffs Harbour, New South Wales.
FURTHER READING: Womersley & Guiry (1994); Boo et al. (2016a); Scott (2017).

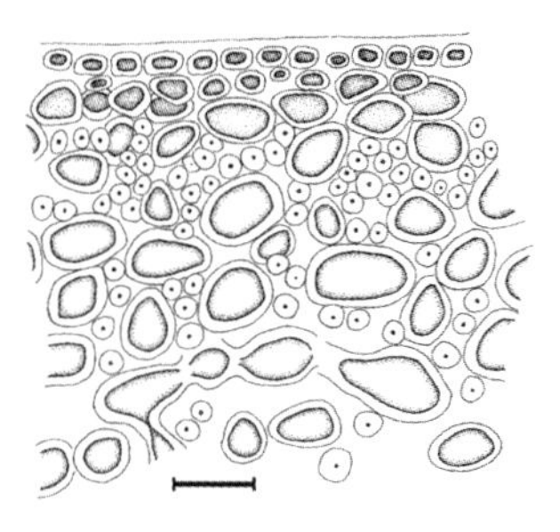

Pterocladia lucida (Cape Peron, WA).

(top) *Pterocladia lucida* – as illustrated by Harvey (1863, pl. 248).

(above) *Pterocladia lucida* – partial section of thallus, with rhizines in the medulla (Rottnest Island, WA). Scale = 20 µm.

Pterocladiella

A genus of three species in Australia, *Pterocladiella* forms dense turfs or carpets on intertidal and shallow subtidal rock. Thalli are cartilaginous, with prostrate axes giving rise to flattened, irregularly branched (*Pterocladiella caerulescens*) or regularly pinnately branched (*Pterocladiella capillacea*) axes. Structurally the thallus is uniaxial and mature branches have a medulla of longitudinal filaments and rhizines, and a pseudoparenchymatous cortex.

Pterocladiella caerulescens

TYPE LOCALITY: Wagap, New Caledonia.
DISTRIBUTION: Widespread in warmer seas.

Pterocladiella capillacea

TYPE LOCALITY: Mediterranean Sea.
DISTRIBUTION: Kalbarri, Western Australia, around southern mainland Australia and Tasmania to Stradbroke Island, Queensland. Widespread in temperate seas and extending into the subtropics.
FURTHER READING: Womersley & Guiry (1994); Santelices & Hommersand (1997); Millar & Freshwater (2005); Boo et al. (2016a); Scott (2017); Huisman et al. (2018a).

Pterocladiella caerulescens (Masthead Island, Qld).

Pterocladiella capillacea (Cape Peron, WA).

(right) *Pterocladiella capillacea* – branch detail (Cape Peron, WA). Scale = 1 mm.

Asparagopsis

A common genus in many parts of the world, particularly in warmer seas, *Asparagopsis* has the appearance of a cluster of dull, pinkish grey foxtails, with each main axis terminating in a feathery tuft. Structurally the plants are uniaxial with a corticating layer that becomes thicker in the main axes. The conspicuous plants are the gametophytes, as the genus has a heteromorphic life history, in which the tetrasporophyte is a small filamentous plant with a totally different form. In the past, the tetrasporophyte was regarded as a separate genus and it was not until the plant was grown in laboratory culture that the two were recognised as phases in the life history of one species. Two species are found in Australia, with *Asparagopsis armata* differing from *Asparagopsis taxiformis* in its less dense tufts and, more importantly, in the production of adventitious branches with reflexed barbels.

Asparagopsis armata

TYPE LOCALITY: Garden Island, Western Australia.

DISTRIBUTION: From the Houtman Abrolhos, Western Australia, around southern mainland Australia and Tasmania to Port Stephens, New South Wales. New Zealand. Mediterranean. Western Europe. Chile.

Asparagopsis taxiformis

TYPE LOCALITY: Alexandria, Egypt.

DISTRIBUTION: From Ellensbrook, Western Australia, around northern Australia to southern Queensland and Lord Howe Island, New South Wales. Also known from the two Gulf regions of South Australia. Cosmopolitan in warmer seas.

FURTHER READING: Zanolla et al. (2014); Andreakis et al. (2016).

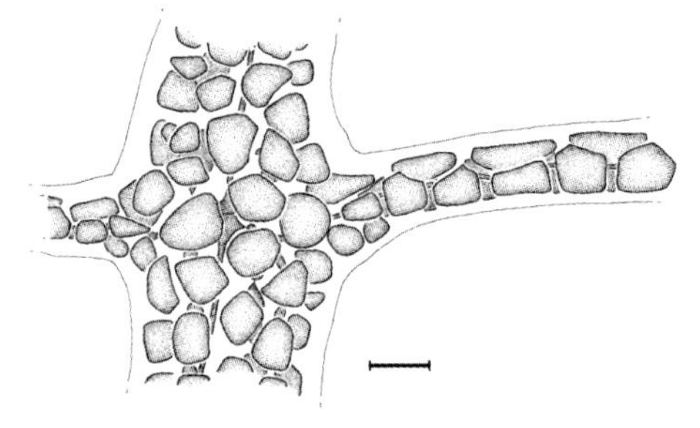

Asparagopsis armata (Cape Peron, WA).

(top) *Asparagopsis armata* – as illustrated by Harvey (1862, pl. 192).

Asparagopsis taxiformis (Cape Peron, WA).

(right) *Asparagopsis taxiformis* – detail of main and lateral axes (Houtman Abrolhos, WA). Scale = 30 µm.

Bonnemaisonia

A genus of about eight species, mostly found in colder seas, with two species on southern Australian coasts. Plants are upright and much-branched with terete to slightly compressed branches bearing branchlets in opposite pairs. Structurally the thallus is uniaxial, with a conspicuous central filament surrounded by a central cavity and then an inner cortex of larger cells bearing an outer layer of small, rounded cells that often form rosettes. In *Bonnemaisonia spinescens*, surface spines are also produced. Species of the genus are known to produce halogenated compounds that inhibit the growth of surface bacteria and reduce fouling. *Bonnemaisonia* is closely related to *Asparagopsis* but can be separated by its distichous rather than radial branching.

Bonnemaisonia spinescens

TYPE LOCALITY: Investigator Strait, South Australia.
DISTRIBUTION: Known from several locations near Adelaide, South Australia, and a single drift plant from Gnarabup, Western Australia.
FURTHER READING: Womersley (1996).

Bonnemaisonia spinescens (Gnarabup, WA).

Delisea

Four species of *Delisea* are known in Australia, mostly confined to the colder southern waters. *Delisea pulchra* grows attached to rock in the subtidal region and is occasionally very common. Thalli are much-branched and have short spines formed in alternating positions on the main axes. Structurally the thallus is uniaxial with the central axis ringed by a layer of rhizoidal cells. The inner cortex is pseudoparenchymatous with large, colourless cells, and grades into a smaller celled outer cortex. *Delisea pulchra* is superficially similar to species of *Phacelocarpus* (see p. 99). Cystocarpic plants are readily distinguished as the cystocarps are borne in a terminal position, as opposed to the lateral cystocarps of *Phacelocarpus*. Positive identification of sterile specimens, however, requires an examination of the internal structure — *Delisea* lacks the inner filamentous medulla that is prominent in *Phacelocarpus*.

Delisea pulchra

TYPE LOCALITY: Australia.
DISTRIBUTION: From Dongara, Western Australia, around southern Australia to Caloundra Queensland. Found in the colder waters of the Southern Ocean, including New Zealand, the subantarctic islands and the Antarctic Peninsula.
FURTHER READING: Womersley (1996); Scott (2017).

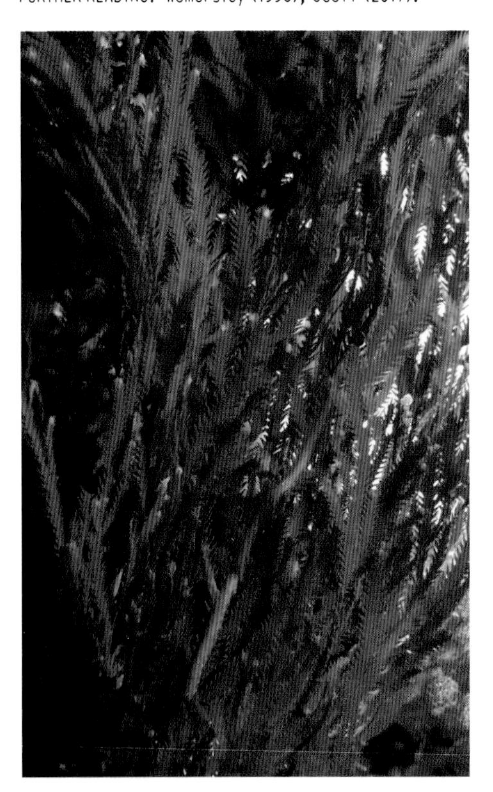

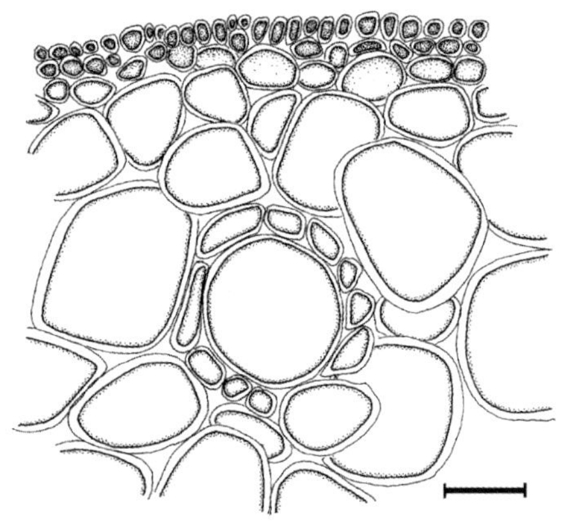

Delisea pulchra (Rottnest Island, WA).

(top) *Delisea pulchra* – as illustrated by Harvey (1858).

(above) *Delisea pulchra* – partial section of thallus showing central axis (Rottnest Island, WA). Scale = 30 µm.

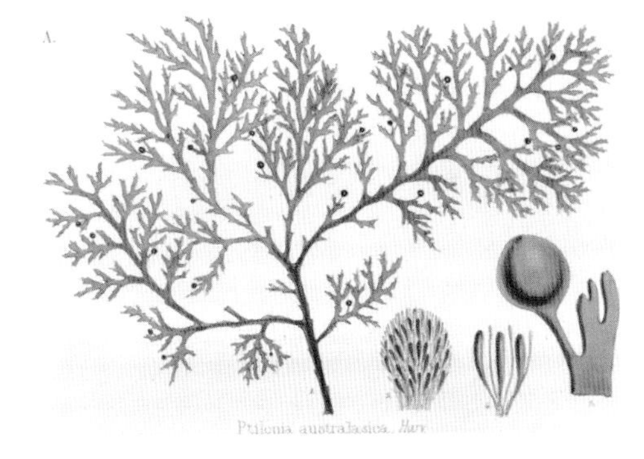

Ptilonia

Ptilonia includes two species in southern Australia, *Ptilonia australasica* and *Ptilonia subulifera* (not illustrated), distinguished by the width of their upper branches (2–4 mm in *P. australasica*, 1.5–1.5 mm in *P. subulifera*). Plants are upright, alternately branched with compressed axes. Structurally the thallus is uniaxial, with a prominent axial filament visible in sections. The outer cortex has conspicuous, usually clear, vesicular cells. *Ptilonia* is similar to *Delisea* (see p. 46), differing in having branches with smooth margins whereas those of *Delisea* have marginal spines.

Ptilonia australasica

TYPE LOCALITY: Georgetown, Tasmania.
DISTRIBUTION: West Island, South Australia, to Walkerville, Victoria, and around Tasmania.
FURTHER READING: Womersley (1996); Scott (2017).

Ptilonia australasica (Mariah Island, Tas; photo G.J. Edgar).

(top) *Ptilonia australasica* – as illustrated by Harvey (1859b, pl. 190A).

Arthrocardia

Arthrocardia is a geniculate coralline genus with two species in Australia, *Arthrocardia wardii* and *Arthrocardia flabellata* ssp. *australica*, both restricted to south-eastern coasts. Thalli have a crustose base bearing several erect, pinnately branched fronds, somewhat similar in appearance to *Corallina* (see p. 49), but the compressed segments tend to be larger and the upper segments less uniformly shaped and often lobed. Structurally the segments contain as many as 40 tiers of medullary cells in *Arthrocardia*, whereas there are 20 or less in *Corallina*. Reproductive conceptacles in *Arthrocardia* are sunken in the apices of segments and there are usually two surmounting branches arising from the fertile segment.

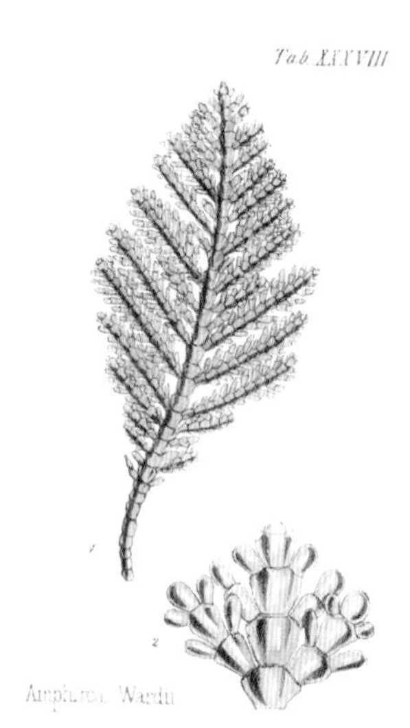

Arthrocardia wardii

TYPE LOCALITY: Port Phillip, Victoria.
DISTRIBUTION: From Pondalowie Bay, South Australia, to Norah Head, New South Wales, and around Tasmania. New Zealand.
FURTHER READING: Womersley & Johansen (1996b); Nelson (2013).

Arthrocardia wardii (Longnose Point, NSW; photo G.J. Edgar).

(top) *Arthrocardia wardii* – as illustrated by Harvey (1847–1849, pl. 38, as *Amphiroa wardii*).

Corallina

Corallina is represented in Australia by the single species *Corallina officinalis*, which is widely distributed in temperate seas. On the south coast of Australia, it can be very common and form turfs near low tide level. Thalli grow to about 6 cm tall with several segmented, pinnately branched fronds arising from a crustose base. Segments are terete in lower portions but become compressed above and broader near the branch apices. Reproductive structures are borne in conceptacles that arise at the apices of lateral branches. *Corallina* is similar in appearance to *Arthrocardia* and the differences are described under that genus (see p. 48).

Corallina officinalis

TYPE LOCALITY: Europe.
DISTRIBUTION: Widespread in temperate seas.
FURTHER READING: Womersley & Johansen (1996b); Nelson (2013).

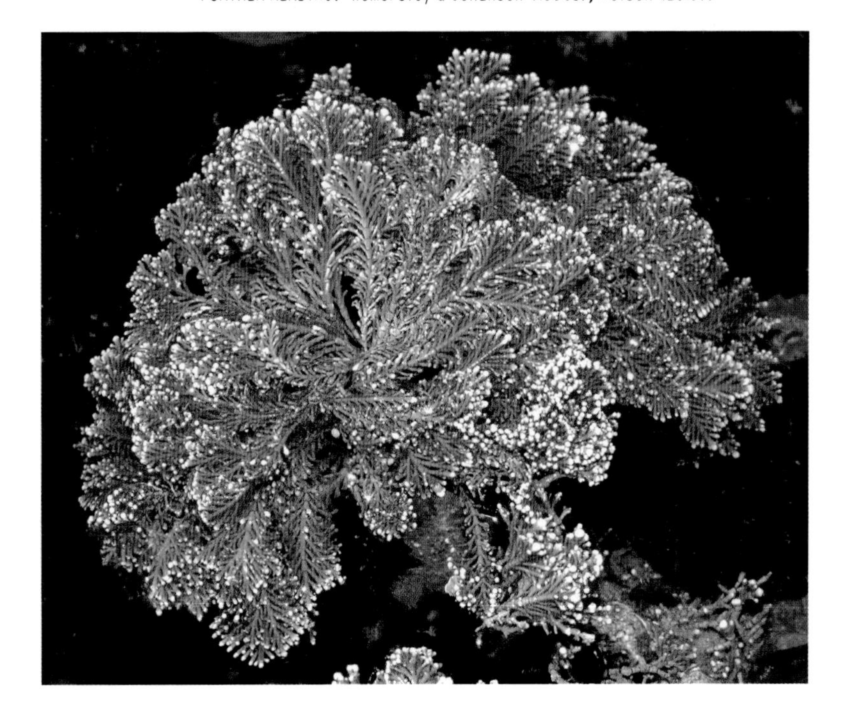

Corallina officinalis (Kingston, SA; photo G.J. Edgar).

Jania

Species of *Jania* are calcified and articulated, generally with terete segments, and distinguished from closely related genera by their dichotomous branching pattern. Structurally the axes have a medulla with cells arranged in tiers: in the calcified segments (intergenicula) a thin cortex is produced and it partially covers the uncalcified joints (genicula). The genus includes several common species, such as *Jania pedunculata* and *Jania micrarthrodia*, which are regularly found as small tufts on seagrasses and larger algae. *Jania pulchella* is unusual in the genus in having lower segments that are flattened and somewhat triangular, while the upper segments are terete and more typical of the genus. *Jania rosea* was previously included in the separate genus *Haliptilon*, but DNA sequence analyses led to it being moved to *Jania*. The species can be an extremely common plant, found in a wide range of habitats. Thalli are regularly epiphytic on stems of the seagrass *Amphibolis* and a variety of larger algae, as well as epilithic on rock. The thallus is calcified and articulated, with the main axes composed of terete (in upper parts somewhat flattened) segments that bear short distichous branches and, in most cases, adventitious branches. These short branches are usually thinner than the main axes and are composed of terete segments. *Jania rosea* is a variable species that displays a wide range of forms. Most are recognisable by the presence of adventitious branches, but where these are absent, the thallus resembles that of *Corallina* (see photo of thallus detail of *Jania rosea*, p. 51). *Jania rosea* can be distinguished by the generally shorter segments in the axes (< 1.0 mm) and smaller number of medullary tiers per segment (three to five as opposed to more than 10). *Jania sagittata* has unusual arrow shaped segments and was previously included in the genus *Cheilosporum*.

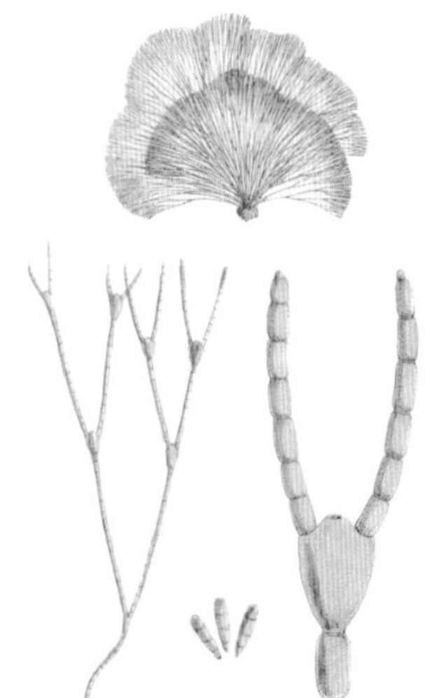

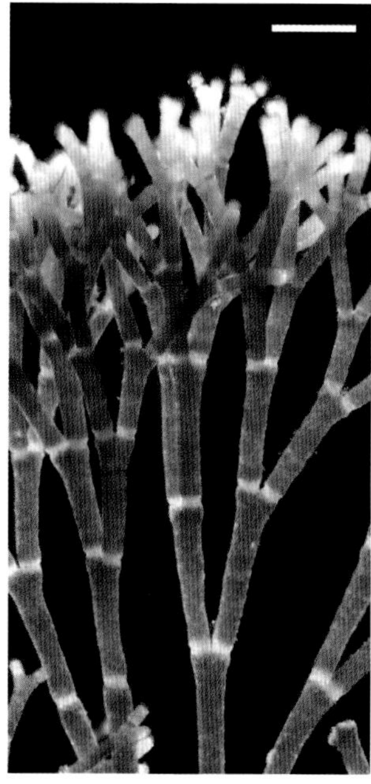

Jania pedunculata

TYPE LOCALITY: Australia.
DISTRIBUTION: Widespread in tropical and subtropical seas.

(top) *Jania micrarthrodia* – as illustrated by Harvey (1863, pl. 251, as *Jania fastigiata*).

(above) *Jania pedunculata* – thallus detail (Cape Peron, WA). Scale = 500 µm.

Jania pedunculata (Long Reef, WA).

Jania pulchella

TYPE LOCALITY: Rottnest Island, Western Australia.
DISTRIBUTION: From the Houtman Abrolhos, Western Australia, around southern Australia to Portland, Victoria.

Jania micrarthrodia

TYPE LOCALITY: Australia.
DISTRIBUTION: From Geraldton, Western Australia, around southern Australia to Walkerville, Victoria, and around Tasmania. New Zealand.

Jania rosea

TYPE LOCALITY: Australia.
DISTRIBUTION: Shark Bay, Western Australia, around southern Australia to Port Denison, Queensland. New Zealand.

Jania sagittata

TYPE LOCALITY: Mauritius.
DISTRIBUTION: Widespread in the southern hemisphere.
FURTHER READING: Johansen & Womersley (1994); Womersley & Johansen (1996b); Nelson (2013).

Jania pulchella (Rottnest Island, WA).

Jania micrarthrodia (Cape Peron, WA).

Jania rosea (Cape Peron, WA).

(inset) *Jania rosea* – detail of distichous thallus (Cape Peron, WA). Scale = 1 mm.

Jania sagittata (Memory Cove, SA; photo G.J. Edgar).

Hydrolithon

Species of *Hydrolithon* form crusts on a wide variety of substrata, from seagrasses to other algae to rocks. *Hydrolithon farinosum* is an extremely common epiphyte, and is often responsible for the pink discolouration on seagrass leaves and other algae. In the photograph here, it is seen on the surface of a *Sargassum*. *Hydrolithon boergesenii* often forms rhodoliths and typically has a rough surface and slight bluish tinge.

Hydrolithon farinosum

TYPE LOCALITY: Mediterranean Sea.
DISTRIBUTION: Virtually cosmopolitan.

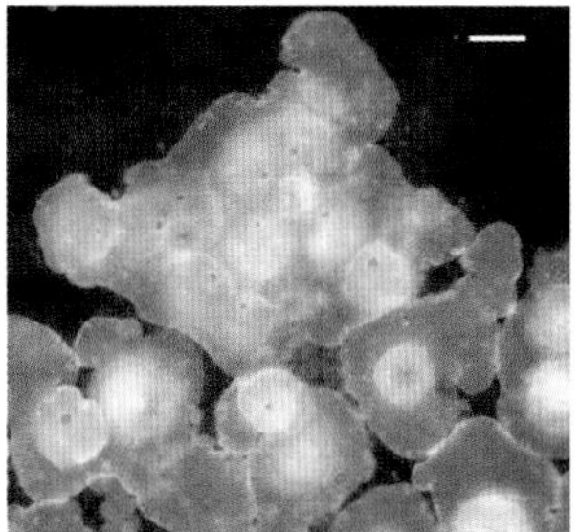

Hydrolithon boergesenii

TYPE LOCALITY: St Croix, U.S. Virgin Islands.
DISTRIBUTION: From the Houtman Abrolhos, Western Australia, around northern Australia to Heron Island, Queensland. Widespread in warmer seas.
FURTHER READING: Adey, Townsend & Boykins (1982); Penrose (1996a); Townsend & Huisman (2018).

Hydrolithon farinosum – growing on *Valonia ventricosa* (Scott Reef, WA).

(above) *Hydrolithon farinosum* – growing on *Sargassum* (Cape Peron, WA). Scale = 100 µm.

Hydrolithon boergesenii (Northwest Island, Qld).

Amphiroa

Amphiroa is a genus of articulated, calcified red algae that bears reproductive structures in conceptacles arising in random locations on the surface of the intergenicula (the calcified region) – not in specific regions as is found in many closely related genera. Most species of *Amphiroa* have flattened branches, and separation of the species is based on their habit and internal structure. *Amphiroa gracilis* and *Amphiroa fragilissima* are unusual in having terete branches. Other distinguishing features are given under the species headings.

Amphiroa crassa

Unique among the Australian species in that genicula never (or rarely) occur at branch points and mature genicula have a calcified collar.

TYPE LOCALITY: Presumed to be Shark Bay, Western Australia.
DISTRIBUTION: In Australia known from Lady Musgrove Island north to Hicks Reef, Queensland, and King and Conway Islands, Western Australia. Also recorded from South America, South-east Asia, some Pacific islands and the Antarctic, but these require verification.

Amphiroa anceps

A common species, *Amphiroa anceps* has dichotomously divided, mostly flattened branches that are generally of equal width throughout the plant. Individual plants vary in branch width and some have terete lower segments.

TYPE LOCALITY: Australia.
DISTRIBUTION: Australia-wide. Widespread in the Indo-West Pacific.

(top) *Amphiroa gracilis* – as illustrated by Harvey (1862, pl. 231).
Amphiroa anceps (Cape Peron, WA).
Amphiroa crassa (King and Conway Islands, WA).
Following page:
Amphiroa foliacea (Cassini Island, WA).
Amphiroa fragillisima (Heron Island, Qld).
Amphiroa gracilis (Cape Peron, WA).
(inset) *Amphiroa gracilis* – detail of thallus (Cape Peron, WA). Scale = 1 mm.
Amphiroa tribulus (Ningaloo, WA).

Amphiroa foliacea

The intergenicula of *Amphiroa foliacea* are highly variable; compressed or flattened intergenicula are always present, but terete intergenicula can also occur.

TYPE LOCALITY: Mariana Islands (Micronesia).
DISTRIBUTION: Widespread in tropical seas.

Amphiroa fragilissima

Amphiroa fragilissima has terete intergenicula and often has expanded collars adjacent to the genicula.

TYPE LOCALITY: West Indies.
DISTRIBUTION: Known from Reveley Island south to Coral Bay, Western Australia, the Gulf of Carpentaria west to Groote Eylandt, Northern Territory, North Reef north to Thursday Island, Queensland, and Lord Howe Island, New South Wales. Widespread in tropical seas.

Amphiroa gracilis

This is a distinctive Western Australian species with terete branches and often with multiple branches at each node.

TYPE LOCALITY: Rottnest Island, Western Australia.
DISTRIBUTION: Shark Bay south to Cape Peron, including the Houtman Abrolhos, Western Australia, and eastwards to Troubridge Hill, Yorke Peninsula, South Australia.

Amphiroa tribulus

Branching in *Amphiroa tribulus* is multiplanar, dichotomous and polychotomous, and the intergenicula vary from terete to compressed to flattened throughout the thallus.

TYPE LOCALITY: West Indies.
DISTRIBUTION: Maret Islands south to the Rowley Shoals, Western Australia, and Hayman Island, Queensland. Elsewhere widespread in tropical seas.
FURTHER READING: Millar (1990); Womersley & Johansen (1996a); Woelkerling et al. (2012); Harvey et al. (2013); Harvey et al. (2018).

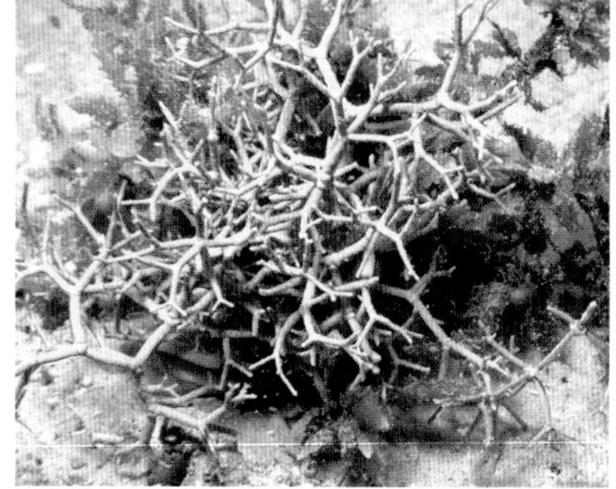

Lithophyllum

Lithophyllum displays a variety of forms, from thin crusts (e.g. *Lithophyllum insipidum*) to much-branched, stony thalli with numerous anastomoses between branches. *Lithophyllum pygmaeum* (not pictured) and *Lithophyllum longense* typically have pointed apices, while those of *Lithophyllum kotschyanum* (not pictured) are blunt. As with all crustose coralline red algae, identification requires an internal examination. *Lithophyllum* is characterised by the presence of secondary pit connections between vegetative cells.

Lithophyllum stictiforme

TYPE LOCALITY: Mediterranean Sea.
DISTRIBUTION: From Eagle Bluff, Shark Bay, Western Australia, around southern Australia and the northern and eastern coasts of Tasmania to Newport, New South Wales.

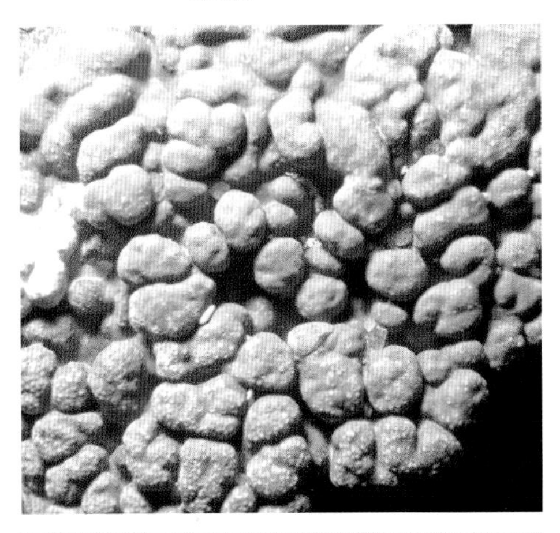

Lithophyllum insipidum

TYPE LOCALITY: Waikiki, O'ahu, Hawaii.
DISTRIBUTION: Known from Coral Bay, Western Australia. Hawaii.

Lithophyllum longense

TYPE LOCALITY: Long Reef, Western Australia.
DISTRIBUTION: In Australia from the Dampier Archipelago and Long Reef, Western Australia.
FURTHER READING: Adey, Townsend & Boykins (1982); Woelkerling (1996b); Townsend & Huisman (2018).

Lithophyllum stictiforme (Rottnest Island, WA).

Lithophyllum insipidum (Coral Bay, WA).

Lithophyllum longense (Long Reef, WA).

Mastophora

Mastophora is yet another genus of coralline algae that displays a variety of forms, from thin crusts on rock to upright branched thalli. *Mastophora rosea*, the species included here, can occur as both forms, but the most obvious is the bluish pink, leafy, dichotomously branched thallus that can be common in the intertidal on tropical reef platforms. Reproductive structures occur in protuberant conceptacles with a single opening.

Mastophora rosea

TYPE LOCALITY: Guam, Mariana Islands.
DISTRIBUTION: Largely restricted to the tropical and temperate western Pacific Ocean. Known from the Dampier Peninsula to the northern Kimberley coast, Western Australia.
FURTHER READING: Keats et al. (2009); Townsend & Huisman (2018).

Mastophora rosea (Cassini Island, WA).

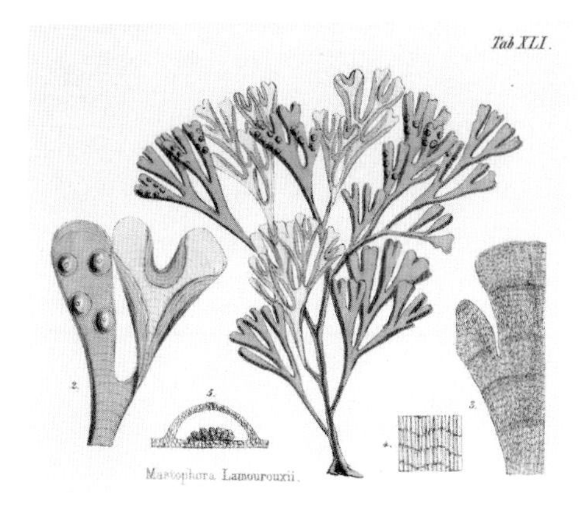

Metamastophora

Metamastophora includes only a single species, *Metamastophora flabellata*, an occasionally common species found growing on rock in subtidal and intertidal habitats. Thalli have erect stipes with flattened, ribbon-like upper branches. The branches have distinct dorsal and ventral surfaces that differ in colour. Structurally the thallus is pseudoparenchymatous. Reproductive structures are borne in raised conceptacles that occur only on the dorsal surface.

Metamastophora flabellata

TYPE LOCALITY: Western Australia.
DISTRIBUTION: Kalbarri, Western Australia, around southern Australia to Waterloo Bay, Wilsons Promontory, Victoria, and Currie, King Island, Tasmania. Southern Africa.
FURTHER READING: Woelkerling (1980); Woelkerling (1996c).

Metamastophora flabellata (Cape Peron, WA).

(top) *Metamastophora flabellata* – as illustrated by Harvey (1849, pl. 41, as *Mastophora lamourouxii*).

(above) *Metamastophora flabellata* – detail showing conceptacles (Rottnest Island, WA).

Metagoniolithon

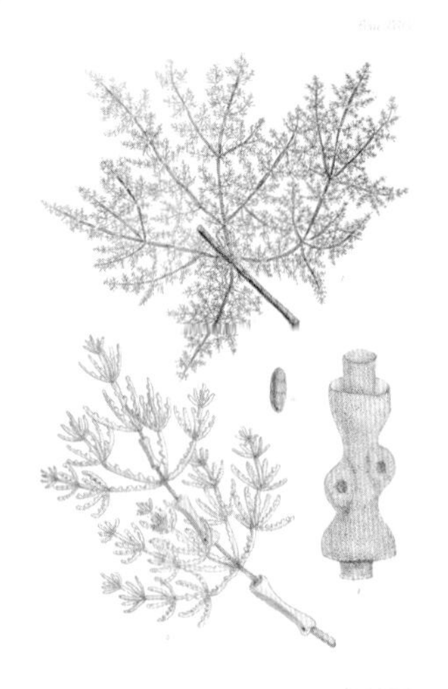

Metagoniolithon is a common genus, endemic to southern and south-western Australia. Thalli are calcified and articulated, with terete axes that are initially dichotomously branched but have secondary production of whorled branches at the nodes. An unusual feature of *Metagoniolithon* is the presence of small mucilaginous caps that form above the apical cells. The combination of these vegetative characteristics allows ready recognition of the genus. Only *Amphiroa gracilis*, which is also calcified and has whorled branches (see p. 54), could possibly be confused with *Metagoniolithon*, but it lacks the mucilaginous caps. Three species of *Metagoniolithon* are known and can be separated based on habitat and features of the branching pattern. *Metagoniolithon radiatum* is a coarse species that grows exclusively on rock. *Metagoniolithon stelliferum* is a common epiphyte on species of the seagrass *Amphibolis* and rarely on other seagrasses and algae. The third species, *Metagoniolithon chara*, also grows epiphytically on seagrasses. It can be distinguished by the production of whorls of uniformly equal length as opposed to their unequal development in *Metagoniolithon stelliferum*.

Metagoniolithon chara

TYPE LOCALITY: Australia.
DISTRIBUTION: From Port Denison, Western Australia, around southern mainland Australia to Waratah Bay, Victoria, and King Island, Bass Strait.

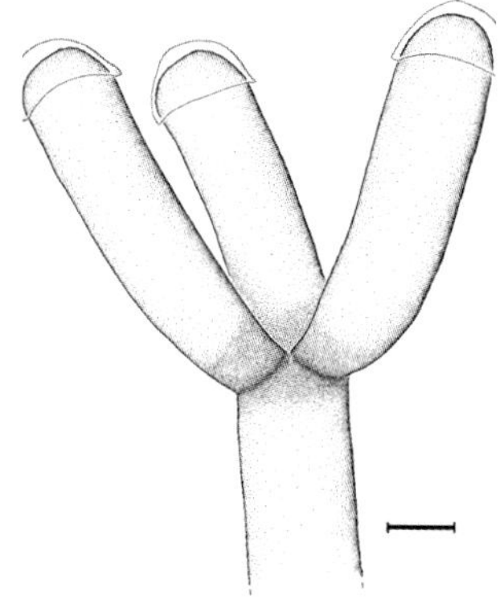

Metagoniolithon chara (Cape Peron, WA).

(above) *Metagoniolithon chara* – apex of thallus showing mucilaginous caps (Jurien Bay, WA). Scale = 200 µm.

(top) *Metagoniolithon stelliferum* – as illustrated by Harvey (1862, pl. 230, as *Amphiroa stelligera*).

Metagoniolithon radiatum

TYPE LOCALITY: Australia.
DISTRIBUTION: From Port Denison, Western Australia, around southern mainland Australia to Cape Patterson, Victoria, and northern Tasmania.

Metagoniolithon stelliferum

TYPE LOCALITY: Australia.
DISTRIBUTION: From Shark Bay, Western Australia, around southern mainland Australia to Refuge Cove, Victoria, and northern Tasmania.
FURTHER READING: Ducker (1979); Womersley & Johansen (1996c).

Metagoniolithon radiatum (Rottnest Island, WA).

Metagoniolithon stelliferum (Cape Peron, WA).

Porolithon

Porolithon is a genus of non-geniculate coralline algae that can form extensive horizontal crusts or, in some species, lumpy or branched thalli. The genus is characterised by its construction in which some inner filaments run parallel to the surface forming a distinct medulla that gives rise to curved filaments that develop into the cortex. Cortical filaments are terminated externally by one to three epithallial cells. Several species occur in Australia but distinguishing between them can be difficult and is best undertaken by molecular analyses.

Porolithon imitatum

TYPE LOCALITY: East Imperieuse Reef, Rowley Shoals, Western Australia.
DISTRIBUTION: From the Dampier Archipelago north to Long Reef, Western Australia.

Porolithon improcerum

TYPE LOCALITY: Montego Bay, Jamaica.
DISTRIBUTION: Known from scattered locations in southern Australia. Bahamas, Caribbean, Jamaica.
FURTHER READING: Penrose (1996a, as *Hydrolithon*); Townsend & Huisman (2018).

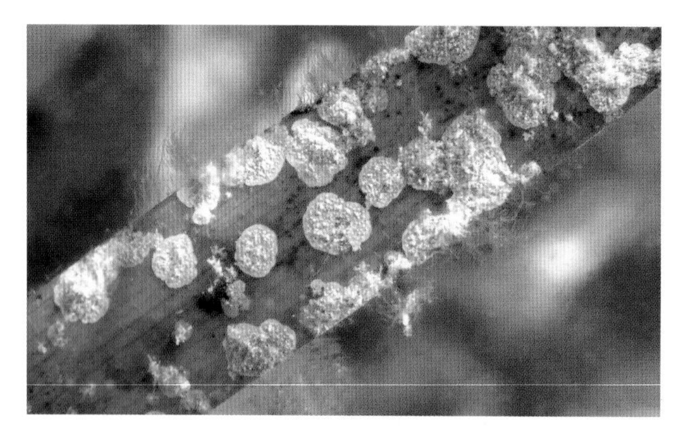

Porolithon imitatum (Imperieuse Reef, Rowley Shoals, WA).

Porolithon improcerum – growing on the seagrass *Posidonia* (Cape Peron, WA).

Neogoniolithon

Like many of the crustose coralline algae, species of *Neogoniolithon* can be encrusting to warty or much branched, with the different growth forms often present in a single species. Distinctive features of the genus include its non-geniculate habit, its monomerous construction, and the cortical filaments terminated by a single rounded surface cell. Some species of *Neogoniolithon* can also occur as rhodoliths, which are unattached, often nodular thalli that can form from broken branches or when thalli grow on all surfaces of small rocks. Several species occur in Australia, the two included here showing the crustose (*Neogoniolithon brassica-florida*) and branched (*Neogoniolithon mamillare*) growth forms.

Neogoniolithon brassica-florida

TYPE LOCALITY: Algoa Bay, Cape Province, South Africa.
DISTRIBUTION: Widespread in temperate and tropical sea.

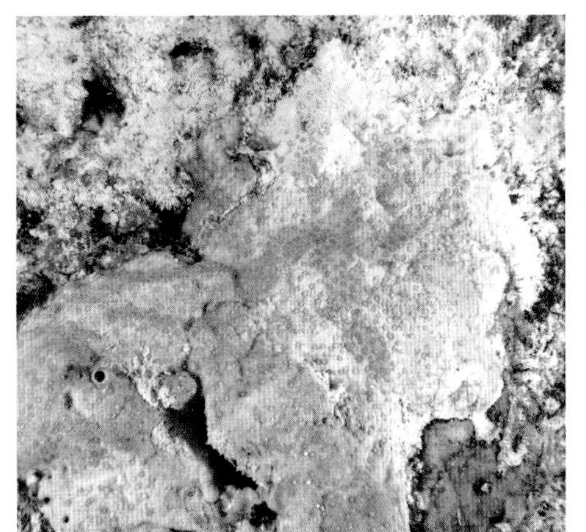

Neogoniolithon mamillare

TYPE LOCALITY: Bahia, Brazil.
DISTRIBUTION: Widespread in tropical seas.
FURTHER READING: Townsend & Huisman (2018).

Neogoniolithon brassica-florida (Ashmore Reef).

Neogoniolithon mamillare (Ashmore Reef).

Spongites

Spongites is a widespread genus, with four species found in southern Australia. Of these, *Spongites hyperellus* is a very distinctive species that forms flat or hemispherical crusts with a dense cover of protuberant, anastomosing branches. The species is largely restricted to Tasmania, where it can be the dominant alga in the lower intertidal zone on coasts subject to strong wave action, attached to rock or invertebrates. It can also occur as unattached rhodoliths.

Spongites hyperellus

TYPE LOCALITY: Western Port, Victoria.
DISTRIBUTION: Known from Western Port, Victoria, and around Tasmania.
FURTHER READING: Penrose (1996b); Scott (2017).

Spongites hyperellus (Safety Cove, Tas; photo M.D. Guiry).

Lithothamnion

Lithothamnion includes numerous species and is represented here by *Lithothamnion proliferum*, a distinctive tropical species. In shallow waters, this species inhabits darker crevices in reefs, and it can usually be recognised by its broad, horizontal protuberances on a smooth surface. Reproductive conceptacles are elliptical in shape with obvious pores.

Lithothamnion proliferum

TYPE LOCALITY: Lumu-Lumu shoal [Pulau Lumulumu], Borneo Bank, Indonesia.
DISTRIBUTION: Known from coral atolls of northern Western Australia and the Great Barrier Reef, Queensland. Widespread in the tropical Indo-Pacific.
FURTHER READING: Keats et al. (1996); Townsend & Huisman (2018).

Lithothamnion proliferum (Rowley Shoals, WA).

Mesophyllum

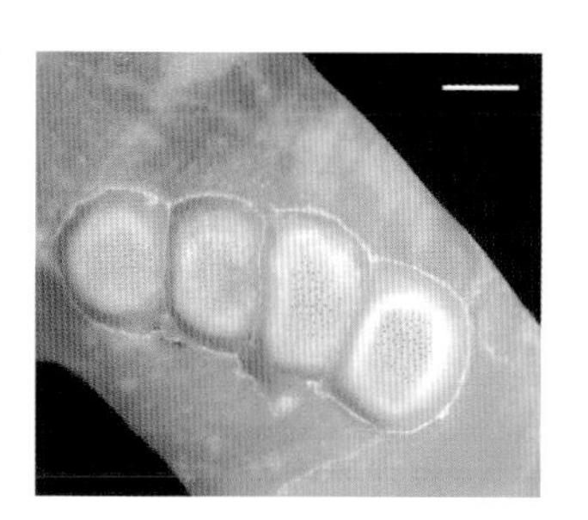

The genus *Mesophyllum*, as is typical of the crustose corallines, displays a broad diversity of forms, from thin crusts to warty to foliose. *Mesophyllum incisum* can occur as most of these, with the discoid form depicted here perhaps one of the more readily recognisable. Plants such as these can be found growing epiphytically on a variety of macroalgae. *Mesophyllum* has uniporate gametangial conceptacles and multiporate tetrasporangial conceptacles.

Mesophyllum engelhartii

TYPE LOCALITY: Cape Jaffa, South Australia.
DISTRIBUTION: From Shark Bay, Western Australia, to Kitty Miller Bay, Phillip Island, Victoria, and the eastern and southern coasts of Tasmania. New Zealand. South Africa. Namibia.

Mesophyllum incisum

TYPE LOCALITY: Bay of Islands, New Zealand.
DISTRIBUTION: From Rottnest Island, Western Australia, around southern mainland Australia and Tasmania to Wilsons Promontory, Victoria. New Zealand.
FURTHER READING: Woelkerling (1996a).

Mesophyllum engelhartii – growing on *Carpopeltis elata* (Cape Peron, WA). Scale = 1 mm.

(top) *Mesophyllum engelhartii* – detail of multiporate conceptacles (Cape Peron, WA). Scale = 250 µm.

Mesophyllum incisum – growing on *Metamastophora flabellata* (Rottnest Island, WA).

Phymatolithon

Phymatolithon is a crustose coralline alga represented in Australia by two species, of which the distinctive southern Australian species *Phymatolithon masonianum* is included here. Plants are generally readily recognised by their loosely attached, pink to gray-pink, plate-like to foliose thalli that can be up to 15 cm broad. The upper surface of the thallus is smooth and often shiny, while the lower surface has numerous perpendicular struts.

Phymatolithon masonianum

TYPE LOCALITY: Ninepin Point, D'Entrecasteaux Channel, Tasmania.
DISTRIBUTION: Cape Buffon, South Australia, to Walkerville, Victoria, and around Tasmania.
FURTHER READING: Wilks & Woelkerling (1994); Woelkerling (1996a); Scott (2017).

Phymatolithon masonianum (Gordon, Tas; photo G.J. Edgar).

Rhizolamellia

Rhizolamellia, including the single species *Rhizolamellia colli*, is a rare genus described originally from specimens collected during a Russian expedition to the Indian Ocean in the 1970s, but not rediscovered until the early 2000s. Plants grow among coral, often in darker crevices in the reef on vertical walls, forming plate-like thalli that often overlap one another and encrust other calcareous objects such as slender corals. The thalli are extremely fragile and easily broken. A distinctive feature of the genus is the production of reproductive conceptacles with an extended tubular pore, as seen in the photo here.

Rhizolamellia colli

TYPE LOCALITY: Fantome Bank, Timor Sea.
DISTRIBUTION: From Ningaloo Reef north to Scott Reef and the Timor Sea.
FURTHER READING: Townsend et al. (2018).

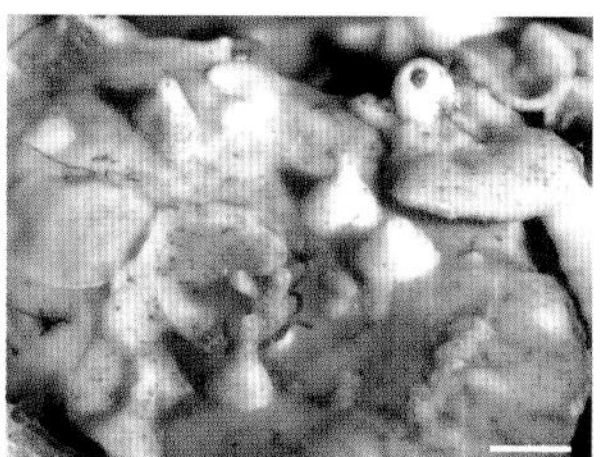

Rhizolamellia colli (Scott Reef, WA).

(above) *Rhizolamellia colli* – surface of thallus showing conceptacles with extended pores (Scott Reef, WA). Scale = 1 mm.

Acrosymphyton

Two species of *Acrosymphyton* are known from Australia: the widespread *Acrosymphyton taylorii* and *Acrosymphyton tenax* from northern New South Wales (not included here). Thalli of *Acrosymphyton taylorii* are very delicate and mucilaginous. They are profusely branched, with the axes often having a banded appearance due to the growth of whorled cortical filaments arising from the central filament. Internally the thallus is rather loosely constructed with the central axial filament bearing initially four lateral filamentous branches. Additional branches can arise at a later stage as the axis matures. The lower cells of carpogonial branches bear clusters of distinctive nutritive cells. *Acrosymphyton* has a lifecycle that includes a microscopic stage, which presumably over-winters until the onset of spring. Thalli can appear in large numbers during early summer and then disappear after a few months.

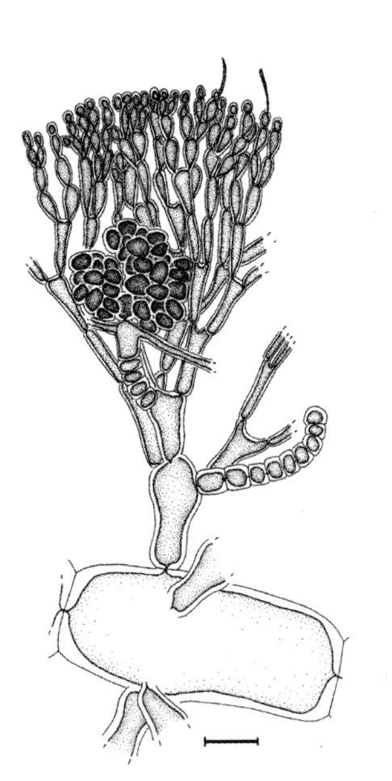

Acrosymphyton taylorii

TYPE LOCALITY: Hauula, Oahu, Hawaiian Islands.
DISTRIBUTION: From Rottnest Island, Western Australia, around northern Australia to Lord Howe Island and Coffs Harbour, New South Wales, with one record from South Australia. Hawaiian Islands.
FURTHER READING: Millar & Kraft (1984); Huisman (2018).

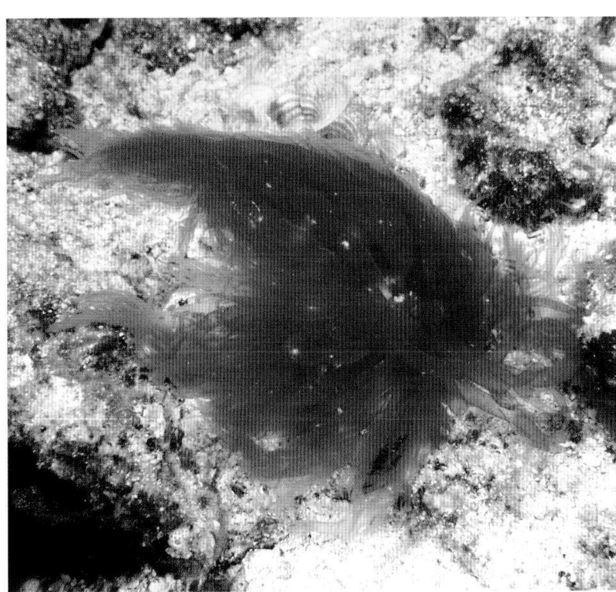

Acrosymphyton taylorii (Ashmore Reef).

(top) *Acrosymphyton taylorii* – internal structure, showing an axial filament, one cortical fascicle and a developing carposporophyte (Houtman Abrolhos, WA). Scale = 30 µm.

(above) *Acrosymphyton taylorii* (Heron Island, Qld).

Amphiplexia

Amphiplexia has a tubular thallus that is irregularly divided and has constrictions at the base of each branch. Structurally the branches are multiaxial, with the interior of the branches mostly hollow and mucilage-filled with the exception of a few scattered medullary filaments. Axial filaments are restricted to the periphery of the central cavity and subtend the two- to three-celled-thick cortical layer. The smaller, outer cortical cells form a distinctive rosette pattern when viewed from the surface of the plant. Cystocarps are protuberant and tetrasporangia are zonately divided and scattered in the cortex. The genus includes two species, of which the relatively rare *Amphiplexia hymenocladioides* is included here.

Amphiplexia hymenocladioides

TYPE LOCALITY: Port Phillip Heads, Victoria.
DISTRIBUTION: From the Houtman Abrolhos, Western Australia, around southern Australia to Port Phillip Heads, Victoria.
FURTHER READING: Kraft (1977a); Kraft & Womersley (1994b).

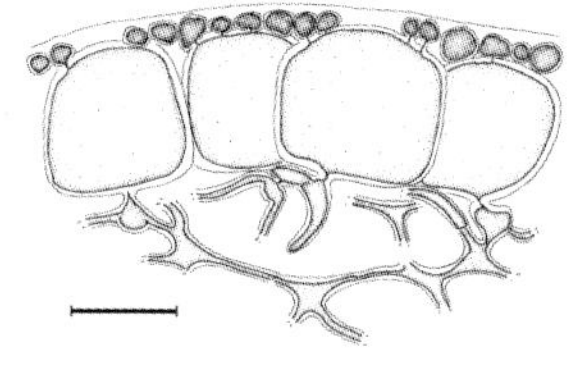

Amphiplexia hymenocladioides (Rottnest Island, WA).

(above) *Amphiplexia hymenocladioides* – section of cortex and medullary filaments (Rottnest Island, WA). Scale = 50 µm.

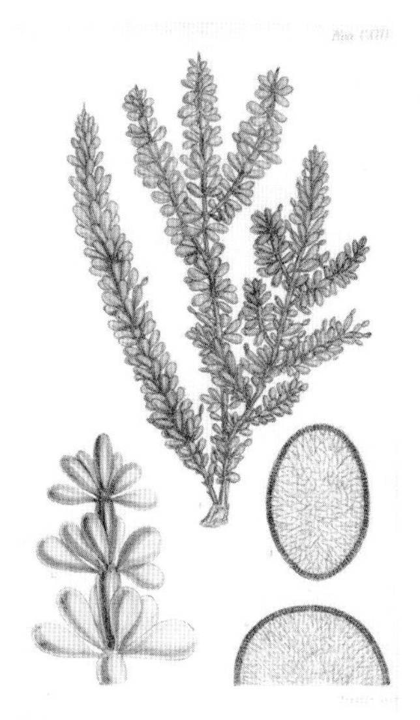

Claviclonium

Thalli of *Claviclonium ovatum*, the only species thus far attributed to the genus, have terete, branched axes that are covered with short, segmented lateral branches. Segments are ovoid or clavate and can be single or in chains of two to three. Structurally the thallus is multiaxial, with a filamentous medulla and a pseudoparenchymatous cortex around two to four cells thick. Cystocarps are embedded in the lateral segments. Tetrasporangia are zonately divided and scattered in the cortex. *Claviclonium ovatum* superficially resembles species of *Botryocladia*, but in the latter the lateral vesicles are never in chains and also have a central cavity rather than a filamentous medulla.

Claviclonium ovatum

TYPE LOCALITY: Australia.
DISTRIBUTION: From the Houtman Abrolhos, Western Australia, around south-western Australia to Eucla, South Australia.
FURTHER READING: Kraft & Min-Thein (1983); Kraft & Womersley (1994b).

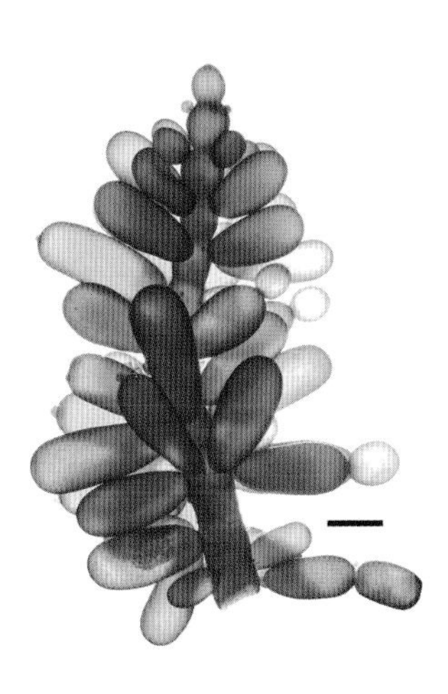

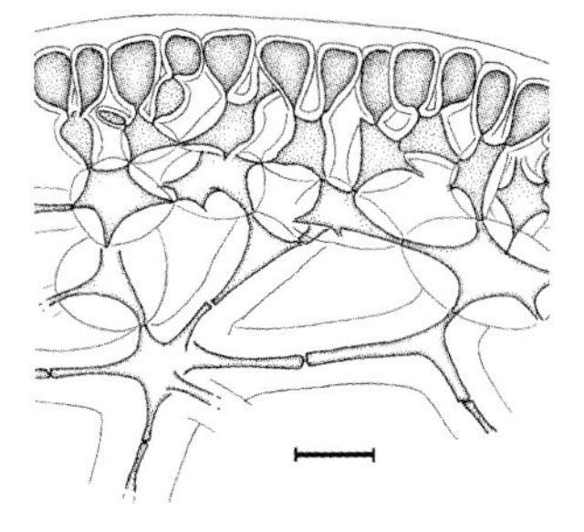

Claviconium ovatum (Carnac Island, WA).

(top) *Claviclonium ovatum* – as illustrated by Harvey (1860, pl. 129, as *Rhabdonia globifera*).

(above) *Claviconium ovatum* – detail of thallus (Cape Peron, WA). Scale = 2 mm.

(right) *Claviconium ovatum* – partial section of lateral branch (Jurien Bay, WA). Scale = 30 µm.

Hennedya

Hennedya includes a single species, *Hennedya crispa*. Thalli are cartilaginous, with a terete stalk grading into flattened branches that are subdichotomously or irregularly branched. Most axes have a conspicuous notch at the apex, a feature that can assist in recognising *Hennedya crispa* in the field. Structurally the thallus is multiaxial and mature branches have a filamentous medulla and pseudoparenchymatous cortex, the latter with a single, inner layer of markedly inflated cells. Cystocarps arise near the apices of fertile thalli and are slightly protuberant. Tetrasporangia arise in a similar position on sporophytes and are zonately divided. In habit, *Hennedya crispa* superficially resembles *Tylotus obtusatus* but the latter has a totally pseudoparenchymatous construction.

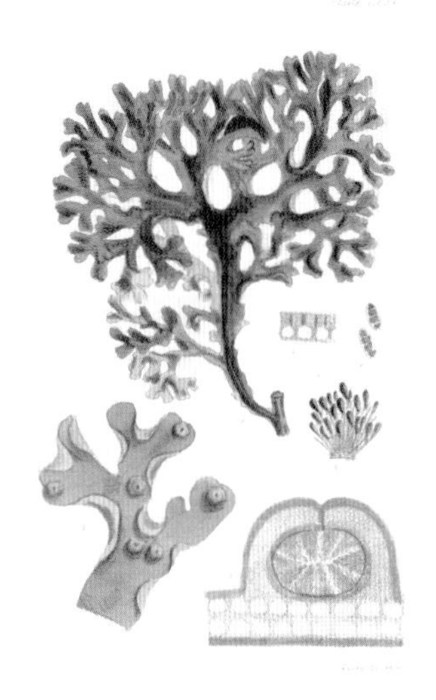

Hennedya crispa

TYPE LOCALITY: Rottnest Island, Western Australia.
DISTRIBUTION: From the Houtman Abrolhos, Western Australia, around southern Australia to Pearson Island, South Australia.
FURTHER READING: Kraft (1977a); Kraft & Womersley (1994b).

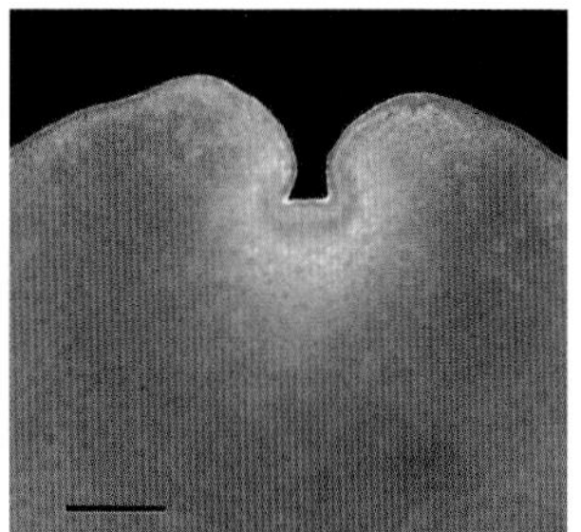

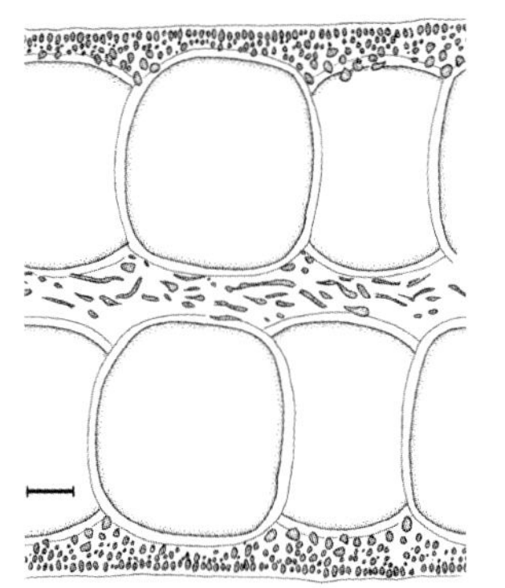

Hennedya crispa (Cape Peron, WA).

(top) *Hennedya crispa* – as illustrated by Harvey (1859a, pl. 75).

(middle) *Hennedya crispa* – branch apex (Cape Peron, WA). Scale = 500 µm.

(bottom) *Hennedya crispa* – section of thallus (Rottnest Island, WA). Scale = 30 µm.

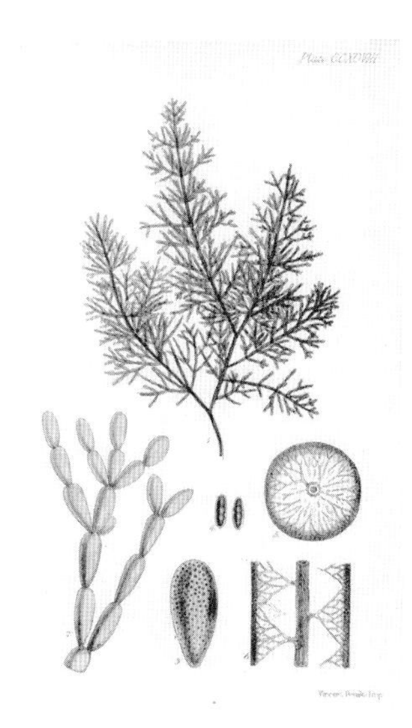

Erythroclonium

Erythroclonium includes five species from southern and western Australia. Thalli are irregularly branched and are distinctly segmented into clavate or slender segments. Structurally the thallus is uniaxial with the central filament remaining conspicuous throughout the plant. Each axial cell bears two branched filaments that traverse the central cavity before forming the outer cortex. Rhizoids also encircle the central filament and in older axes will often fill the inner cavity. The species of *Erythroclonium* can be separated by the size and shape of the segments and the presence of short branches on lower axes. Segments of *Erythroclonium muelleri* are generally less than 5 mm long, whereas those of *Erythroclonium sonderi* are elongate and can be up to 20 mm long.

Erythroclonium muelleri

TYPE LOCALITY: Lefevre Peninsula, South Australia.
DISTRIBUTION: Cape Peron, Western Australia, around southern mainland Australia to Port Phillip, Victoria, and around Tasmania.

Erythroclonium sonderi

TYPE LOCALITY: Fremantle, Western Australia.
DISTRIBUTION: From the Houtman Abrolhos, Western Australia, around southern mainland Australia to Robe, South Australia and northern Tasmania.
FURTHER READING: Womersley (1994); Huisman (2018).

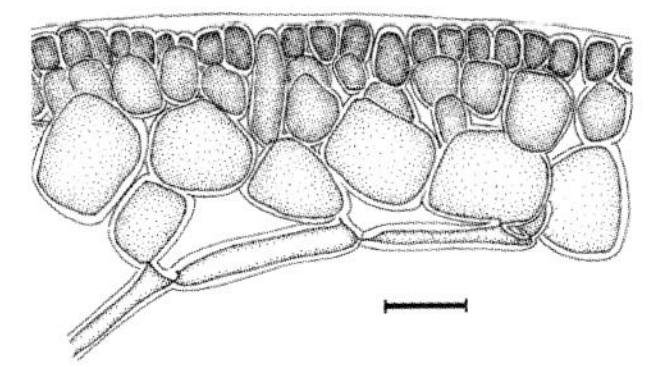

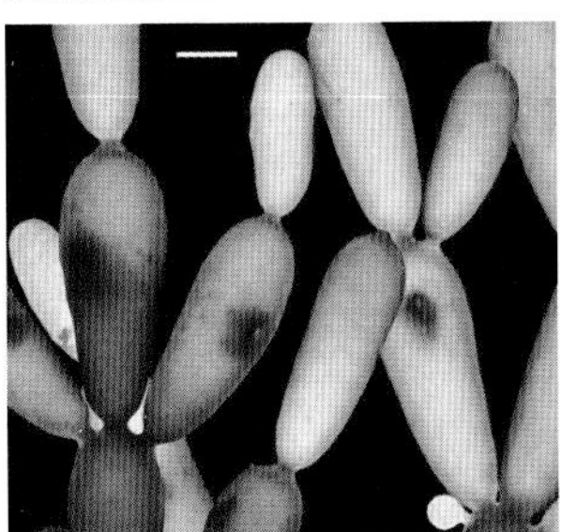

Erythroclonium muelleri (Cape Peron, WA).

(left) *Erythroclonium muelleri* – branch detail (Cape Peron, WA). Scale = 1mm.

(top) *Erythroclonium muelleri* – as illustrated by Harvey (1863, pl. 298).

Erythroclonium sonderi (Houtman Abrolhos, WA).

(above) *Erythroclonium sonderi* – partial section of cortex (Houtman Abrolhos, WA). Scale = 30μm.

Schmitzia

Schmitzia includes uniaxial, mucilaginous species somewhat similar in appearance to *Acrosymphyton* (p. 67), but differing in lacking a banding pattern and details of reproductive structures. In *Schmitzia*, the carpogonial branches are three cells long and lack lateral branches, whereas those of *Acrosymphyton* can be over 10 cells long and are pinnately branched. In addition, the gonimoblast in *Schmitzia* arises from connecting filaments some distance from the junction with the auxiliary cell, whereas in *Acrosymphyton* they arise adjacent to the auxiliary cell. Two species of *Schmitzia* have been recorded in Australia: *Schmitzia japonica* from the Coffs Harbour region, New South Wales, and the Great Barrier Reef species included here, which has not been formally described but has been mentioned previously in publications as *Schmitzia falcata*.

Schmitzia 'falcata'

TYPE LOCALITY: Undescribed species.
DISTRIBUTION: Known from the Capricorn Group of the Great Barrier Reef.
FURTHER READING: Millar (1990).

Schmitzia 'falcata' (Sykes Reef, Qld).

Corynocystis

A genus with only a single species, *Corynocystis prostrata* has a creeping thallus that is deep red and segmented, with flattened branches. Structurally the thallus is multiaxial, with a filamentous medulla and a cortex of narrow, widely separated, branched filaments. Interspersed between the cortical filaments are longitudinally aligned, thick-walled rhizines. A distinctive feature of *Corynocystis* is the production of hemispherical cystocarps at the tips of branches. Since it was first described in 1999, numerous records of this distinctive species have appeared, suggesting that it is reasonably common in the Indo-Pacific region. *Corynocystis prostrata* almost invariably occurs in shaded recesses in reef walls.

Corynocystis prostrata

TYPE LOCALITY: Bulusan, Sorsogon province, Philippines.
DISTRIBUTION: Known from the Rowley Shoals, Western Australia, Ashmore Reef, and the southern Great Barrier Reef, Queensland; otherwise apparently widespread in the Indo-West Pacific.
FURTHER READING: Kraft et al. (1999); Huisman et al. (2009); Huisman (2018).

Corynocystis prostrata (Lamont Reef, Qld).

Craspedocarpus

Craspedocarpus includes eight species, of which four occur in southern Australia. Plants are flattened and irregularly to subdichotomously branched from the margins. Structurally the thallus is uniaxial and, when viewed under the microscope, the central axial filament can often be seen from the surface, especially when stained. The surface of the thallus generally has a pattern of rosettes of smaller cells surrounding the larger inner cells. Tetrasporangia are zonately divided and cystocarps are protuberant. In *Craspedocarpus venosus*, the cystocarps occur on branch margins, but in other species of the genus they can be scattered.

Craspedocarpus venosus

TYPE LOCALITY: Australia.
DISTRIBUTION: Fremantle, Western Australia, to Western Port, Victoria, and Flinders Island, Bass Strait.
FURTHER READING: Min-Thein & Womersley (1976); Womersley (1994).

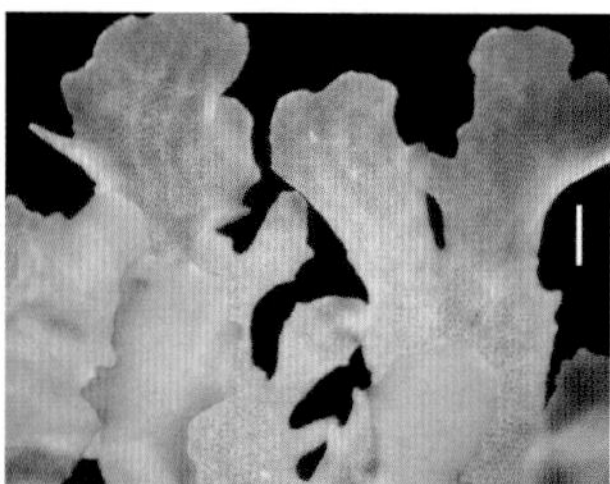

Craspedocarpus venosus (Cape Peron, WA).

(left) *Craspedocarpus venosus* – branch detail (Cape Peron, WA). Scale = 1 mm.

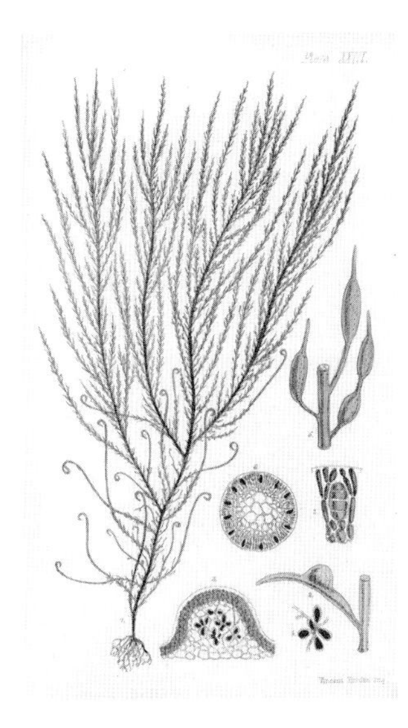

Hypnea

The genus *Hypnea* includes a number of widespread and extremely common species. Thalli are profusely branched, with most axes bearing numerous short, often spine-like, lateral branches. Hooked apices are common in some species and serve to anchor the thallus to any available substratum. Structurally the thallus is uniaxial and the central filament is apparent in sections taken from all parts of the plant. The medulla is composed of large, colourless cells and is bordered by a small-celled cortex. Cystocarps are protuberant and tetrasporangia are zonately divided and formed in raised patches on the short lateral branches. The species of *Hypnea* included here are common in Australia but many others also occur.

Hypnea charoides

Hypnea charoides can be recognised by its often dense cover of short spinous branches. The species is often epiphytic on seagrasses such as *Amphibolis*.

TYPE LOCALITY: Australia
DISTRIBUTION: Kimberley coast, Western Australia, to Cape Jaffa, South Australia, and northern Tasmania.

Hypnea charoides (Cape Peron, WA).
(inset) *Hypnea charoides* – branch detail (Cape Peron, WA). Scale = 1 mm.
(top) *Hypnea ramentacea* – as illustrated by Harvey (1858, pl. 23, as *Hypnea episcopalis*).

Hypnea corona

Hypnea corona grows on rock in the intertidal and shallow subtidal. It produces short, star-shaped branches that can be shed and serve to propagate the alga.

TYPE LOCALITY: Cape Peron, Western Australia.
DISTRIBUTION: Widespread in tropical and subtropical regions, south to at least the Perth region in Western Australia. Introduced into the Mediterranean where it is now widespread.

Hypnea filiformis

Hypnea filiformis is common in the Perth region where it occurs on rock in the intertidal and shallow subtidal. It can be distinguished from other *Hypnea* species in having branchlets that are elongate and slightly constricted at the base, and tetrasporangia occurring in cortical sori that are only slightly raised and not in obvious nemathecia.

TYPE LOCALITY: Garden Island, Western Australia.
DISTRIBUTION: Barrow Island, Western Australia, to Nora Creina, South Australia.

Hypnea musciformis

Hypnea musciformis has hook-like appendages that occur at the tips of regular branches. The species can be extremely common and is an invasive pest in the Hawaiian Islands. Although reported to be widely distributed, molecular analyses are indicating that it might represent a species complex.

TYPE LOCALITY: Trieste, Italy.
DISTRIBUTION: Widespread in most warm seas.

Hypnea pannosa

Hypnea pannosa is a tropical species that is rarely found as far south as Rottnest Island. Elsewhere it can be very common and is often found sprawling between coral on intertidal and shallow reefs. The species has a distinctive pale blue to purple colouration when seen underwater.

TYPE LOCALITY: San Agustín, Oaxaca, Mexico.
DISTRIBUTION: From Rottnest Island, Western Australia, around northern Australia to Queensland. Widespread in tropical seas.

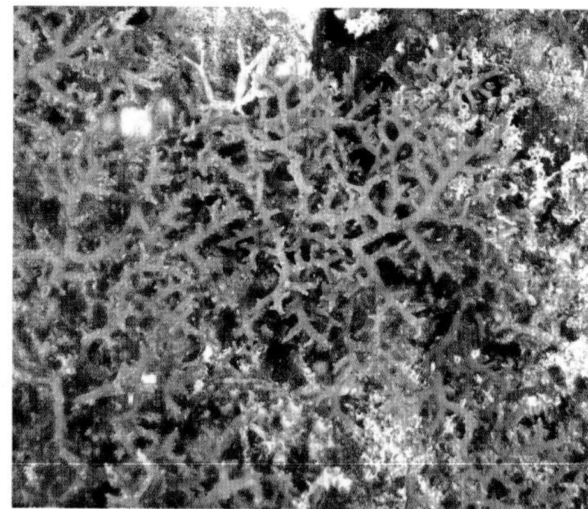

Hypnea ramentacea

One of several species with hook-like structures, those of *Hypnea ramentacea* differ from most others in being formed at the ends of specialised elongate branches that do not bear other lateral branches.

TYPE LOCALITY: Australia.
DISTRIBUTION: Port Denison, Western Australia, around southern Australia to Walkerville, Victoria, and northern Tasmania.

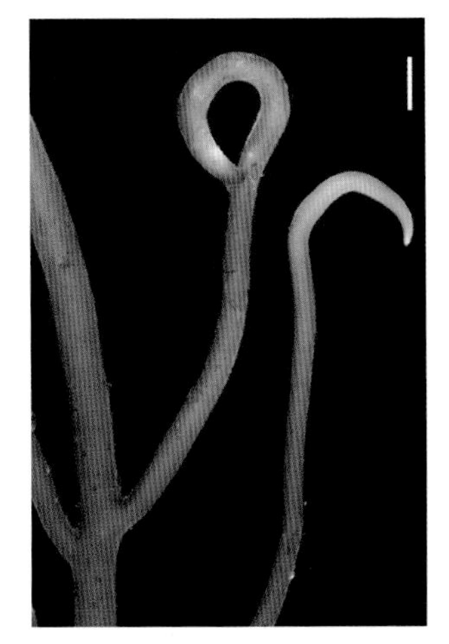

Hypnea valentiae

Hypnea valentiae can vary in colour from red to yellow. The species occurs in similar habitats to *Hypnea charoides*, but does not have the dense spinous branches of that species.

TYPE LOCALITY: Red Sea.
DISTRIBUTION: Widely distributed in tropical and subtropical seas.
FURTHER READING: Womersley (1994); Huisman (2018); Huisman et al. (2021).

Opposite page:
Hypnea corona (Cape Peron, WA).
(inset) *Hypnea corona* – with dark-red stellate propagules (Cape Peron, WA). Scale = 1 mm.
Hypnea filiformis (Cape Peron, WA).
(inset) *Hypnea filiformis* – partial section of thallus (Carnac Island, WA). Scale = 30 µm.
Hypnea musciformis (Cape Peron, WA).
(inset) *Hypnea musciformis* – upper branch with terminal hook (Cape Peron, WA). Scale = 1 mm.
Hypnea pannosa (Cassini Island, WA).

Hypnea ramentacea (Cape Peron, WA).

(above) *Hypnea ramentacea* – specialised branch with terminal hook (Cape Peron, WA). Scale = 1 mm.

Hypnea valentiae (Cape Peron, WA).

Hypneocolax

Hypneocolax is a small parasite of various species of *Hypnea*, in Australia most commonly on *Hypnea ramentacea* (as in the photo). Plants form white to pale red, lobed growths to a couple of millimetres tall. While the genus is regarded as a parasite because of its lack of pigmentation, some thalli can be red and presumably are able to photosynthesise. Plants can substantially reduce the growth rate of the host *Hypnea*, by as much as 70 per cent. The genus is taxonomically closely related to the host and produces similar reproductive structures.

Hypneocolax stellaris f. *orientalis*

TYPE LOCALITY: Dobo, Pulao Wokam, Kepulauan Aru (Aru Islands), Indonesia.
DISTRIBUTION: Garden Island, Western Australia, to Robe, South Australia. Tropical to temperate Indo-Pacific.
FURTHER READING: Mshigeni (1976); Womersley (1994); Huisman et al. (2007).

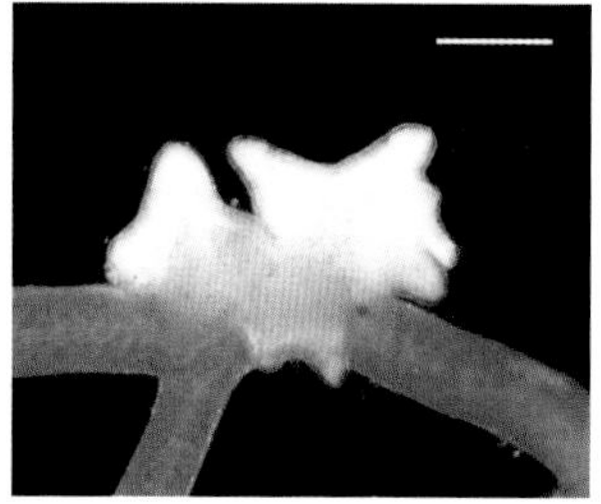

Hypneocolax stellaris f. *orientalis* – growing on *Hypnea ramentacea* (indicated by arrows). (Cape Peron, WA).

(above) *Hypneocolax stellaris* f. *orientalis* – detail of thallus (Cape Peron, WA). Scale = 500 µm.

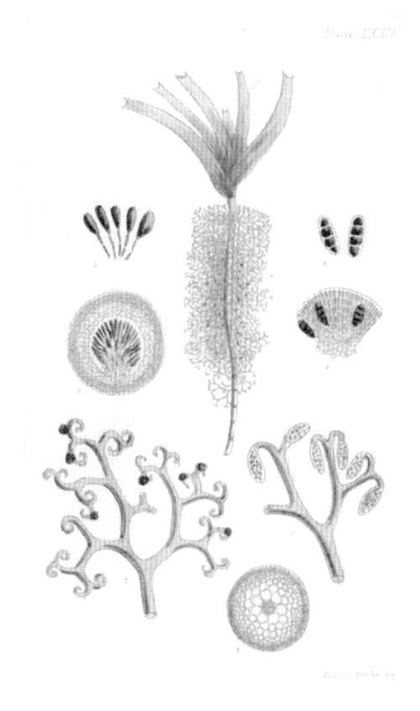

Dicranema

Seaweeds can vary substantially in their habitat requirements and their ability to settle and grow on different surfaces. Some only grow on rock, others only in sand, and yet others are almost always epiphytic, which means that they grow on other marine plants. One of this latter group is *Dicranema revolutum*, which is extremely selective to the point of growing only on the woody stems of the seagrass *Amphibolis*. In this habitat, it can be very common. *Dicranema revolutum* is quite firm and retains its shape even when detached and washed up on the beach, where accumulations of this unusual alga are common. It is superficially similar in appearance to another epiphyte that also grows exclusively on the stems of *Amphibolis*, *Mychodea pusilla*, but that species is less regularly branched and has pointed branch tips, whereas *Dicranema revolutum* has rounded branch tips. Plants are a red-brown colour and grow to a height of about 5 cm. The branches are regularly forked and all grow to a similar length, such that individual plants often have a spherical outline. Structurally the plants are composed of densely packed cells surrounding a central filamentous core.

Dicranema revolutum

TYPE LOCALITY: Freycinet Harbour, Shark Bay, Western Australia.
DISTRIBUTION: Shark Bay, Western Australia, to Walkerville, Victoria, and Flinders Island, Bass Strait.
FURTHER READING: Kraft (1977b); Kraft & Womersley (1994a).

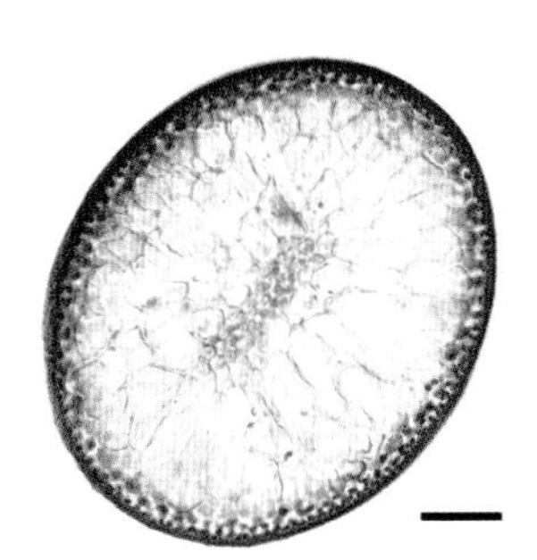

Dicranema revolutum, growing on *Amphibolis antarctica* (Cape Peron, WA).

(top) *Dicranema revolutum* – as illustrated by Harvey (1859a, pl. 74).

(above) *Dicranema revolutum* – section of thallus (Cape Peron, WA). Scale = 100 µm.

Tylotus

The genus *Tylotus* includes two species, of which only *Tylotus obtusatus* is known in Australia. Thalli are cartilaginous, irregularly subdichotomously branched with flattened axes that are 5–12 mm broad. Attachment to the substratum is via a number of elongate processes that arise from the undersurface of the thallus near the base. Upper branches become secondarily attached in a similar manner. Structurally the thallus is entirely pseudoparenchymatous, with a large-celled medulla grading to a small-celled cortex. Cystocarps are protuberant and borne on the under surface. Tetrasporangia are zonately divided and occur in raised patches on the upper surface. *Tylotus obtusatus* is generally a dark-red colour, turning almost black when dried, although some plants can have a yellow tinge.

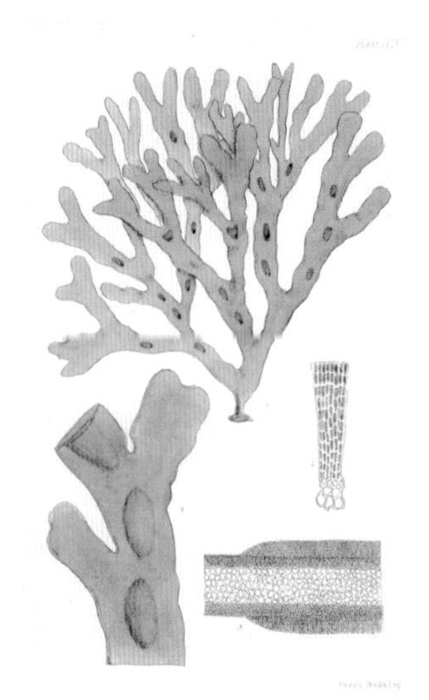

Tylotus obtusatus

TYPE LOCALITY: Western Australia.
DISTRIBUTION: From Champion Bay, Western Australia, around southern Australia to Inverloch, Victoria.
FURTHER READING: Kraft (1977b); Kraft & Womersley (1994a).

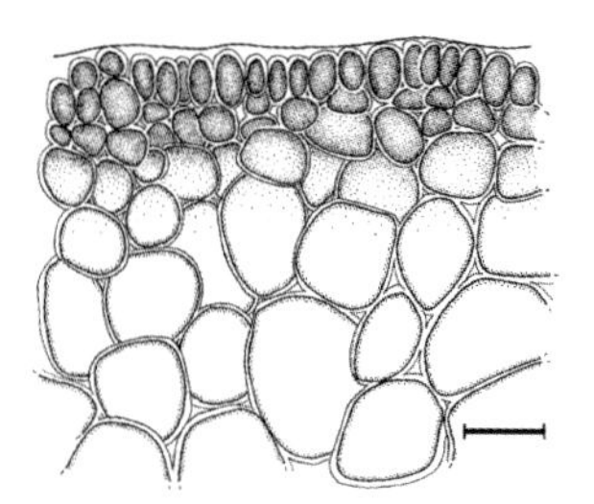

Tylotus obtusatus – luxuriant plant with a yellow colour (Cape Peron, WA).

(top) *Tylotus obtusatus* – as illustrated by Harvey (1862, pl. 210, as *Curdiea obtusata*).

(middle) *Tylotus obtusatus* (Cape Peron, WA).

(above) *Tylotus obtusatus* – section of medulla and cortex (Eglinton Rocks, WA). Scale = 30 µm.

Dudresnaya

Four species of *Dudresnaya* are known from Australia. *Dudresnaya capricornica* occurs in warmer waters, where it grows on rock or coral rubble. Thalli are soft and irregularly radially branched, with terete axes that have a filamentous appearance. Structurally the thallus is uniaxial, with each axial cell producing five to seven lateral filaments that branch and eventually form a filamentous outer cortex. Cystocarps are loosely constructed and are borne in the inner cortex near the axial filament. Tetrasporangia are zonately divided and are borne terminally or laterally on cortical filaments. The remaining species are *Dudresnaya australis* (not pictured), a southern species that is invariably an epiphyte, commonly on the seagrass *Amphibolis*, *Dudresnaya hawaiiensis*, a species closely related and similar in appearance to *Dudresnaya capricornica*, and *Dudresnaya barrowensis*, known only from Barrow Island in Western Australia.

Dudresnaya capricornica

TYPE LOCALITY: One Tree Island, Great Barrier Reef, Queensland.

DISTRIBUTION: The Houtman Abrolhos and Rottnest Island, Western Australia. Great Barrier Reef, Queensland. Lord Howe Island, New South Wales.

Dudresnaya hawaiiensis

TYPE LOCALITY: Kāne'ohe Bay, O'ahu, Hawaiian Islands.

DISTRIBUTION: Widespread in warmer waters of the Indo-Pacific. South Africa.

FURTHER READING: Robins & Kraft (1985); Huisman (2018).

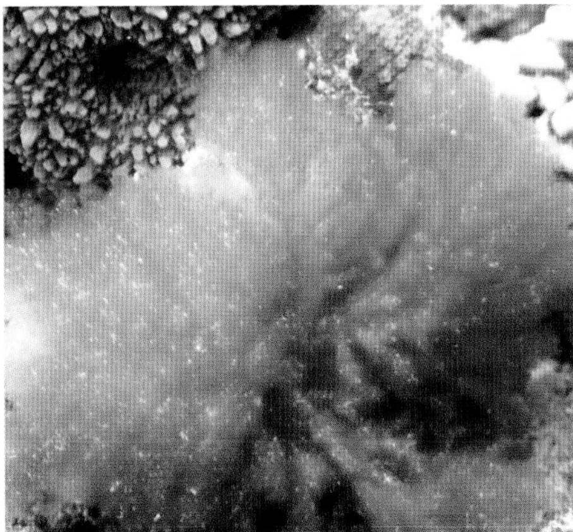

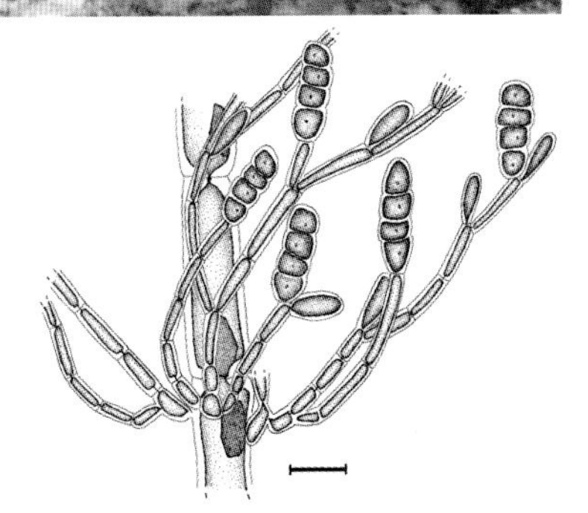

Dudresnaya capricornica (Heron Island, Qld).

(left) *Dudresnaya capricornica* – detail of thallus structure with zonate tetrasporangia (Houtman Abrolhos, WA). Scale = 20µm.

Dudresnaya hawaiiensis (Sykes Reef, Qld).

Gibsmithia

Undeniably one of the more unusual of the red algae, *Gibsmithia* has (apparently) perennial cartilaginous stalks that bear an apical cluster of soft, gelatinous branches. In section, the stalks are densely filamentous and appear to have annular growth rings. Structurally, the gelatinous branches are multiaxial and the filaments are loosely associated in the gelatinous matrix. Secondary fusions and lateral connections between filaments are common. Cystocarps, cruciately divided tetrasporangia and spermatangia are formed in the gelatinous branches. Four species of *Gibsmithia* are known in Australia, with *Gibsmithia indopacifica*, *Gibsmithia larkumii* and *Gibsmithia womersleyi* included here. The first is known only from tropical regions and can be distinguished by its cortical filaments extending beyond the gelatinous matrix, giving the thallus a furry appearance. *Gibsmithia larkumii* also has a tropical distribution, but has a smooth surface. *Gibsmithia womersleyi* is apparently restricted to the colder waters of southern Australia and the thallus has a smooth surface.

Gibsmithia indopacifica

TYPE LOCALITY: Paliton Wall (off Paliton beach), Siquijor, Philippines.
DISTRIBUTION: Widespread in tropical waters of the Indo-West Pacific; in Australia known from the Great Barrier Reef, Queensland, and from the Ningaloo Reef north to Echuca Shoal, Western Australia.

Gibsmithia indopacifica (Ashmore Reef).

Gibsmithia larkumii

TYPE LOCALITY: 'The Keyhole', One Tree Island, Southern Great Barrier Reef, Queensland.
DISTRIBUTION: Possibly widespread in the tropical Indo-West Pacific.

Gibsmithia womersleyi

TYPE LOCALITY: Esperance, Western Australia.
DISTRIBUTION: Augusta to Esperance, Western Australia.
FURTHER READING: Kraft (1986b); Schils & Coppejans (2002); Gabriel et al. (2016), (2017).

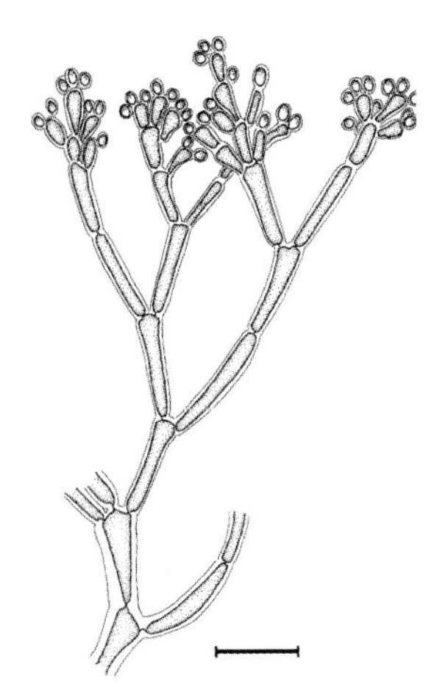

Gibsmithia larkumii (Ashmore Reef).

Gibsmithia womersleyi (Augusta, WA).

(above) *Gibsmithia womersleyi* – cortical filaments with apical spermatangia (Augusta, WA). Scale = 20µm.

Kraftia

Kraftia includes only a single species, *Kraftia dichotoma*, which grows exclusively on the stems of the seagrass *Amphibolis*, often in a lower position under the shade of the canopy. Plants grow to a few centimetres tall and have a soft texture. They are subdichotomously branched, with slightly compressed axes. Structurally both the medulla and cortex are filamentous, with the cortical filaments dichotomously or trichotomously branched. Cystocarps are embedded in the thallus. *Kraftia* has an unusual life history, as the tetrasporophytes produce a basal crust that wraps around the seagrass stem and produces terminal, zonately divided tetrasporangia.

Kraftia dichotoma

TYPE LOCALITY: Victor Harbor, South Australia.
DISTRIBUTION: Port Denison, Western Australia, to Walkerville, Victoria.
FURTHER READING: Shepley & Womersley (1983); Womersley (1994).

Kraftia dichotoma (Cape Peron, WA). Scale = 2 mm.

Rhodopeltis

The genus *Rhodopeltis* was first described by W.H. Harvey in 1863 for a small filamentous plant that he believed to be an independent species growing epiphytically on what was then known as *Amphiroa australis*. It was later discovered that this 'species' was in fact the reproductive branches of the host plant, which itself was only remotely related to true *Amphiroa*. As a consequence, the name *Rhodopeltis* was adopted. Thalli of *Rhodopeltis* are calcified with segmented axes and can easily be confused with the unrelated *Amphiroa* (p. 53) or *Dichotomaria* (p. 28). The flattened, somewhat circular segments are characteristic of *Rhodopeltis*. Two species are known from Australia. *Rhodopeltis australis* has round- to oval-shaped segments that have smooth surfaces. The second species, *Rhodopeltis borealis*, was originally described from southern Japan and Taiwan and has only recently been recognised as occurring in Australia. This species is apparently restricted to the south-west of Western Australia and is common on subtidal rock walls at Rottnest Island. *Rhodopeltis borealis* can be confused with *Amphiroa anceps*, but can usually be recognised by the broadening of the mid-portion of the segments so that they are oval in shape.

Rhodopeltis australis

TYPE LOCALITY: Rottnest Island, Western Australia.
DISTRIBUTION: From Cottesloe and Rottnest Island, Western Australia, around southern Australia to Point Roadknight, Victoria.

Rhodopeltis borealis

TYPE LOCALITY: Ryukyu-retto, Japan. Ryusensui, Kotosho, Taiwan.
DISTRIBUTION: From the Houtman Abrolhos to Rottnest Island, Western Australia. Taiwan. Southern Japan. Philippines.
FURTHER READING: Nozawa (1970); Womersley (1994).

Rhodopeltis australis (Rottnest Island, WA).

Rhodopeltis borealis (Rottnest Island, WA).

(top) *Rhodopeltis australis* – as illustrated by Harvey (1859a, pl. 77, as *Amphiroa australis*).

Chondracanthus

Chondracanthus is represented in Australia by the single species *Chondracanthus acicularis*. The species is widespread worldwide, but has only been recorded from a few disjunct locations in Australia and might have been introduced. Plants grow on intertidal and shallow subtidal rock, or jetty pylons, and are often entangled with other algae. The thallus is wiry and cartilaginous, with terete or compressed, curved axes that are sharply pointed and often have distinctive white spots. Structurally the thallus is multiaxial, with a filamentous medulla and a cortex of anticlinal files of cells.

Chondracanthus acicularis

TYPE LOCALITY: Adriatic Sea.
DISTRIBUTION: Known from Cape Peron and Albany, Western Australia. Widespread in warm to temperate seas.
FURTHER READING: Coppejans et al. (2009).

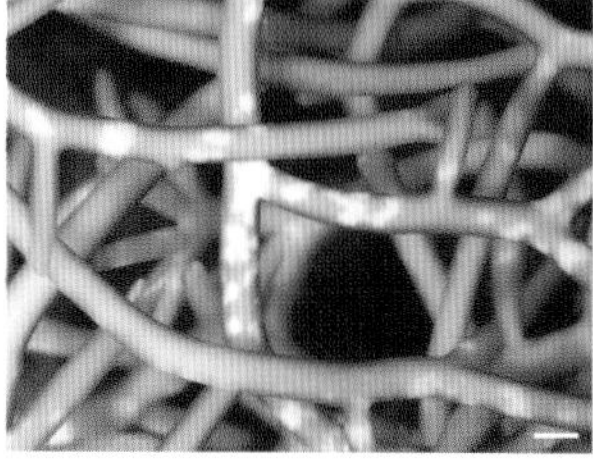

Chondracanthus acicularis (Albany, WA).

(top) *Chondracanthus acicularis* – closer view of thallus (Cape Peron, WA). Scale = 1 mm.

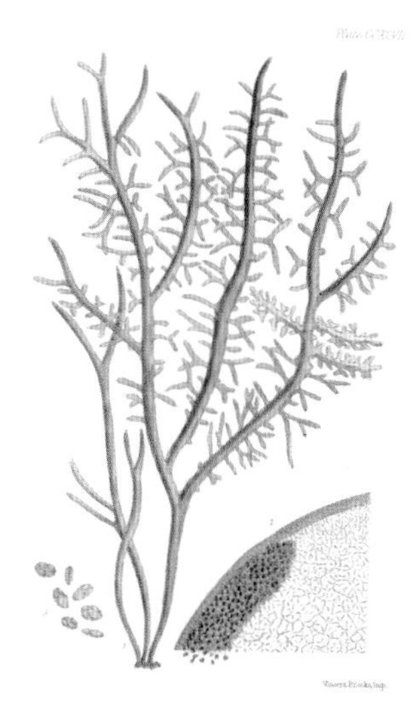

Gigartina

Species of *Gigartina* vary greatly in their external appearance, from flattened and almost foliose to terete, densely branched and cartilaginous. The genus displays its greatest diversity and abundance in the colder waters of southern Australia, with only *Gigartina disticha* found on the west coast. Structurally *Gigartina* is multiaxial and both medulla and cortex are filamentous. The filaments of the cortex are aligned anticlinally and are subdichotomously divided. Cystocarps are borne in short lateral branches and occupy much of the branch, giving the impression of a stalked cystocarp. Tetrasporangia are formed in short chains and are densely aggregated into clusters near the boundary of the medulla and cortex. *Gigartina disticha* is a distinctive species, characterised by cartilaginous axes that are terete to slightly compressed and bear short distichous lateral branches. *Gigartina recurva* is found only on the east coast of Tasmania where it is common near low tide level on steeply sloping rock. It has clumped axes that are largely unilaterally branched.

Gigartina disticha

TYPE LOCALITY: Busselton, Western Australia.
DISTRIBUTION: Jurien Bay, Western Australia, around southern Australia to the Fitzroy River mouth, Victoria.

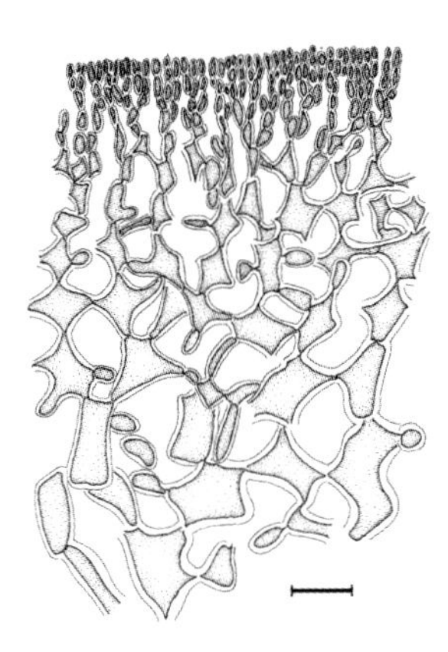

Gigartina disticha (Two Rocks, WA).

(top) *Gigartina disticha* – as illustrated by Harvey (1863, pl. 297).

(above) *Gigartina disticha* – partial section of thallus (Jurien Bay, WA). Scale = 30 µm.

Gigartina recurva

TYPE LOCALITY: Bicheno, Tasmania.
DISTRIBUTION: East coast of Tasmania.
FURTHER READING: Edyvane & Womersley (1994).

Gigartina recurva
(Taroona, Tas; photo G.J. Edgar).

Austrokallymenia

The genus *Austrokallymenia* was erected following a taxonomic study incorporating DNA analyses that indicated that several southern hemisphere species referred to as *Kallymenia* were not closely related to the Mediterranean type species. Two new species, *Austrokallymenia rebeccae* (not pictured) and *Austrokallymenia roensis*, were also described for recently discovered entities from Lord Howe Island and Rottnest Island, respectively. Plants are blade-like and can be solidly membranous (as in *Austrokallymenia roensis*, included here) or have regular circular perforations, depending on the species. Structurally the medulla is composed of sparse filaments and stellate cells with elongate, slender arms, grading to a cortex of several layers of cells. Carposporophytes are embedded in the medulla and distributed widely across blades, composed of clusters of carposporangia separated by sterile filaments. Tetrasporangia are scattered in the cortex, and are cruciately or irregularly zonately divided.

Austrokallymenia roensis

TYPE LOCALITY: Roe Reef, Rottnest Island, Western Australia.
DISTRIBUTION: Known only from Rottnest Island.
FURTHER READING: Saunders et al. (2017).

Austrokallymenia roensis (Rottnest Island, WA).

Callophyllis

Thalli of *Callophyllis* are much divided with flattened axes branched in one plane. *Callophyllis rangiferina* is a widespread species found throughout temperate Australian waters. Branch width is extremely variable and is apparently dependent on the level of water movement. Structurally the thallus has a pseudoparenchymatous medulla with intermingled rhizoidal filaments, grading into a small-celled cortex. Cystocarps are formed near the apices and are generally embedded in the thallus, although they occasionally become protuberant. Tetrasporangia are cruciately divided and found in the outer cortex. *Callophyllis lambertii* is a more robust species and lacks the lanceolate apices found in *Callophyllis rangiferina*.

Callophyllis lambertii

TYPE LOCALITY: Australia.
DISTRIBUTION: Head of the Great Australian Bight, South Australia, to Walkerville, Victoria, and around Tasmania.

Callophyllis rangiferina

TYPE LOCALITY: Kent Islands, Bass Strait.
DISTRIBUTION: From the Houtman Abrolhos, Western Australia, around southern mainland Australia and Tasmania to Tathra, New South Wales.
FURTHER READING: Womersley (1994).

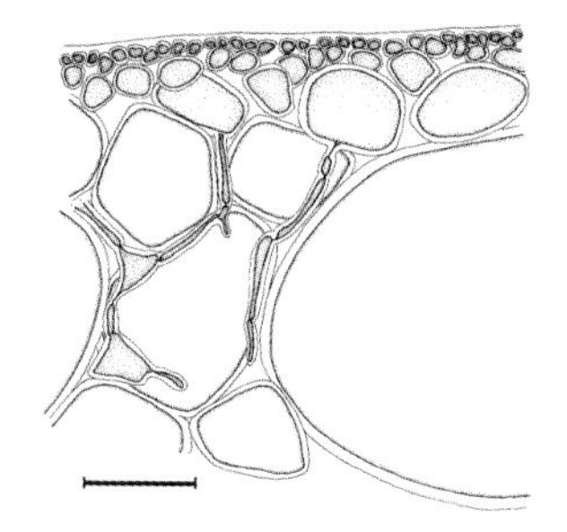

Callophyllis lambertii (Gordon, Tas; photo G.J. Edgar).

Callophyllis rangiferina (Houtman Abrolhos, WA).

(right) *Callophyllis rangiferina* – partial section of thallus (Houtman Abrolhos, WA). Scale = 50 µm.

Glaphyrymenia

Glaphyrymenia includes the single species *Glaphyrymenia pustulosa*. Thalli are foliose and generally unbranched. Structurally the thallus is almost entirely filamentous, with a medulla of scattered filaments and a thin cortex. Cystocarps are immersed in the thallus and tetrasporangia are cruciately divided and scattered in the cortex. The habit of *Glaphyrymenia pustulosa* is similar to that of some species of *Austrokallymenia* (see p. 89), but the former can be readily identified by its totally filamentous construction. In *Austrokallymenia* a pseudoparenchymatous cortex borders the filamentous medulla.

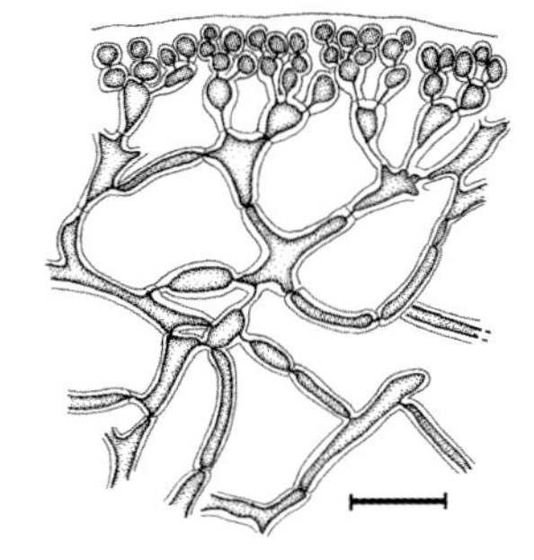

Glaphyrymenia pustulosa

TYPE LOCALITY: Port Phillip, Victoria.
DISTRIBUTION: From the Houtman Abrolhos, Western Australia, around southern mainland Australia to Port Phillip, Victoria, and south-eastern Tasmania. New Zealand.
FURTHER READING: Womersley & Norris (1971); Womersley (1994).

Glaphyrymenia pustulosa (Rottnest Island, WA).

(top) *Glaphyrymenia pustulosa* – partial section of thallus (Rottnest Island, WA). Scale = 20 µm.

Leiomenia

The genus *Leiomenia* includes species with foliose thalli that are generally unbranched. The surface of the thallus is smooth, and in some species, such as *Leiomenia cribrosa* and *Leiomenia lacunata*, the thallus is perforated to some degree. Structurally the thallus is composed of a filamentous medulla with interspersed larger cells and a pseudoparenchymatous cortex. Cystocarps are immersed in the thallus and tetrasporangia are cruciately divided and located in the outer cortex. The primarily southern Australian *Leiomenia cribrosa* was formerly included in *Kallymenia* and is a distinctive species that is readily recognised, although it might be confused with *Austrokallymenia cribrogloea*, another perforated species with a similar appearance.

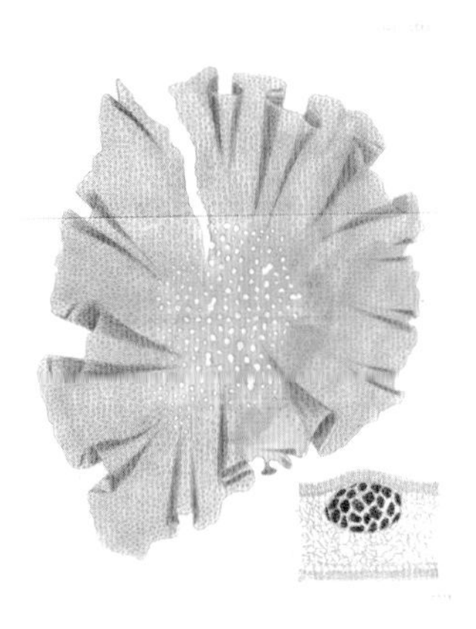

Leiomenia cribrosa

TYPE LOCALITY: Fremantle, Western Australia.
DISTRIBUTION: Houtman Abrolhos, Western Australia, around southern mainland Australia to Flinders, Victoria, and around Tasmania.

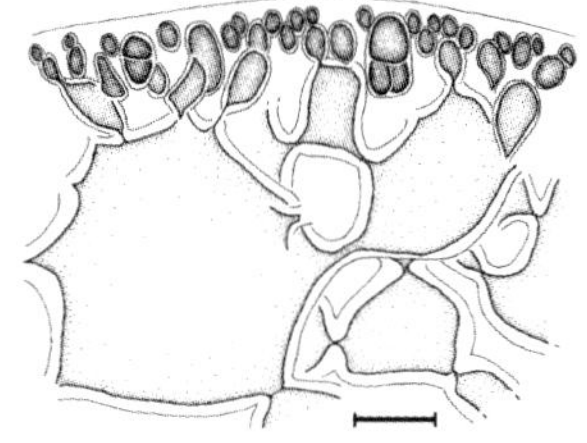

Leiomenia cribrosa (Rottnest Island, WA).

(top) *Leiomenia cribrosa* – as illustrated by Harvey (1859a, pl. 73, as *Kallymenia cribrosa*).

(right) *Leiomenia cribrosa* – partial section of cortex and large outer medullary cells (Houtman Abrolhos, WA). Scale = 30 µm.

Leiomenia imbricata (Hibernia Reef, WA).

Leiomenia lacunata (Ashmore Reef).

Leiomenia imbricata

TYPE LOCALITY: Hibernia Reef, Western Australia.
DISTRIBUTION: Known from the type locality (circa 42 km northeast of Ashmore Reef) and the Cocos (Keeling) Islands, Australia.

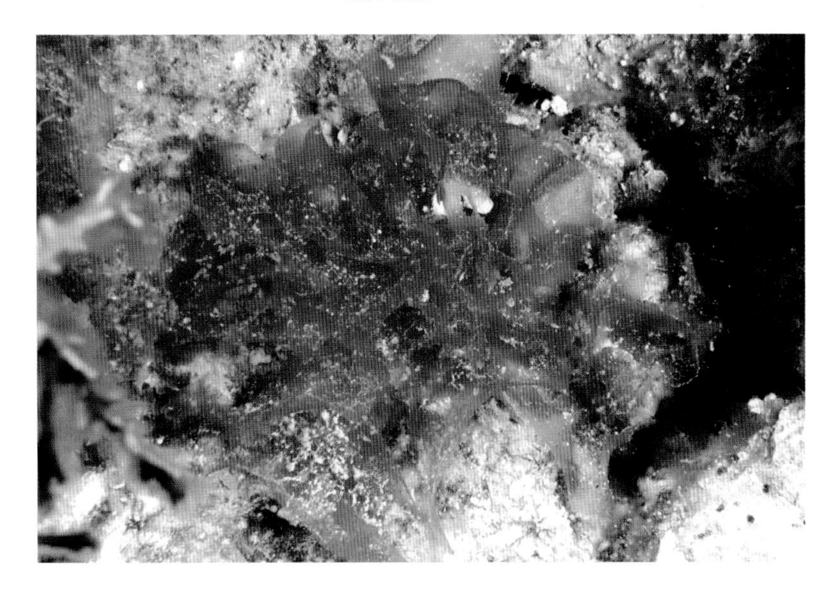

Leiomenia lacunata

TYPE LOCALITY: Ashmore Reef, northern channel, Australia.
DISTRIBUTION: Known from Ashmore Reef, Coral Bay, Ningaloo Reef and the Montebello Islands, Western Australia.
FURTHER READING: Womersley (1994, as *Kallymenia*); Saunders et al. (2017).

Rhytimenia

Rhytimenia is based on the rare *Rhytimenia maculata*, which was originally described as a species of *Kallymenia* in 1928 but not seen again for over 80 years. The species was rediscovered at Ashmore Reef, growing on rock on vertical reef walls at 12–15 m depth. Plants are bladelike and a light red colour, with a distinctive wrinkled surface. They can grow to 25 cm tall, but are gelatinous and very delicate, fragmenting easily if not handled carefully. Structurally the thallus has a medulla of branched filaments, arranged in all directions, forming an interconnected web, often with cells forming cross shapes with extended arms, but rarely with stellate cells. The cortex is relatively thin, with only three or four cell layers. Cystocarps are large and hemispherical, often protuberant, with numerous clusters of carposporangia mixed with filaments. Tetrasporangia are scattered in the outer cortex and are cruciately or irregularly divided.

Rhytimenia maculata

TYPE LOCALITY: Pulau Sebangatan, Makassar Strait, Indonesia.
DISTRIBUTION: Known only from the Indonesian type locality and from Ashmore Reef.
FURTHER READING: Huisman et al. (2016); Huisman & Saunders (2018a).

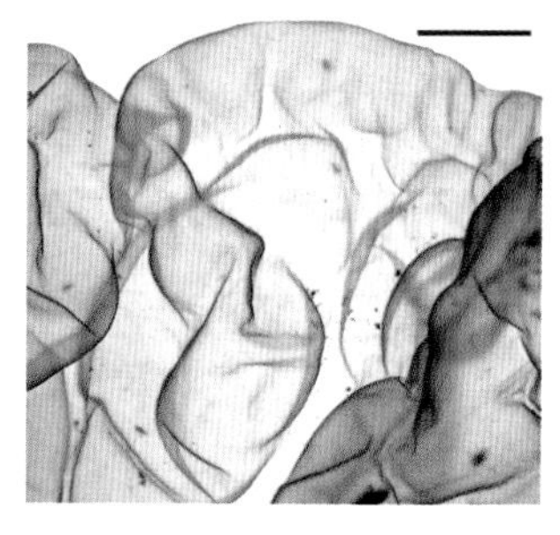

Rhytimenia maculata (Ashmore Reef).

(above) *Rhytimenia maculata* – wrinkled surface of thallus (Ashmore Reef). Scale = 1 mm.

Stauromenia

Two species of *Stauromenia* are known: one from New Zealand and *Stauromenia lacerata* from southern and south-western Australia. The species was previously included in *Thamnophyllis*, but based on molecular and morphological studies that genus is now regarded as being restricted to southern Africa. Thalli of *Stauromenia lacerata* are foliose and divided into lobes. Structurally they have a medulla of large cells mixed with short filaments and a pseudoparenchymatous cortex. Cystocarps are slightly protuberant. A distinctive feature of the species is the presence of darkly staining ganglionic cells that are aligned in series – in stained material these are generally visible from the surface of the thallus.

Stauromenia lacerata

TYPE LOCALITY: Great Taylor Bay, Bruny Island, Tasmania.
DISTRIBUTION: From the Perth region, Western Australia, around south-western Australia to Gulf St. Vincent, South Australia, and the east coast of Tasmania.
FURTHER READING: Womersley (1994, as *Thamnophyllis*); D'Archino et al. (2012); Saunders et al. (2017).

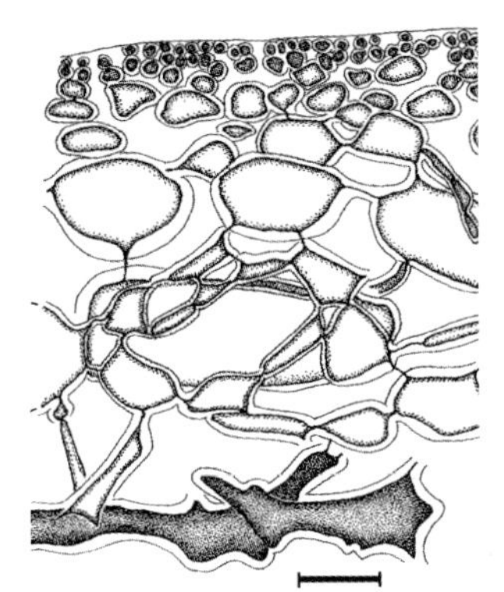

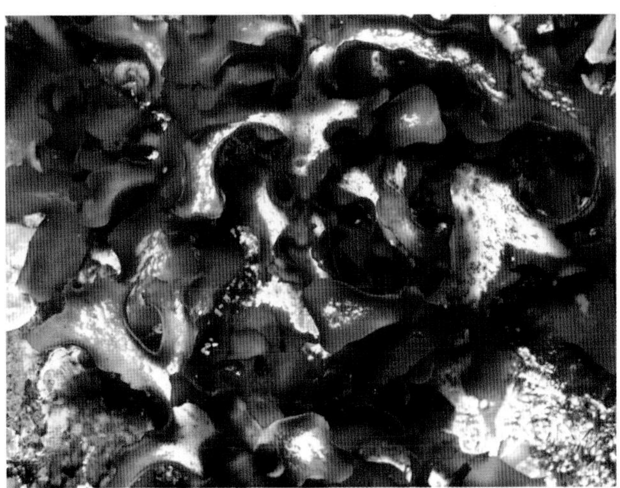

Stauromenia lacerata (Rottnest Island, WA).

(above) *Stauromenia lacerata* – section of medulla and cortex (Rottnest Island, WA). Scale = 30 µm..

Mychodea

Mychodea includes some 20 species. The genus (and the family Mychodeaceae) is wholly endemic to Australia, where most species occur on the southern coast. Depending on the species, plants can have terete, compressed, or conspicuously flattened branches. Structurally the thallus is uniaxial, with the central axial filament conspicuous near the apices but becoming indistinct in older parts of the thallus. Several species have a medulla with conspicuously enlarged cells forming a ring around the central core of filaments, although this is not always obvious in species with flattened branches. Cystocarps are protuberant and tetrasporangia are scattered in the cortex and zonately divided. *Mychodea marginifera* has flattened branches and grows on a variety of substrata, including rock, jetty pylons and seagrasses. *Mychodea disticha* has compressed branches with marginal laterals. The third species included here, *Mychodea pusilla*, is unusual in that it grows exclusively on the woody stems of *Amphibolis antarctica*, where it forms spherical clumps and has terete branches.

Mychodea marginifera

TYPE LOCALITY: Port Phillip Heads, Victoria.
DISTRIBUTION: Geraldton, Western Australia, around southern Australia to Western Port, Victoria, and around Tasmania.

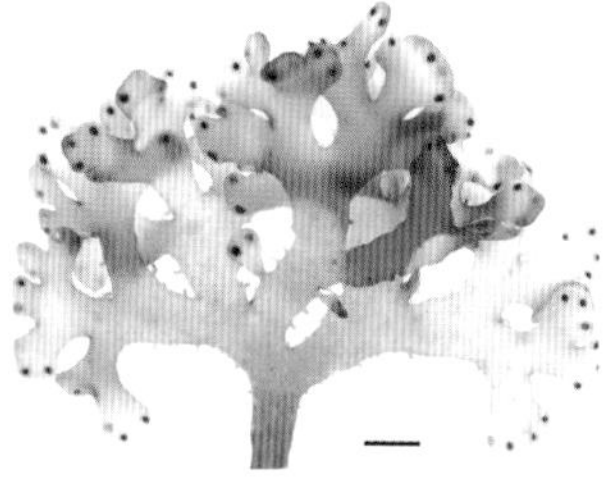

Mychodea marginifera (Cape Peron, WA).

(above) *Mychodea marginifera* – branch detail (Cape Peron, WA). Scale = 3 mm.

(top) *Mychodea pusilla* – as illustrated by Harvey (1863, pl. 266, as *Acanthococcus pusillus*).

Mychodea pusilla

TYPE LOCALITY: King George's Sound, Western Australia.
DISTRIBUTION: Geraldton, Western Australia, around southern Australia to Walkerville, Victoria, and around Tasmania.

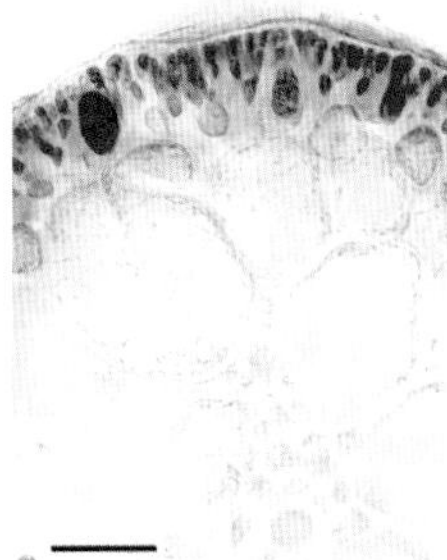

Mychodea disticha

TYPE LOCALITY: East coast of Tasmania.
DISTRIBUTION: Port Denison, Western Australia, to Walkerville, Victoria, and around Tasmania.
FURTHER READING: Kraft (1978); Kraft & Womersley (1994c); Kraft & Saunders (2017); Scott (2017).

Mychodea pusilla (Cape Peron, WA)

(middle, left) *Mychodea pusilla* – branch detail (Cape Peron, WA). Scale = 1 mm.

(middle, right) *Mychodea pusilla* – section of thallus with tetrasporangia (Cape Peron, WA). Scale = 50 µm.

Mychodea disticha (Cape Peron, WA).

Nizymenia

Nizymenia includes three species, mostly restricted to southern and south-western Australia. Depending on the species, thalli are either flattened and branched from the margins (*Nizymenia australis* and *Nizymenia conferta*) or terete and subdichotomously branched (*Nizymenia furcata*). Structurally the thallus is uniaxial, with each axial cell bearing three or four lateral filaments that traverse the medulla and form a cellular cortex. A dense, rhizoidal medulla develops around the axial filament.

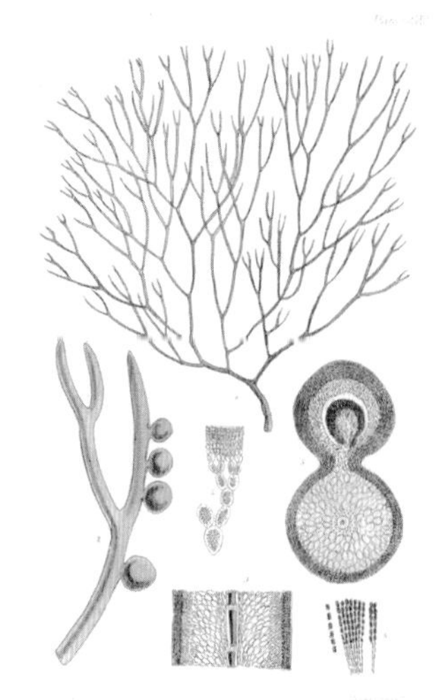

Nizymenia conferta

TYPE LOCALITY: Australia.
DISTRIBUTION: Geraldton, Western Australia, around southern mainland Australia to Port Phillip Heads, Victoria, and north-west Tasmania.

Nizymenia furcata

TYPE LOCALITY: Port Phillip Heads, Australia.
DISTRIBUTION: Rottnest Island, Western Australia, around southern mainland Australia to Port Phillip Heads, Victoria.
FURTHER READING: Womersley (1994); Chiovitti et al. (1995).

Nizymenia conferta (Rottnest Island, WA).

(left) *Nizymenia conferta* – partial section of thallus (Hamelin Bay, WA). Scale = 30 µm.

Nizymenia furcata (Rottnest Island, WA).

(top) *Nizymenia furcata* – as illustrated by Harvey (1862, pl. 215, as *Heringia furcata*).

Phacelocarpus

Thalli of *Phacelocarpus* are generally readily recognised by their terete to compressed axes bearing alternate lateral spines. Structurally the thallus is uniaxial, with each axial cell bearing four lateral filaments, the first-produced forming one of the marginal spines. All lateral filaments ultimately form a pseudoparenchymatous cortex. Rhizoids are produced from the lower cells of lateral filaments and surround the central axial filament. Reproductive structures arise on the spinous branches or on short stalks that develop between them. Five species of *Phacelocarpus* can be found in Australia, separated by features of their branching pattern and position of reproductive structures. The genus is superficially similar to *Delisea pulchra* (see p. 46), but can be separated by its filamentous medulla and lateral cystocarps. *Delisea pulchra* has a pseudoparenchymatous medulla and cystocarps are borne near the branch apices.

Phacelocarpus apodus

TYPE LOCALITY: Southern Australia.
DISTRIBUTION: From Perth, Western Australia, around southern mainland Australia and Tasmania to Broulee, New South Wales.
FURTHER READING: Womersley (1994).

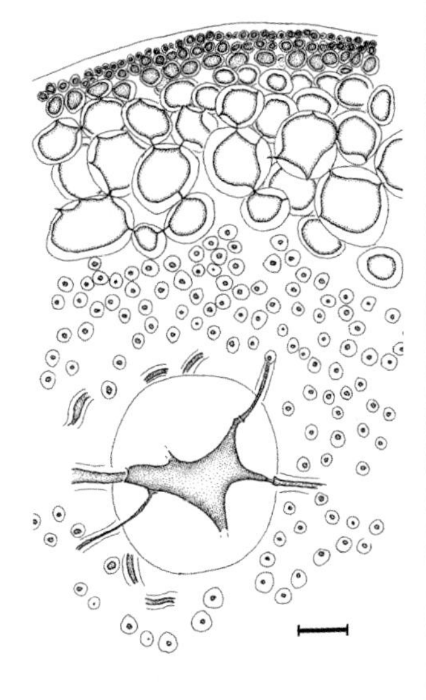

Phacelocarpus apodus (Rottnest Island, WA).

(above) *Phacelocarpus apodus* – partial section of thallus (Rottnest Island, WA). Scale = 30 µm.

Stenogramma

Stenogramma is a genus of three species in Australia, including the widespread *Stenogramma interruptum*, originally described from Spain and known from numerous temperate locations worldwide. Australian plants can reach 15 cm tall and have flattened, subdichotomously divided, linear blades, with smooth, rarely proliferous, margins. Structurally the thallus has a medulla of large cells grading into a smaller celled cortex. The genus is unique in having, in female gametophytes, procarps and carposporophytes arranged along a central line, giving the appearance of an interrupted midrib (hence the species epithet 'interruptum'). Tetrasporangia are cruciately divided and arise in scattered nemathecia on the fronds.

Stenogramma interruptum

TYPE LOCALITY: Cadiz, Spain.
DISTRIBUTION: Nuyts Reef, South Australia, to Crawfish Rock, Western Port Bay, Victoria, and south-east Tasmania; Arrawarra, New South Wales. New Zealand. Widespread in the cold temperate northern hemisphere.
FURTHER READING: Millar (1990); Lewis & Womersley (1994).

Stenogramma interruptum (Bicheno, Tas; photo G.J. Edgar).

(top) *Stenogramma interruptum* – as illustrated by Harvey (1862, pl. 220, as *Stenogramme interrupta*).

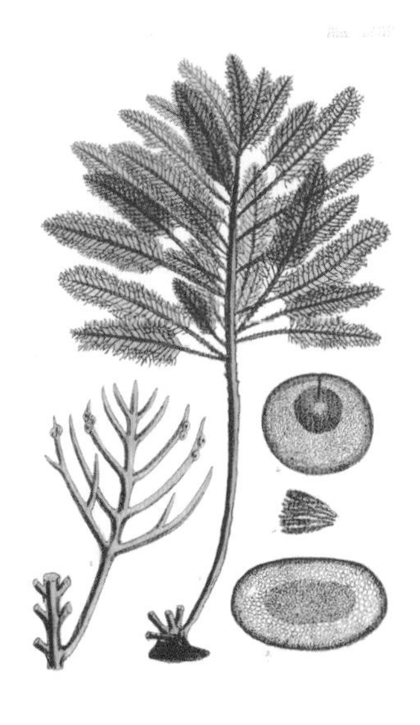

Callophycus

Species of *Callophycus* are commonly encountered in Australian subtidal habitats. They are generally robust, cartilaginous plants with flattened branches that in some species have a distinct midrib. Branching is pinnate and serrations can be present on the branch margins. Structurally the thallus is multiaxial. The medulla is entirely filamentous and is clearly demarcated from the pseudoparenchymatous cortex. Of the four species included here, *Callophycus serratus* is restricted to warmer waters and has serrated margins. *Callophycus harveyanus* and *Callophycus oppositifolius* have more temperate distributions and have smooth margins. They can be separated by their appearance – *Callophycus oppositifolius* has slender lateral branches on a distinctly broader main axis, while the axes of *Callophycus harveyanus* are more uniform in width. *Callophycus dorsiferus* has a closely bipinnate habit, the smaller branches with marginal serrations.

Callophycus dorsiferus

TYPE LOCALITY: Australia.
DISTRIBUTION: From Port Denison to Hamelin Bay, Western Australia.

Callophycus harveyanus

TYPE LOCALITY: Fremantle, Western Australia.
DISTRIBUTION: From Port Denison, Western Australia, around south-western Australia to Eucla, South Australia.

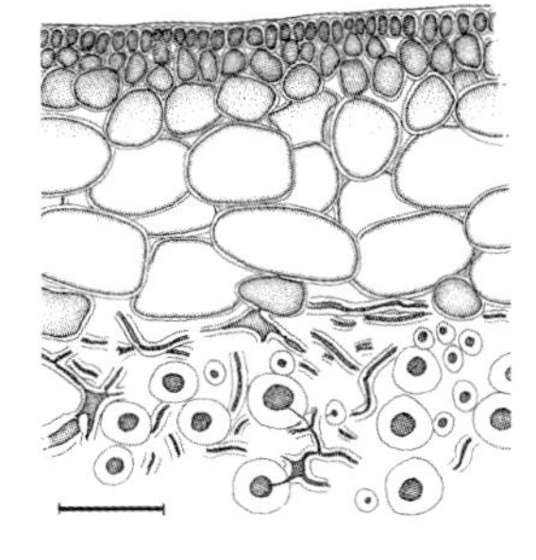

(top) *Callophycus oppositifolius* – as illustrated by Harvey (1862, pl. 187, as *Thysanocladia oppositifolia*).

Callophycus dorsiferus (Canal Rocks, WA; photo G.J. Edgar).

Callophycus harveyanus (Cape Peron, WA).

(far right) *Callophycus harveyanus* – detail of upper branches (Cape Peron, WA). Scale = 5 mm.

(right) *Callophycus harveyanus* – partial section of thallus (Rottnest Island, WA). Scale = 50 µm.

Callophycus oppositifolius

TYPE LOCALITY: Australia.
DISTRIBUTION: From the Houtman Abrolhos, Western Australia, around south-western Australia to Kangaroo Island and southern Yorke Peninsula, South Australia.

Callophycus serratus

TYPE LOCALITY: Tonga.
DISTRIBUTION: Known from the Houtman Abrolhos, Western Australia, and Cape York, Queensland. Widespread in warmer waters of the Indo-Pacific.
FURTHER READING: Kraft (1984b); Womersley (1994).

Callophycus oppositifolius (Rottnest Island, WA).

Callophycus serratus (Houtman Abrolhos, WA).

Contarinia

Contarinia is known in Australia by two rare species, of which *Contarinia pacifica* is included here. Plants are prostrate, with flattened blades growing closely to the substratum. Structurally the thallus can be uniaxial or multiaxial, depending on the species, with horizontally growing filaments giving rise to assurgent filaments on both the upper and lower surfaces. Vesicular cells occur at the tips of filaments. Tetrasporangia are formed in raised patches and are irregularly cruciately to irregularly zonately divided.

Contarinia pacifica

TYPE LOCALITY: Easter Island, south-eastern Pacific Ocean.
DISTRIBUTION: In Australia known only from the Rowley Shoals, Western Australia. Easter Island, south-eastern Pacific Ocean.
FURTHER READING: Dixon & Huisman (2018).

Contarinia pacifica (Rowley Shoals, WA).

Portieria

Portieria was previously thought to be represented in Australia by the single species *Portieria hornemannii*, which was regarded as an extremely common plant in tropical and subtropical waters of the Indian and western Pacific Oceans. However, studies have shown that several cryptic species may be present, these being genetically distinguishable but with no discernible morphological differences. For convenience, and until the taxonomy is settled, a broadly defined *Portieria hornemannii* is recognised here. Thalli are profusely branched, with flattened, narrow axes that arise in an alternate-distichous or subdichotomous pattern. Apices of most axes are strongly incurved. Structurally the thallus is uniaxial, with each axial cell producing from one to four lateral branches. Species of *Portieria* usually occur in shallow waters and are a conspicuous element on tropical reef flats.

Portieria hornemannii

TYPE LOCALITY: Red Sea.
DISTRIBUTION: From the Houtman Abrolhos, Western Australia, around northern Australia to southern New South Wales. Warmer waters of the Indo-Pacific.
FURTHER READING: Millar (1990); Payo et al. (2013); Dixon & Huisman (2018); Leliaert et al. (2018).

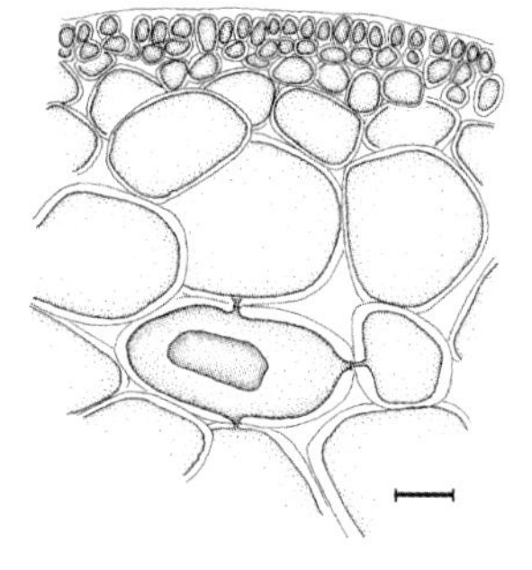

Portieria hornemannii (Heron Island, Qld).

(above) *Portieria hornemannii* – partial section of thallus (Ningaloo Reef, WA). Scale = 30 µm.

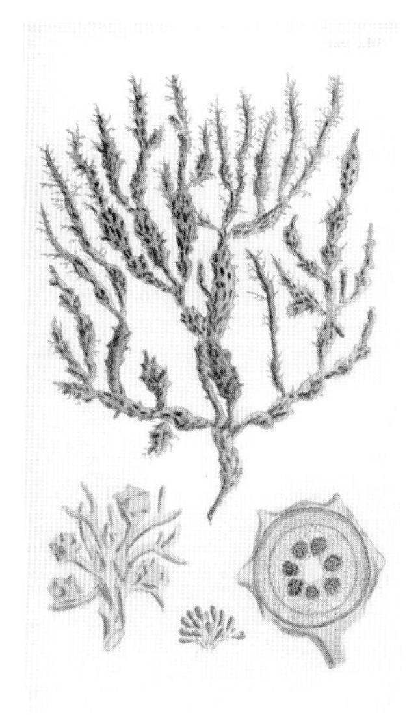

Betaphycus

The genus *Betaphycus* includes one of the more distinctive of the Australian algae, *Betaphycus speciosus*, a common species in the south-west where it is generally found attached to rock in high energy zones. Thalli are generally coarse and cartilaginous, with numerous short lateral spines. Structurally the thallus is multiaxial and has a filamentous medulla bordered by a thick cortex of stellate cells. Species of *Betaphycus* were once included in the closely related genus *Eucheuma*. *Betaphycus speciosus* was the 'jelly weed' of the early Swan River Colonists, who collected drift specimens that were then boiled to extract the carrageenan.

Betaphycus speciosus

TYPE LOCALITY: Western Australia.
DISTRIBUTION: Known from the northern Kimberley coast to Perth, Western Australia. Mauritius. Madagascar. Réunion. New Caledonia.
FURTHER READING: Dumilag et al. (2014); Huisman (2018).

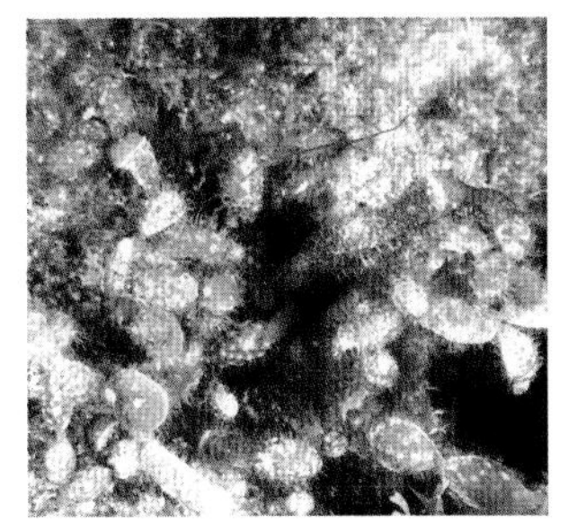

Betaphycus speciosus – a greenish plant from the tropics (Long Reef, WA).

(top) *Betaphycus speciosus* – as illustrated by Harvey (1859a, pl. 64, as *Eucheuma speciosum*).

(above) *Betaphycus speciosus* (Coral Bay, WA).

Eucheuma

The genus *Eucheuma* is at its most diverse in the tropics, where several species are known to occur. Thalli are generally coarse and cartilaginous, often with short lateral spines. Structurally the thallus is multiaxial and has a filamentous medulla bordered by a thick cortex of stellate cells. *Eucheuma denticulatum* is restricted to tropical regions. In many countries, *Eucheuma* is farmed for carrageenan, a cell-wall component that is used as a thickening agent in a variety of foodstuffs and in other applications.

Eucheuma denticulatum

TYPE LOCALITY: Supposedly Cape of Good Hope, South Africa.
DISTRIBUTION: From the Houtman Abrolhos (rarely) and North West Cape region, Western Australia, around northern Australia to Queensland. Warmer waters of the Indo-Pacific.
FURTHER READING: Kraft (1972); Cribb (1983); Zuccarello et al. (2006); Huisman (2018).

Eucheuma denticulatum (Long Reef, WA).

Meristotheca

Two species of *Meristotheca* have been recorded from Australian waters: the widespread *Meristotheca papulosa* and *Meristotheca procumbens* from northern New South Wales and Lord Howe Island (not included here). Thalli of *Meristotheca papulosa* are generally flattened and blade-like, but can display a wide range of branching patterns, from profusely branched in an alternate or dichotomous manner, to mostly undivided with marginal proliferations. Living thalli often have a mottled appearance. Structurally the thallus is multiaxial, with a filamentous medulla (often with mixed stellate cells) grading into a pseudoparenchymatous cortex.

Meristotheca papulosa

TYPE LOCALITY: Hodeida, Yemen.
DISTRIBUTION: From Rottnest Island, Western Australia, around northern Australia to Lord Howe Island, New South Wales. Warmer waters of the Indo-Pacific.
FURTHER READING: Gabrielson & Kraft (1984).

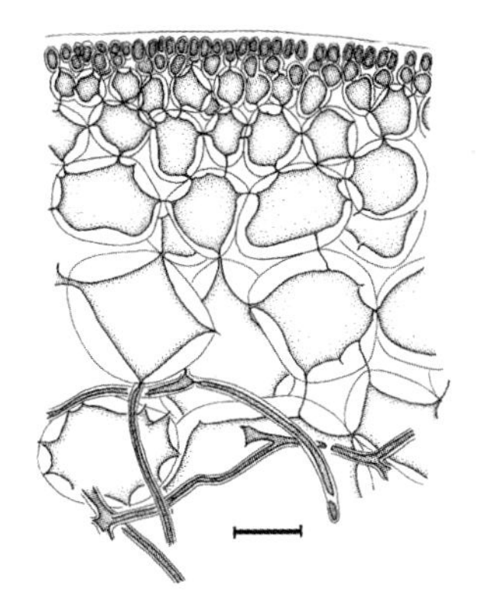

Meristotheca papulosa (Bateman Bay, WA).

(above) *Meristotheca papulosa* – partial section of thallus (Houtman Abrolhos, WA). Scale = 30µm.

Mimica

A genus separated from the closely related *Eucheuma*, based on the lack of an axial core and DNA sequence analyses. Only the single species *Mimica arnoldii* is known from Australia, an unusual species that closely mimics the appearance of *Acropora* corals, a characteristic that has presumably evolved to deter herbivorous fishes. The genus name was coined to allude to this characteristic.

Mimica arnoldii

TYPE LOCALITY: Gisser Island (Pulau Geser), near Seram, Indonesia
DISTRIBUTION: Warmer waters of the Indo-Pacific.
FURTHER READING: Richards & Huisman (2014); Huisman (2018, as Eucheuma); Santiañez & Wynne (2020).

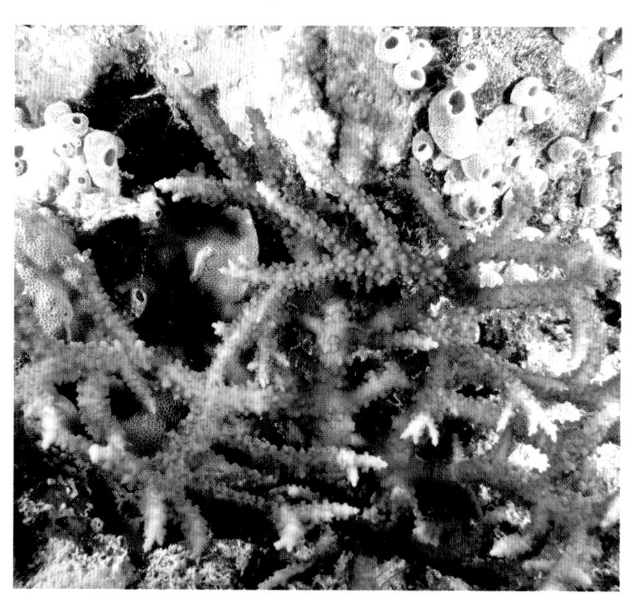

Mimica arnoldii (Ashmore Reef).

The coral *Acropora abrotanoides*, mimicked by the seaweed *Mimica arnoldii* (Ashmore Reef).

Sarconema

Sarconema includes only three species, with *Sarconema filiforme* and *Sarconema scinaioides* found in Australia. Thalli are cartilaginous with terete axes that are regularly dichotomously branched. Structurally the thallus is multiaxial and has a narrow medulla of longitudinal filaments and a pseudoparenchymatous cortex. Cystocarps are embedded in the axes and cause a noticeable swelling. Tetrasporangia are scattered and zonately divided. *Sarconema filiforme* is similar in habit to *Gracilaria cliftonii* (p. 113) but differs in the regularity of its branching and in the presence of a filamentous medulla. All species of *Gracilaria* have a pseudoparenchymatous medulla. *Sarconema scinaioides* is typically smaller than *Sarconema filiforme*, but with thicker axes. In the Perth region of Western Australia, the typical dark red, elongate form of *Sarconema filiforme* is common in subtidal habits. *Sarconema scinaioides* is also present, but tends to be pale and occurs on reef crests in the shallow subtidal.

Sarconema filiforme

TYPE LOCALITY: Western Australia.

DISTRIBUTION: From Hamelin Bay, Western Australia, around northern Australia to Jervis Bay, New South Wales, and rare occurrences in southern Australia. Warmer waters of the Indo-Pacific.

Sarconema scinaioides

TYPE LOCALITY: In the vicinity of Karachi, Pakistan.

DISTRIBUTION: From the tropics south to the Perth region, Western Australia. Apparently widespread in the Indian Ocean, introduced in the Mediterranean.

FURTHER READING: Papenfuss & Edelstein (1974); Womersley (1994).

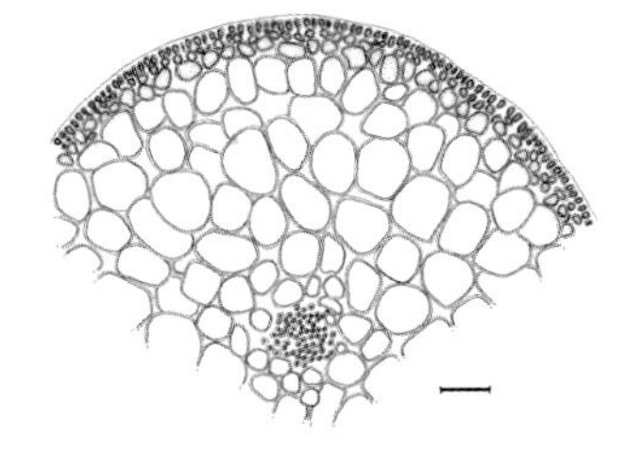

Sarconema filiforme (Cape Peron, WA).

(left) *Sarconema filiforme* – partial section of thallus (Rottnest Island, WA). Scale = 100 µm.

Sarconema scinaioides (Cape Peron, WA).

Solieria

Thalli of *Solieria* have terete to slightly compressed axes that are generally irregularly branched and have a distinct constriction at the base of lateral branches. Structurally the thallus is multiaxial, and mature axes have a laxly filamentous medulla and a pseudoparenchymatous cortex. Cystocarps are embedded in the axes and cause a noticeable swelling when mature. Tetrasporangia are scattered and zonately divided. Several species of *Solieria* are known from Australia, with *Solieria robusta* the most commonly encountered.

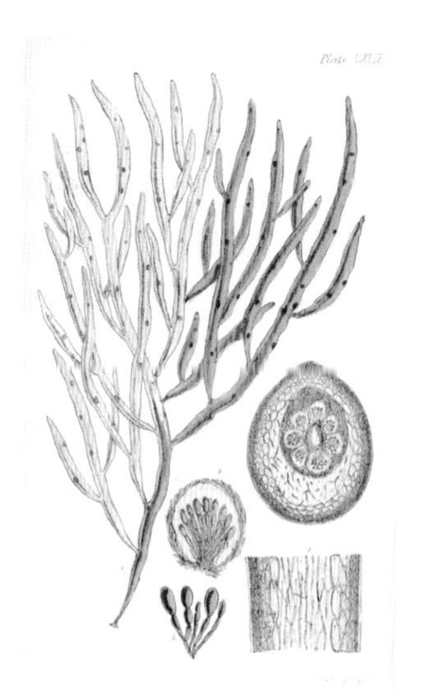

Solieria robusta

TYPE LOCALITY: Australia.
DISTRIBUTION: Widespread in the Indian and western Pacific Oceans.
FURTHER READING: Gabrielson & Kraft (1984); Womersley (1994); Scott (2017); Huisman (2018).

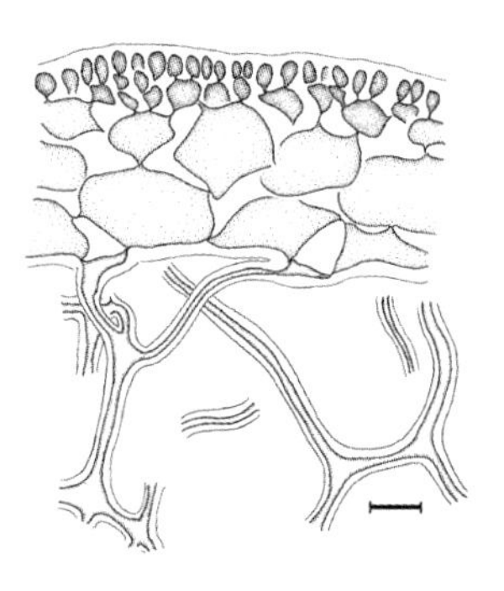

Solieria robusta (Cape Peron, WA).

(top) *Soliera robusta* – as illustrated by Harvey (1860, pl. 149, as *Solieria australis*).

(above) *Solieria robusta* – partial section of thallus (Houtman Abrolhos, WA). Scale = 30µm.

(left) *Solieria robusta* – cystocarpic specimen (Carnac Island, WA).

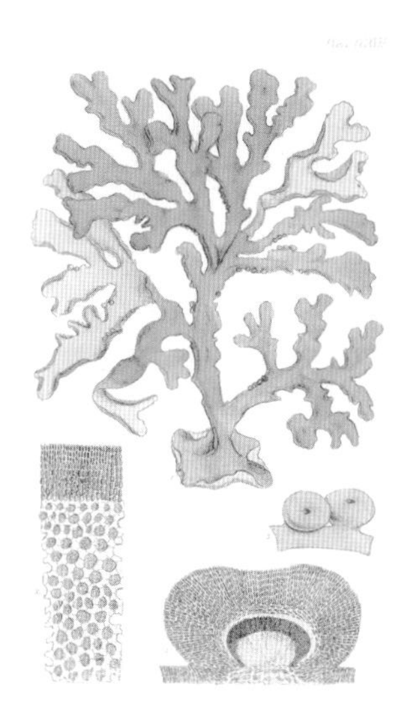

Curdiea

Curdiea includes several species found in Australian waters, of which *Curdiea obesa* and *Curdiea irvineae* are included here. Both are thick, coarse plants with compressed branches. Structurally the thallus is multiaxial, with a medulla of large, colourless cells grading into a cortex of smaller, pigmented cells that are often in anticlinal rows. Cystocarps are proberant and are usually restricted to the thallus margins. Tetrasporangia are cruciately divided and grouped into fertile patches. *Curdiea obesa* and *Curdiea irvineae* are both partially decumbent and can be separated by the narrower, linear branches of the latter. Harvey (1862) described *Curdiea obesa* (as *Sarcocladia obesa*) as a 'clumsy-looking alga' and commented, 'It is not a plant likely, by its beauty, to attract any but a botanist'.

Curdiea irvineae

TYPE LOCALITY: Geographe Bay, Western Australia.
DISTRIBUTION: From Seabird to Geographe Bay, Western Australia.

Curdiea obesa

TYPE LOCALITY: Rottnest Island, Western Australia.
DISTRIBUTION: From the Houtman Abrolhos, Western Australia, around southern Australia to Nora Creina, South Australia.
FURTHER READING: Womersley (1996).

Curdiea irvineae (Map Reef, WA).

Curdiea obesa (Cape Peron, WA).

(left) *Curdiea obesa* – partial section showing medulla and dorsal cortex (Rottnest Island, WA). Scale = 30 µm.

(top) *Curdiea obesa* – as illustrated by Harvey (1862, pl. 217, as *Sarcocladia obesa*).

Gracilaria

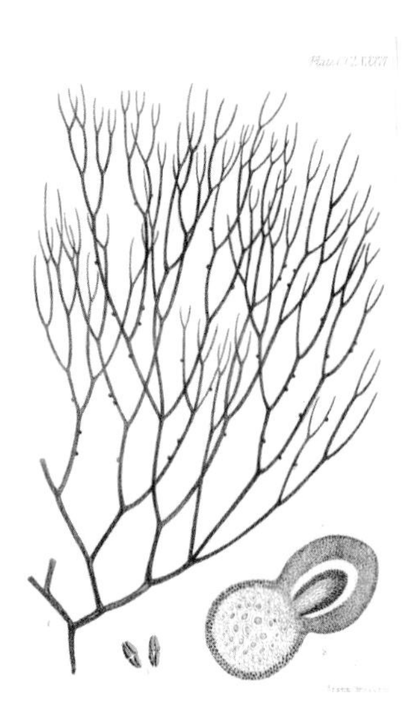

Species of *Gracilaria* come in a wide variety of forms. Some are terete and regularly dichotomously branched, while others have a flattened thallus with lateral branches. Others again have very coarse, cartilaginous branches with numerous protuberances. All share the common vegetative feature of a pseudoparenchymatous medulla, without an easily recognisable central axis. This feature is surprisingly rare in the red algae and the only taxa likely to be confused with species of *Gracilaria* are some members of the Rhodymeniaceae. Reproductive material can be required to positively identify *Gracilaria*. It is characterised by a cystocarp in which the base of the gonimoblast is broad and covers the floor of the cystocarp cavity, while in the Rhodymeniaceae, the gonimoblast is borne on a central stalk. The species of *Gracilaria* are separated on details of their habit, cross-section (whether there is an abrupt or gradual transition in cell size between the medulla and cortex), and arrangement of spermatangia. The last-mentioned feature is only rarely observed and hence many identifications can only be tentative.

Gracilaria blodgettii

TYPE LOCALITY: Key West, Florida.
DISTRIBUTION: South-western Australia. Possibly China and Japan. USA.

Gracilaria canaliculata

TYPE LOCALITY: Wagap, New Caledonia.
DISTRIBUTION: From the Houtman Abrolhos, Western Australia, around northern Australia. Indo-Pacific.

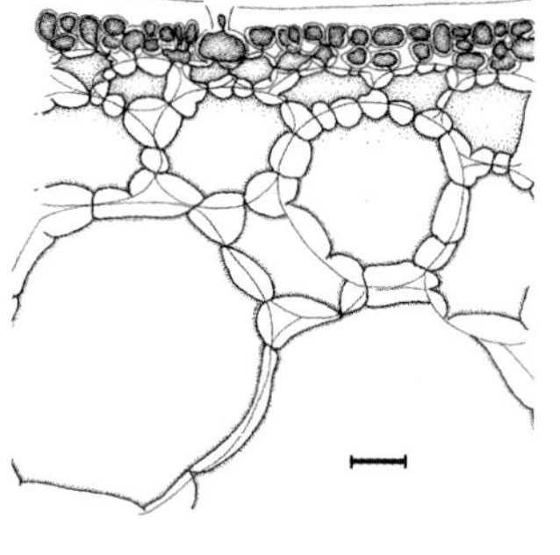

Gracilaria cliftonii

TYPE LOCALITY: Fremantle, Western Australia.
DISTRIBUTION: From the Perth region, Western Australia, around southern mainland Australia and Tasmania to Walkerville, Victoria.

Gracilaria flagelliformis

TYPE LOCALITY: Western Australia.
DISTRIBUTION: From Geraldton to Geographe Bay, Western Australia.

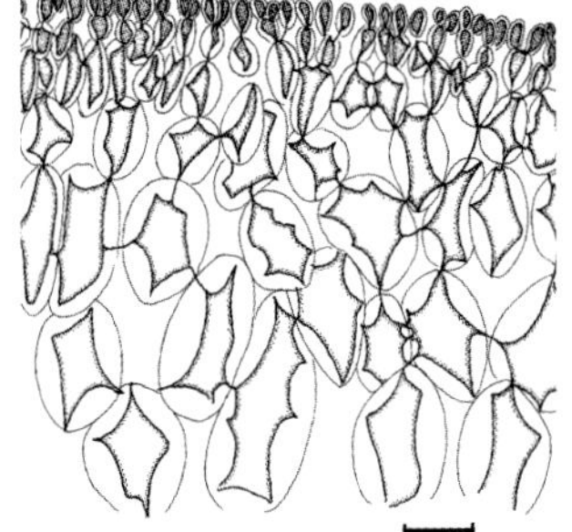

Gracilaria salicornia

TYPE LOCALITY: Manila Bay, Philippines.
DISTRIBUTION: Widespread in warmer waters of the Indian and Pacific Oceans.

Opposite page:

Gracilaria blodgettii (Perth, WA).

Gracilaria canaliculata (Ashmore Reef).

(bottom right) *Gracilaria canaliculata* – partial section of thallus (Houtman Abrolhos, WA). Scale – 30 µm.

(opposite top) *Gracilaria cliftonii* – as illustrated by Harvey (1863, pl. 286, as *Gracilaria furcellata*).

Gracilaria cliftonii (Map Reef, WA).

Gracilaria flagelliformis (Seabird, WA).

Gracilaria flagelliformis – partial section of thallus (Jurien Bay, WA). Scale = 30 µm.

Gracilaria salicornia (Barrow Island, WA).

Gracilaria textorii

TYPE LOCALITY: Japan.
DISTRIBUTION: Apparently widespread in the Indo-Pacific.

Gracilaria vieillardii

TYPE LOCALITY: Wagap, New Caledonia.
DISTRIBUTION: Warmer waters of the Indian and western Pacific Oceans.

Gracilaria eucheumatoides

TYPE LOCALITY: Ryukyu Islands, Japan.
DISTRIBUTION: North West Cape region, Western Australia, and northern Australia. Warmer waters of the Indo-Pacific.

Gracilaria textorii (Houtman Abrolhos, WA).

Gracilaria vieillardii (James Price Point, WA).

Gracilaria eucheumatoides (Montebello Islands, WA).

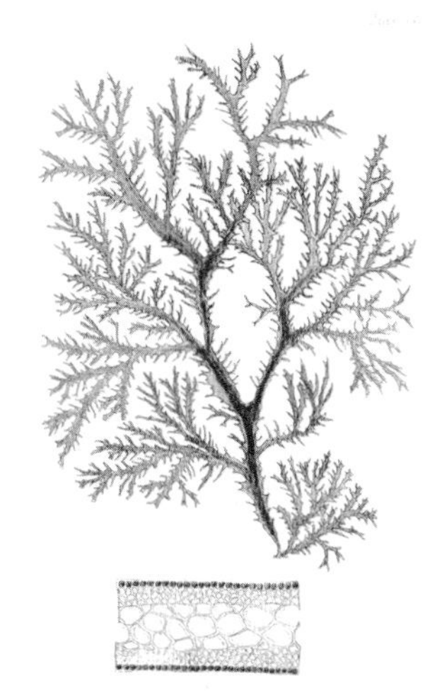

Gracilaria preissiana

TYPE LOCALITY: Western Australia.
DISTRIBUTION: From Geraldton and the Houtman Abrolhos to King George Sound, Western Australia.

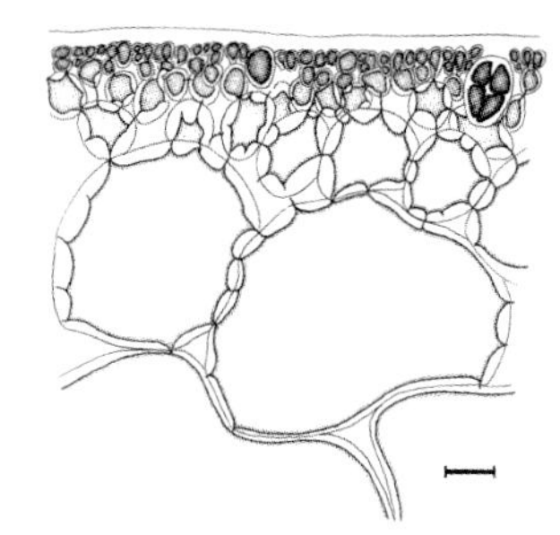

Gracilaria urvillei

TYPE LOCALITY: Near Insel Toud [Warrior Islet], Torres Strait, Queensland.
DISTRIBUTION: Known from northern Australia, southern Java, New Guinea, Singapore and the China Sea.
FURTHER READING: Withell et al. (1994); Womersley (1996); Millar & Xia (1997); Gurgel & Fredericq (2004); Lyra et al. (2021); Gurgel et al. (2018).; Huisman (2018); Lyra et al. (2021).

Gracilaria preissiana (Cape Peron, WA).

(top) *Gracilaria preissiana* - as illustrated by Harvey (1859a, as *Calliblepharis preissiana*).

(above) *Gracilaria preissiana* – partial section of thallus with cruciate tetrasporangia (Houtman Abrolhos, WA). Scale = 30 µm.

Gracilaria urvillei (James Price Point, WA).

Melanthalia

Melanthalia includes four species, three of which are endemic to southern Australia and New Zealand, and a fourth which is found only in New Caledonia. Plants are readily recognised by their consistent subdichotomous branching, with linear, compressed branches, each with a terminal cap of smaller cells. Structurally the thallus is multiaxial and pseudoparenchymatous throughout. Cystocarps are protuberant and arise on branch margins, with a thick pericarp. Tetrasporangia are cruciately divided and occur in shallow nemathecia below branch apices. The three Australian species are similar in structure, differing primarily in their branch widths and degree of compression. Branches in *Melanthalia obtusata* are 2–4 mm broad and strongly compressed; those of *Melanthalia abscissa* are similar but only 0.7–1.3 mm broad. Both species are found on high energy coasts.

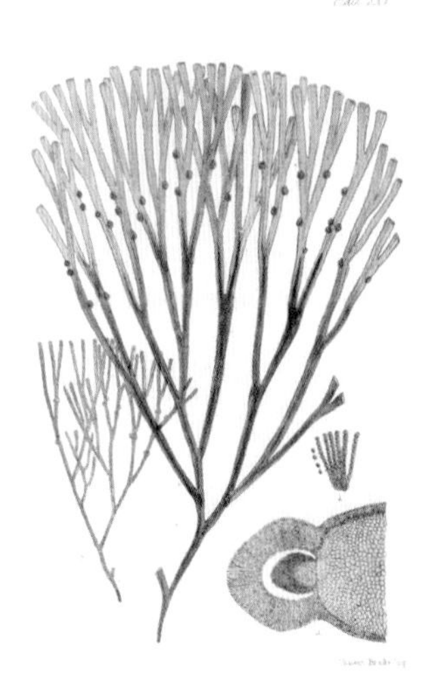

Melanthalia abscissa

TYPE LOCALITY: New Zealand.

DISTRIBUTION: Wedge Island, South Australia, to Wilsons Promontory, Victoria, and around Tasmania.

Melanthalia obtusata

TYPE LOCALITY: South-east Tasmania.

DISTRIBUTION: Pennington Bay, Kangaroo Island, South Australia, to Port Phillip Heads, Victoria, and Deal Island, Bass Strait, and around Tasmania.

FURTHER READING: Womersley (1996); Edgar (2008); Nelson (2013); Nelson et al. (2013); Scott (2017).

Melanthalia abscissa (Hogan Island, Tas; photo G.J. Edgar).

Melanthalia obtusata (Ile de Golfe, Tas; photo G.J. Edgar).

(top) *Melanthalia obtusata* – as illustrated by Harvey (1858, pl. 25). The smaller plant on the left (figure 2) is *Melanthalia abscissa*.

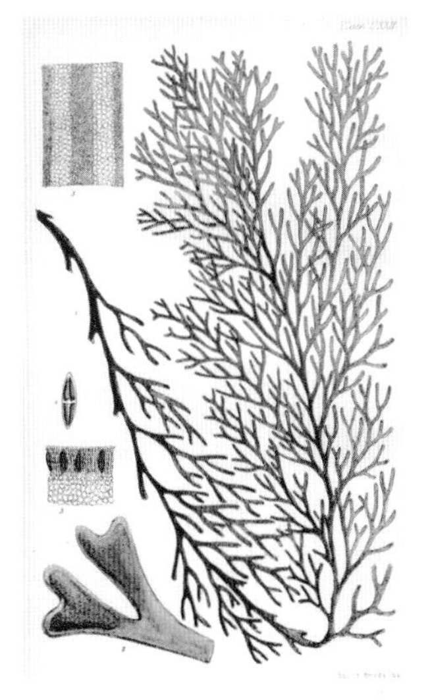

Carpopeltis

Thalli of *Carpopeltis* are cartilaginous, with flattened axes that are alternately branched. Structurally the thallus is multiaxial, and mature axes are composed of a narrow filamentous medulla that grades into a broad pseudoparenchymatous cortex. Cystocarps are immersed in the thallus and tetrasporangia are cruciately divided and arise in slightly raised patches. Several species of *Carpopeltis* are known in Australia. *Carpopeltis elata* is a large plant with narrow branches (mostly 1–2 mm broad) that are flexuous. Two other species occur in Australia: *Carpopeltis spongeaplexa* has lower axes that are invariably covered with a layer of sponge, while *Carpopeltis phyllophora* can be distinguished by its broader (to 3–5 mm), sponge-free axes.

Carpopeltis elata

TYPE LOCALITY: Rottnest Island, Western Australia.
DISTRIBUTION: From Geraldton south to the Recherche Archipelago, Western Australia.
FURTHER READING: Womersley & Lewis (1994).

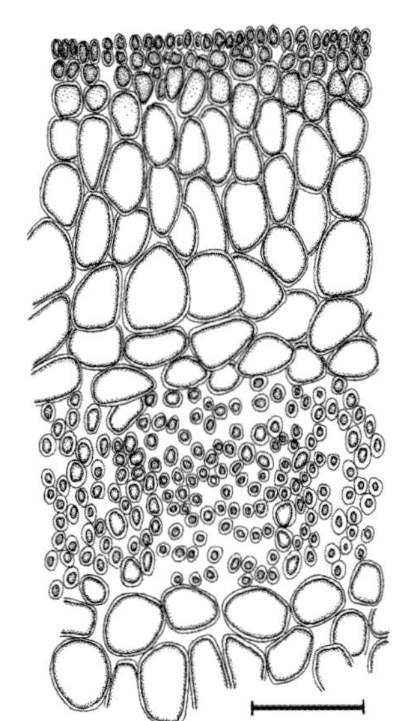

Carpopeltis phyllophora

TYPE LOCALITY: Port Arthur, Tasmania.
DISTRIBUTION: Geraldton, Western Australia, to Phillip Island, Victoria, and around Tasmania.

Carpopeltis elata (Cape Peron, WA).

(above) *Carpopeltis elata* – partial section of thallus (Rottnest Island, WA). Scale = 50 µm.

(top) *Carpopeltis elata* – as illustrated by Harvey (1860, pl. 122, as *Acropeltis elata*).

Carpopeltis phyllophora (Cape Peron, WA).

Codiophyllum

Codiophyllum is a genus of two species, with only *Codiophyllum flabelliforme* known in Australia. Thalli have a substantial woody stalk bearing fan-shaped blades composed of a meshwork of several layers of narrow branches. The thallus is generally covered by an encrusting sponge that obscures the algal branches. Structurally the axes have a filamentous medulla with occasional refractive ganglionic cells and a pseudoparenchymatous cortex four to eight cells thick. Reproductive structures are borne in specialised small branches that arise near the upper margins of the thallus and are clear of the sponge layer. *Codiophyllum flabelliforme* is often difficult to recognise as an alga because of its sponge coating. It is usually found on rocky reefs, growing on vertical walls.

Codiophyllum flabelliforme

TYPE LOCALITY: Western Australia.
DISTRIBUTION: From the Houtman Abrolhos to King George Sound, Western Australia.
FURTHER READING: Scott et al. (1984); Womersley & Lewis (1994).

Codiophyllum flabelliforme (Rottnest Island, WA).

(top) *Codiophyllum flabelliforme* – as illustrated by Harvey (1859a, pl. 113, as *Thamnoclonium flabelliforme*).

(above) *Codiophyllum flabelliforme* – fertile branches (Rottnest Island, WA). Scale = 1 mm.

Corynomorpha

Corynomorpha is a genus of two species: *Corynomorpha prismatica* from the Indian Ocean and *Corynomorpha clavata* (not shown) from tropical America. Both are rare species. Plants are club-shaped and cartilaginous, and grow in clusters. The Australian thalli are unbranched, but branched specimens have been reported from other areas. Structurally the thallus is multiaxial, with a medulla composed of elongate filaments and stellate cells, grading abruptly into a pseudoparenchymatous cortex. Reproductive structures are borne in slightly swollen apical regions, generally with a constriction between the fertile region and the lower thallus.

Corynomorpha prismatica

TYPE LOCALITY: India.
DISTRIBUTION: Known from Barrow Island and the Montebello Islands, Western Australia, and Torres Strait, Queensland. Warmer seas of the Indo-Pacific.
FURTHER READING: Desikachary et al. (1990); Wynne (1995); Huisman (2018).

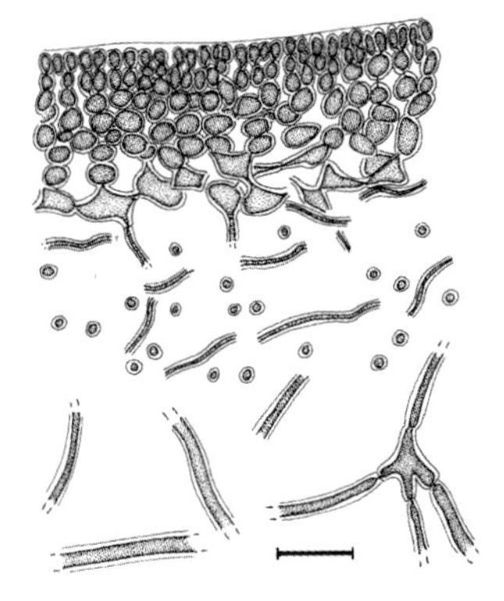

Corynomorpha prismatica (Montebello Islands, WA).

(above) *Corynomorpha prismatica* – partial section of thallus (Montebello Islands, WA). Scale = 20µm.

Cryptonemia

Cryptonemia includes some 30 species worldwide, with several found in southern and western Australian waters. Thalli are foliose (with lobed or ruffled margins) or regularly branched, depending on the species. Most species have a distinct cartilaginous stalk. Structurally the thallus is multiaxial, with a filamentous medulla and pseudoparenchymatous cortex, the latter often with an outer layer of anticlinal filaments. Refractive ganglioid cells are common in the medulla. Cystocarps are immersed in the thallus and tetrasporangia are scattered and cruciately divided. *Cryptonemia kallymenioides* is a distinctive species that grows on undercut rock walls, generally in partially shaded positions. The species is instantly recognisable by the presence of thick, woody stalks (these are apparently perennial), which, in section, show distinct growth rings. The frond is generally red with a conspicuous pink or yellow mottling. *Cryptonemia undulata* has, as the name suggests, strongly undulate upper branches.

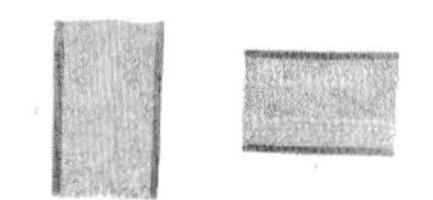

Cryptonemia kallymenioides

TYPE LOCALITY: Fremantle, Western Australia.
DISTRIBUTION: From the Houtman Abrolhos to Hamelin Bay, Western Australia, and Cannan Reefs, South Australia.

Cryptonemia undulata

TYPE LOCALITY: Port Phillip, Victoria.
DISTRIBUTION: Geraldton, Western Australia, to Western Port, Victoria. India.
FURTHER READING: Scott et al. (1982); Womersley & Lewis (1994).

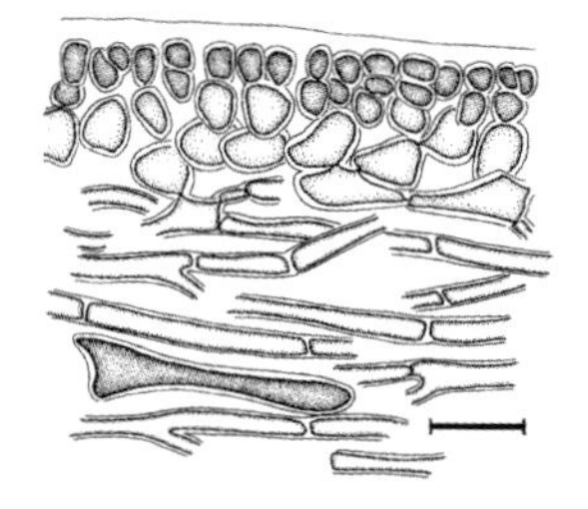

Cryptonemia kallymenioides (Rottnest Island, WA).

(left) *Cryptonemia kallymenioides* – partial section of frond (Rottnest Island, WA). Scale = 20 µm.

Cryptonemia undulata (Cape Peron, WA).

(top) *Cryptonemia undulata* – as illustrated by Harvey (1862, pl. 205).

Grateloupia

Species *Grateloupia* occur in many countries and the genus is a popular food in Asian and other cuisine. In Japan, it is known as *mukadenori* and the Hawaiians call it *limu huluhuluwaena*. Preparation for the table is minimal as plants are mostly eaten fresh. Some washing is usually necessary to remove sand, and the plants are then chopped into small pieces and lightly salted. They can then be used as a garnish in fish dishes or added to cooked beef at serving time. *Grateloupia subpectinata* can be common and often grows on intertidal reefs, where it can be found in rock pools with its base covered in sand. Plants are soft and slightly slippery to touch. They often grow in clusters and each plant is generally bushy with a pyramidal outline, although there is much variation. The stems are flattened and lateral branches arise in one plane. Internally the branches have a loose construction of sparse filaments. *Grateloupia imbricata* is seemingly an introduction to Australia, where it is currently known only from the Perth region of Western Australia. Plants typically occur on rock in the intertidal and shallow subtidal, generally on exposed coasts. They can often be yellow in colour, particularly in the upper branches. Branching is subdichotomous or irregular and branches are broader than those of *Grateloupia subpectinata*.

Grateloupia imbricata

TYPE LOCALITY: Shimoda, Japan.
DISTRIBUTION: Known from Spain, Canary Islands, the Azores, China, Japan, Korea and Taiwan. In Australia from Cottesloe to Cape Peron, Western Australia.

Grateloupia subpectinata

TYPE LOCALITY: Japan.
DISTRIBUTION: Widespread in temperate seas. This species is thought to have been introduced into harbour areas in Australia, New Zealand, Europe and Britain.
FURTHER READING: Nelson et al. (2013); Scott (2017).

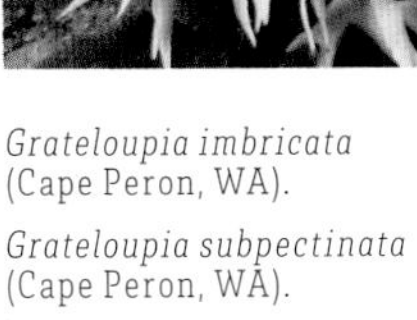

Grateloupia imbricata (Cape Peron, WA).

Grateloupia subpectinata (Cape Peron, WA).

Halymenia

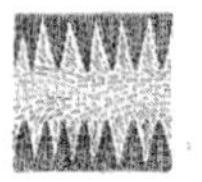

Thalli of *Halymenia* are foliose or irregularly to pinnately branched with lobed or ruffled margins. Structurally the thallus is multiaxial, with a filamentous medulla grading into an inner cortex of stellate cells and a peripheral layer of anticlinal filaments. Refractive ganglioid cells are common in the medulla. A distinctive feature of *Halymenia* is the presence, in the medulla, of numerous transversely aligned filaments connecting the cortical layers. Cystocarps are immersed. Several species occur in Australia and identifying them can be difficult. The following is a selection, with distinguishing features given under the species names.

Halymenia bullosa

Halymenia bullosa was previously included as the single species in the genus *Epiphloea*, but recent studies have moved it to *Halymenia*. Plants have a large, fleshy blade borne on a thick, rigid stipe. The blades are generally mottled in appearance and when mature often have numerous corrugations and depressions. Structurally the thallus is multiaxial, with a filamentous medulla that grades into a pseudoparenchymatous cortex of stellate cells. The outer cortical layer is composed of elongate, anticlinal cells. *Halymenia bullosa* is a distinctive species, recognizable by its unusual habit.

TYPE LOCALITY: Fremantle, Western Australia.
DISTRIBUTION: From Hamelin Bay to the North West Cape region, Western Australia, Lord Howe Island, New South Wales, and Norfolk Island.

Halymenia durvillei

Halymenia durvillei is a robust, richly branched species when mature, often with surface proliferations.

TYPE LOCALITY: Port Praslin, Papua New Guinea.
DISTRIBUTION: Widespread in the Indo-West Pacific.

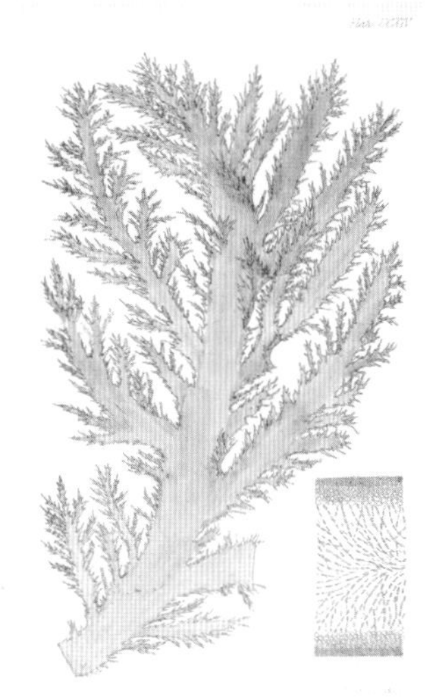

Halymenia harveyana

Halymenia harveyana was previously included as a subspecies of *Halymenia floresii*, but is now recognised as an independent species. It is pinnately branched and similar in appearance to *Halymenia durvillei*, but is more delicate and mostly lacks surface proliferations.

TYPE LOCALITY: Port Phillip Heads, Victoria.
DISTRIBUTION: Jurien Bay, Western Australia, around southern mainland Australia to Walkerville, Victoria.

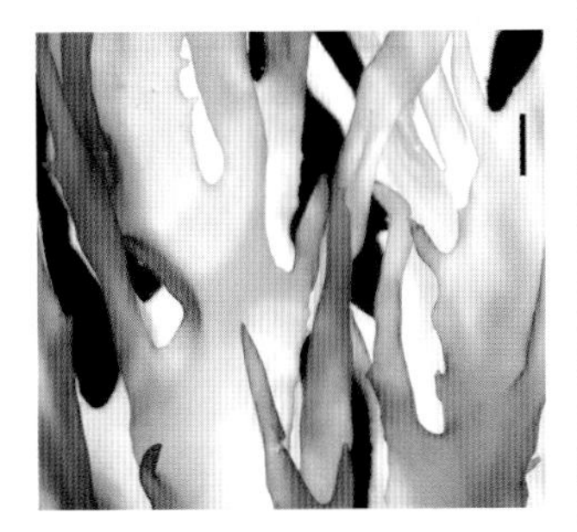

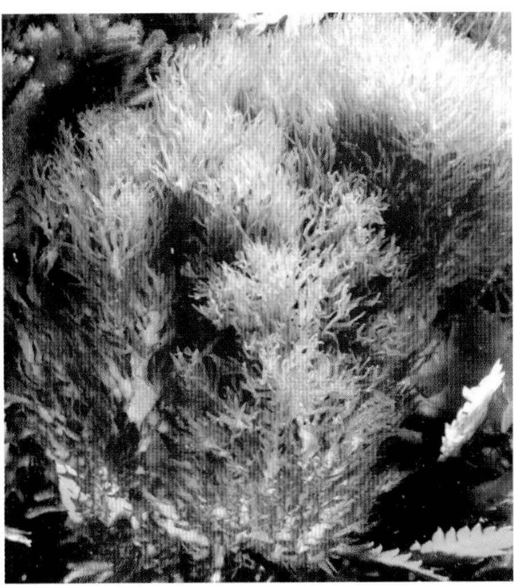

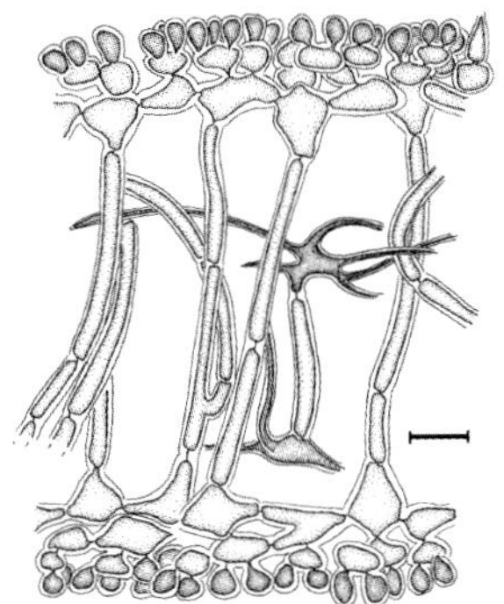

Halymenia maculata

This species is generally recognisable by its bullate blades, giving it the appearance of a marine cabbage.

TYPE LOCALITY: Mauritius.
DISTRIBUTION: Widespread in tropical areas of the Indo-West Pacific.

Opposite page:

Halymenia bullosa (Rottnest Island, WA).

(top) *Halymenia bullosa* – as illustrated by Harvey (1863, pl. 277, as *Schizymenia bullosa*).

Halymenia durvillei (Heron Island, Qld).

Halymenia harveyana (Cape Peron, WA).

(top) *Halymenia harveyana* – as illustrated by Harvey (1862, pl. 214, as *Halymenia floresia*).

(middle) *Halymenia harveyana* – detail of thallus (Cape Peron, WA). Scale = 1 mm.

(above) *Halymenia harveyana* – section of thallus (Rottnest Island, WA). Scale = 30 µm.

Halymenia maculata (Cassini Island, WA).

Halymenia malaysiana

Plants of *Halymenia malaysiana* are foliose, with a pale pink to red colour and an often mottled appearance.

TYPE LOCALITY: Pulau Merambong, Johor, southern Peninsular Malaysia.
DISTRIBUTION: In Australia known from Broome, Long Reef and Cassini Island, Western Australia. Malaysia. Philippines. Indonesia.
FURTHER READING: Womersley & Lewis (1994); Hernández-Kantún et al. (2012); Tan et al. (2015); Huisman & De Clerck (2018).

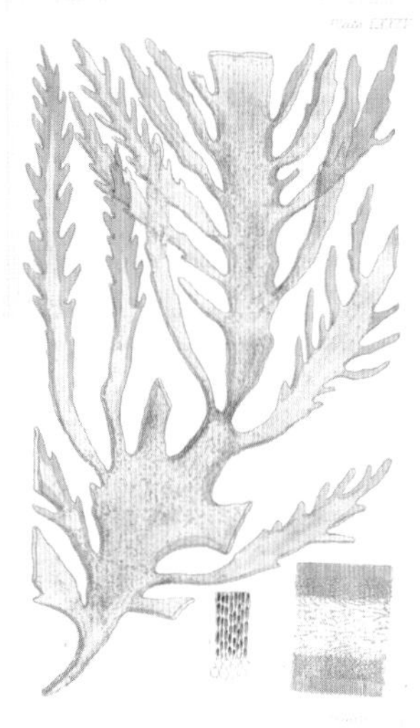

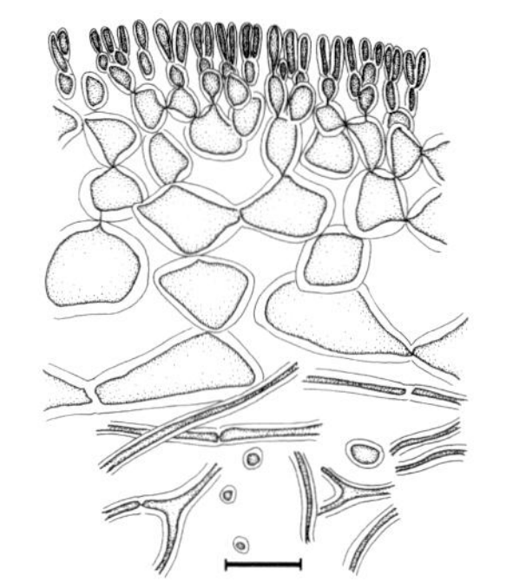

Halymenia ulvoidea

Another species previously placed in its own genus *Gelinaria*, but recently moved to *Halymenia*. Plants can grow to 75 cm in height and are often a bright yellow colour when viewed underwater. *Halymenia ulvoidea* is superficially similar to *Halymenia harveyana*, but is generally a coarser and larger plant.

TYPE LOCALITY: Western Australia.
DISTRIBUTION: From the Houtman Abrolhos, Western Australia, around southern mainland Australia to Walkerville Victoria, and northern Tasmania.
FURTHER READING: Scott (2017); Rodríguez-Prieto et al. (2022).

Halymenia malaysiana (Mavis Reef, WA).

Halymenia ulvoidea (Cape Peron, WA).

(top) *Halymenia ulvoidea* – as illustrated by Harvey (1859a, pl. 85, as *Gelinaria ulvoidea*).

(above) *Halymenia ulvoidea* – partial section of thallus (Rottnest Island). Scale = 20 µm.

Howella

A genus of three species from geographically distant localities; the type species *Howella gorgoniarum* is known only from deep waters off Bermuda and St. Croix, western Atlantic Ocean, and *Howella latifrons* is known from the Indian Ocean coast of South Africa. The Australian species *Howella lemanniana* was previously included in *Thamnoclonium* but was removed to *Howella* based DNA sequence analyses. *Howella lemanniana* is a rare species, known only from a few collections from the Perth region.

Howella lemanniana

TYPE LOCALITY: Fremantle.
DISTRIBUTION: Warmer waters of the Indo-Pacific.
FURTHER READING: Schneider et al. (2019).

Howella lemanniana – as illustrated by Harvey (1859a, pl. 114, as *Thamnoclonium lemannianum*).

Polyopes

Polyopes is a widespread genus of about 11 species, with some species considered to be invasive. It is characterised by a densely branched habit, subdichotomous branching, and slightly raised tetrasporangial nemathecia near branch ends. Structurally the medulla is filamentous, with some stellate medullary cells and inner cortical cells, and the cortex is composed of progressively smaller cells. Carpogonial branches and auxiliary cells occur in ampullae, with compact, immersed cystocarps. Tetrasporangia are cruciately divided. The genus includes several species in Australia, of which *Polyopes tasmanicus* is included here. This species is similar in appearance to some forms of *Grateloupia subpectinata* (see p. 121), which differ in being more robust and cartilaginous, in the abundance of spindle-shaped branches, and in details of reproductive structures.

Polyopes tasmanicus

TYPE LOCALITY: Taroona, Tasmania.
DISTRIBUTION: Eastern Tasmania.
FURTHER READING: Womersley & Lewis (1994, as *Grateloupia tasmanica*); Kawaguchi et al. (2002).

Polyopes tasmanicus (Cressy Beach, Tas; photo M.D. Guiry).

Spongophloea

Spongophloea is a genus of three species, with *Spongophloea tissotii* known from the west coast of Australia north of Shark Bay. On first view, this species looks nothing like an alga, as it has an obligate sponge coating that obscures the algal tissue. It is only when plants are fertile that their true affinities become obvious, as reproductive structures are borne in small mushroom-shaped leaflets that protrude through the sponge tissue. In addition to the sponge tissue, a unicellular green alga is present in the coating. Structurally the thallus has a pseudoparenchymatous medulla and cortex, with the outer cortex producing numerous short protuberances as well as distinctive filaments of spherical cells.

Spongophloea tissotii

TYPE LOCALITY: Kai Islands, Indonesia.
DISTRIBUTION: Aru Archipelago, Indonesia. Thursday Island and Dunk Island in Queensland, as well as Shark Bay, the Dampier Archipelago, and the vicinity of Broome, Western Australia.
FURTHER READING: Huisman (2010); Huisman et al. (2011); Huisman & De Clerck (2018).

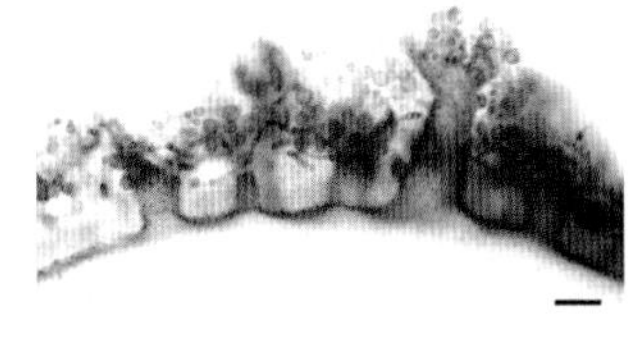

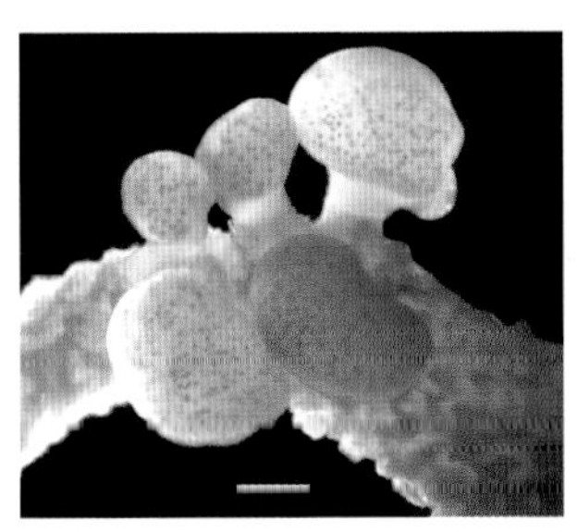

Spongophloea tissotii (James Price Point, WA).

(top) *Spongophloea tissotii* – partial section of thallus (Shark Bay, WA). Scale = 100 µm.

(above) *Spongophloea tissotii* – reproductive leaflets (Shark Bay, WA). Scale = 1 mm.

Thamnoclonium

Thamnoclonium is represented in Australia by the single species *Thamnoclonium dichotomum*, which is commonly found in southern waters. Thalli of *Thamnoclonium* have flattened or terete branches that are covered by numerous small protuberances and then coated with a thin layer of sponge tissue. Structurally the thallus has a filamentous medulla and a pseudoparenchymatous cortex – in older branches a secondary cortex develops and growth rings are evident. Reproductive structures are borne in small leaflets that arise near the apices and are free of the sponge cover.

Thamnoclonium dichotomum

TYPE LOCALITY: Australia.
DISTRIBUTION: From Perth region, Western Australia, around southern mainland Australia and Tasmania to northern New South Wales.

Thamnoclonium dichotomum (Albany, WA).

(top) *Thamnoclonium dichotomum* – as illustrated by Harvey (1863, pl. 293, as *Thamnoclonium hirsutum*).

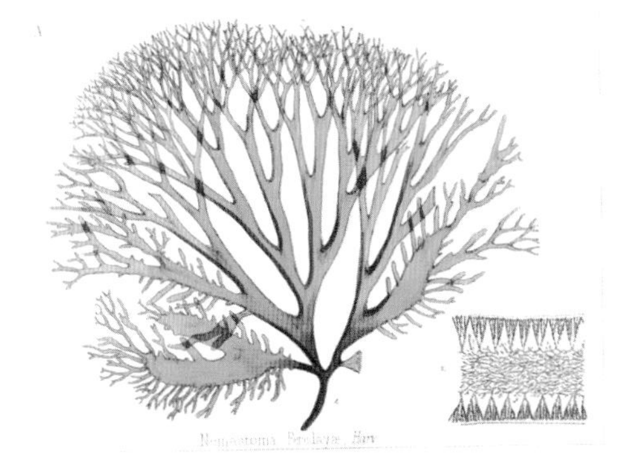

Tsengia

Tsengia includes about 10 species and displays considerable morphological variation, with some species foliose and unbranched while others, such as *Tsengia feredayae* and *Tsengia flammea* included here, are regularly branched with terete to slightly compressed axes. Depending on the species, plants can be firm or soft and mucilaginous. *Tsengia* is well-represented in the colder waters of southern Australia, where three species are known to occur, including *Tsengia feredayae*, but *Tsengia flammea* is a tropical species currently known only from Ashmore Reef. It is named for the orange-red colour of the thallus with conspicuous orange branch tips. Structurally the thallus is multiaxial, with a medulla of entangled filaments grading to a cortex of branched fascicles.

Tsengia feredayae

TYPE LOCALITY: Georgetown, Tasmania.

DISTRIBUTION: Nuyts Reef, South Australia, to Walkerville, Victoria, and the north-east and south coasts of Tasmania. New Zealand (North Island).

Tsengia flammea

TYPE LOCALITY: Ashmore Reef, Australia.

DISTRIBUTION: Known only from the type locality, where it is epilithic in the subtidal.

FURTHER READING: Womersley & Kraft (1994); Scott (2017); Huisman & Saunders (2018b).

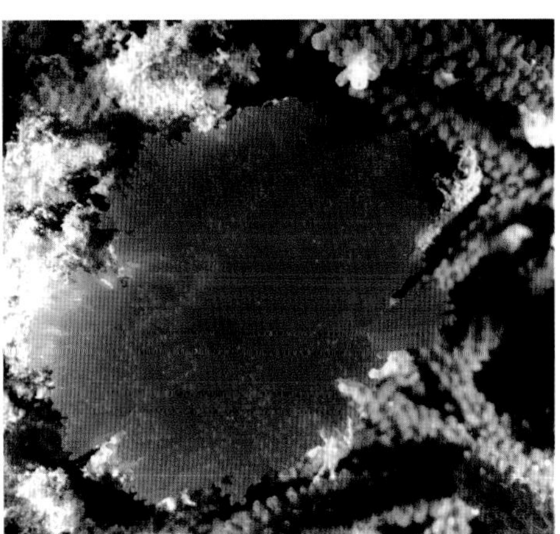

Tsengia feredayae (Tinderbox, Tas; photo G.J. Edgar).

(top) *Tsengia feredayae* – as illustrated by Harvey (1859b, pl. 195, as *Nemastoma feredayae*).

Tsengia flammea (Ashmore Reef).

Cryptocallis

The name *Cryptocallis* means 'concealed beauty' and refers to the habit of its only species, *Cryptocallis dixoniorum*, which grows in the dark recesses of reef undercuts and remains hidden from the casual observer. Plants are blade-like and prostrate, with a cartilaginous texture and a mottled red colour. Structurally the thallus is multiaxial, with a filamentous inner medulla and an outer medulla with three or four layers of ellipsoidal, ovoid or subspherical cells, grading to a three- or four-layered cortex of pigmented cells. The genus has unusual communal cystocarps that occur within an extended pericarp. Tetrasporangia are formed in raised patches and are irregularly cruciately divided.

Cryptocallis dixoniorum

TYPE LOCALITY: Brue Reef, Kimberley, Western Australia.
DISTRIBUTION: Known from the Kimberley coast of north-western Australia. Northern Philippines.
FURTHER READING: Huisman & Saunders (2018c).

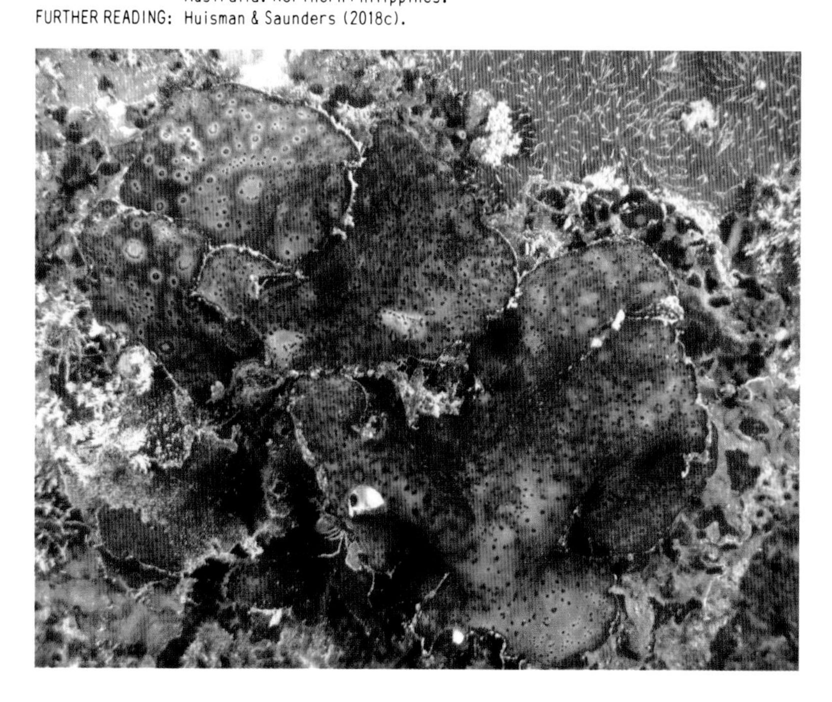

Cryptocallis dixoniorum (Brue Reef, WA).

Sebdenia

Species of *Sebdenia* display a range of morphologies, from strongly compressed and bladelike, to dichotomously divided and terete, as in *Sebdenia polydactyla*, the only species found in Australian seas. Thalli of *Sebdenia polydactyla* are soft and slightly mucilaginous. Structurally they are composed of an inner medulla of loosely entwined filaments that grade into an outer medulla of ganglioid cells and eventually a pseudoparenchymatous cortex. Tetrasporangia are cruciately divided and cystocarps are embedded in the thallus. A distinctive feature of the species is the presence of vesicular cells on the medullary filaments. *Sebdenia polydactyla* is similar in appearance to a species of *Scinaia* (see p. 34), but structurally it lacks the cortical utricles (inflated hyaline cells) of that genus. The secondary photo below is of an as-yet undescribed species from Heron Island.

Sebdenia polydactyla

TYPE LOCALITY: Port Okha, Gujarat, India.

DISTRIBUTION: From Cape Peron, Western Australia, around northern Australia to northern New South Wales and Lord Howe Island. Widespread in warmer seas of the Indian Ocean.

FURTHER READING: Huisman & Saunders (2018c).

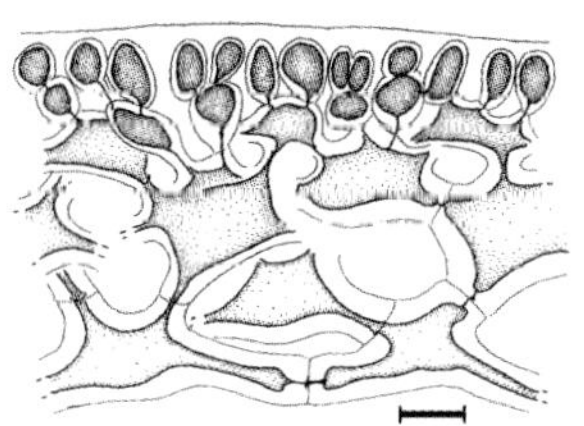

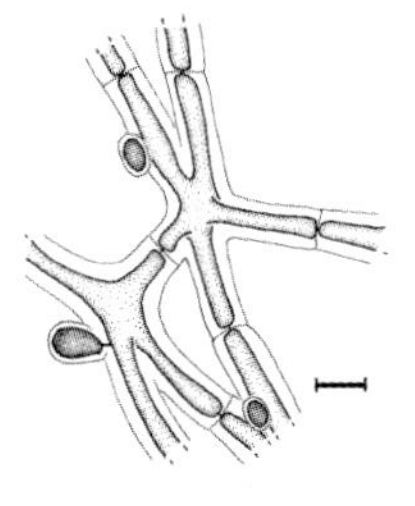

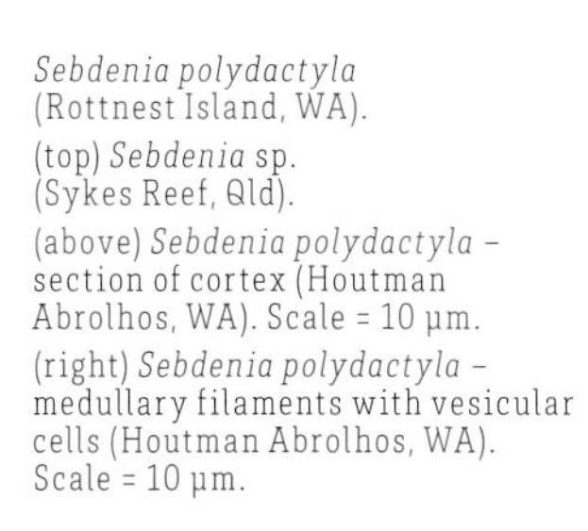

Sebdenia polydactyla (Rottnest Island, WA).

(top) *Sebdenia* sp. (Sykes Reef, Qld).

(above) *Sebdenia polydactyla* – section of cortex (Houtman Abrolhos, WA). Scale = 10 µm.

(right) *Sebdenia polydactyla* – medullary filaments with vesicular cells (Houtman Abrolhos, WA). Scale = 10 µm.

Predaea

Species of *Predaea* have gelatinous thalli that are irregularly branched or lobed. Structurally the thallus is multiaxial and the periphery is composed of dichotomously divided filaments forming an outer cortex that can be smooth or furry, depending on the species. A distinguishing feature of the genus is the presence of clusters of small cells borne on cells either side of the auxiliary cell. Several species are known in Australia, mostly from tropical and subtropical regions. They are distinguished primarily by structural features, including the shapes of the cortical filaments (cylindrical in *Predaea weldii*, ellipsoidal to subspherical in *Predaea laciniosa* and *Predaea sophieae*). External features can also assist in identification: *Predaea laciniosa* has a characteristic ruffled appearance and *Predaea weldii* often has iridescent orange tips.

Predaea laciniosa

TYPE LOCALITY: Coral Gardens, Heron Island, Queensland.

DISTRIBUTION: From the Houtman Abrolhos, Western Australia, around northern Australia to the southern Great Barrier Reef. Widely distributed in tropical seas.

Predaea sophieae

TYPE LOCALITY: Albert Reef, southern Kimberley, Western Australia.

DISTRIBUTION: Known from Albert Reef and Beagle Reef, southern Kimberley, Western Australia.

Predaea weldii

TYPE LOCALITY:	Kāne'ohe Bay, O'ahu , Hawaiian Islands.
DISTRIBUTION:	From the Houtman Abrolhos, Western Australia, around northern Australia to Lord Howe Island, New South Wales. Warmer waters of the Indo-Pacific.
FURTHER READING:	Kraft (1984a); Millar & Guiry (1989); Saunders & Kraft (2002); Schils & Coppejans (2002); Saunders et al. (2005); Huisman (2018).

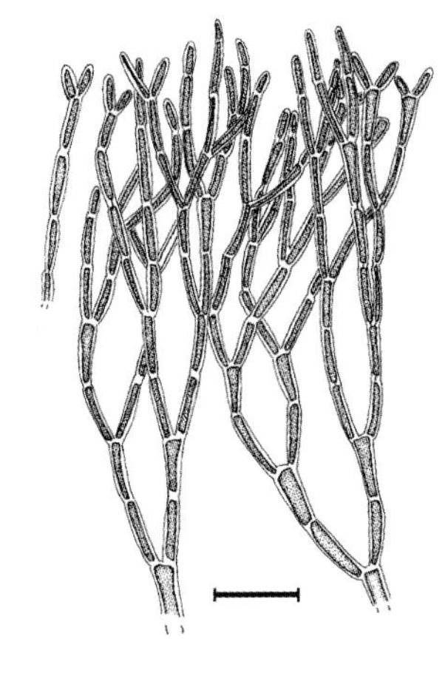

Predaea laciniosa (Heron Island, Qld).

Predaea sophieae (Albert Reef, WA).

Predaea weldii (Ashmore Reef).

(above) *Predaea weldii* – cortical filaments (Houtman Abrolhos, WA). Scale = 20 µm.

Platoma

Platoma is represented in Australia by four species, of which *Platoma cyclocolpum* is included here. Thalli are flattened and somewhat fleshy, irregularly subdichotomously branched or with broad blades with marginal proliferations. Structurally the thallus is multiaxial, with a medulla of loosely interwoven filaments that grade into a cortex of branched, anticlinal filaments. A distinctive feature of *Platoma cyclocolpum* is the presence of large vesicular cells in intercalary positions in cortical filaments. *Platoma* sp. is an as-yet unnamed species from Heron Island in the Great Barrier Reef.

Platoma cyclocolpum

TYPE LOCALITY: Teneriffe, Canary Islands.
DISTRIBUTION: From the Houtman Abrolhos south to Cape Bouvard, Western Australia. Canary Islands.
FURTHER READING: Masuda & Guiry (1995); Huisman (1999); Gabriel et al. (2010).

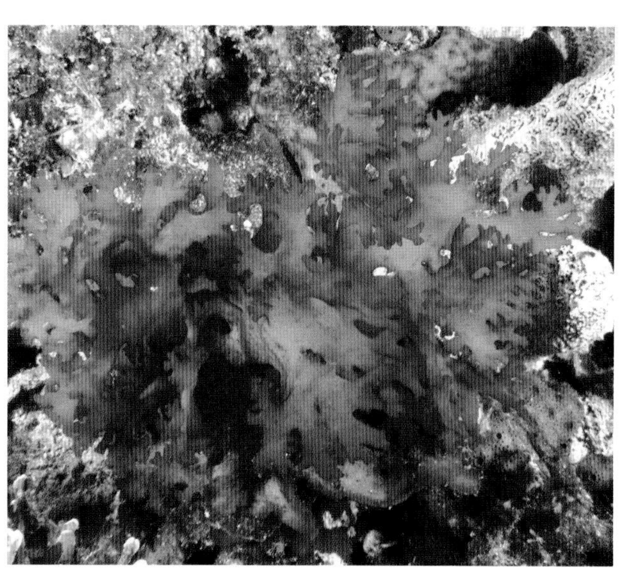

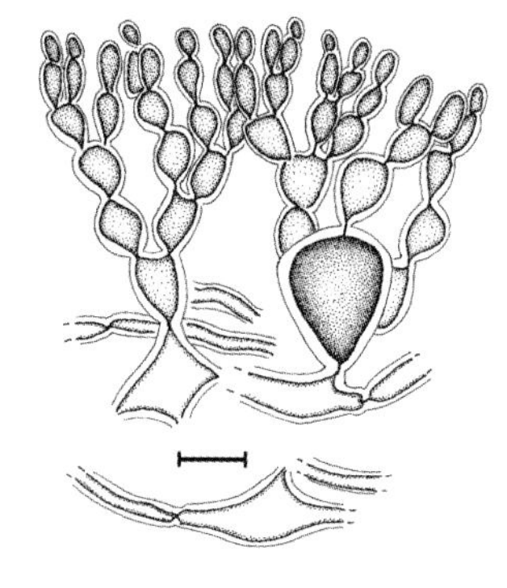

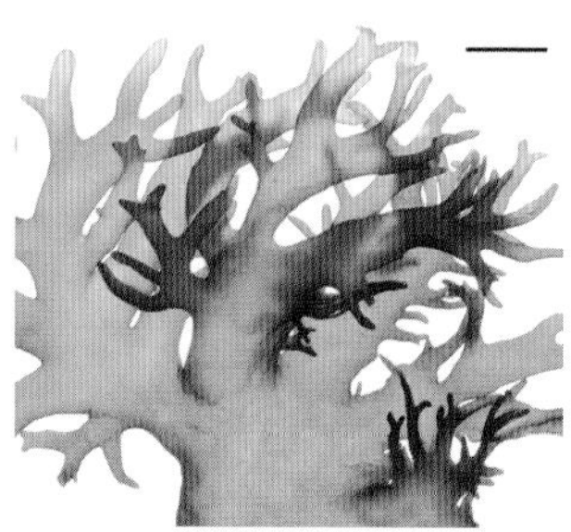

Platoma cyclocolpum (Rottnest Island, WA).

(above) *Platoma cyclocolpum* – detail of medulla and cortex with intercalary vesicular cell (Rottnest Island, WA). Scale = 20 µm.

(left) *Platoma cyclocolpum* (Cape Peron, WA). Scale = 1 cm.

(far left) *Platoma* sp. (Sykes Reef, Qld).

Schizymenia

Schizymenia is represented in Australia by the single species *Schizymenia dubyi*, which can be locally common in intertidal and shallow pools. Plants are foliose, medium- to dark-red-brown, and slightly mucilaginous, and can grow to 60 cm tall. Structurally they have a medulla of entangled filaments and a cortex of discrete branch systems, usually (but not always) with vesicular cells in the outer cortex. *Schizymenia dubyi* has a heteromorphic life history; the conspicuous plant is the gametophyte, and the tetrasporophyte is a small crust that produces zonate tetrasporangia, although these have not yet been observed in Australia. The species is regarded as invasive in some countries, possibly introduced by hull fouling.

Schizymenia dubyi

TYPE LOCALITY: Cherbourg, France.
DISTRIBUTION: Ocean Reef to Cape Peron, Western Australia, and Pondalowie Bay, South Australia, to Point Lonsdale, Victoria. Widespread in most seas.
FURTHER READING: Womersley & Kraft (1994); Gabriel et al. (2011); Ramirez et al. (2012).

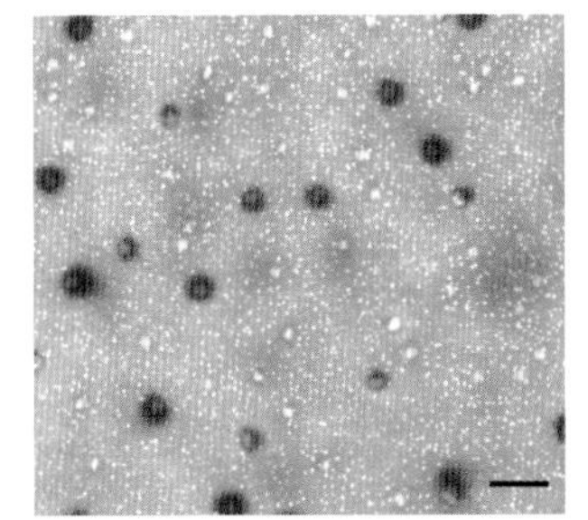

Schizymenia dubyi (Trigg, WA).

(above) *Schizymenia dubyi* – surface view of thallus. The pale spots are vesicular cells (Cape Peron, WA). Scale = 500 µm.

Titanophora

Titanophora includes eight species, mostly restricted to warmer and tropical seas. Thalli are flattened and blade-like – usually with some marginal proliferations or irregularly subdichotomously branched. A distinctive feature of the genus is the light calcification of the medullary region. This gives the thallus a whitish appearance in fresh specimens. Structurally the thallus is multiaxial, with a medulla of loosely interwoven filaments that grade into a cortex of anticlinal branched filaments. In most species, including *Titanophora pikeana*, large vesicular cells are present in the cortex. *Titanophora pikeana* is generally found associated with coral reefs, where it grows in protected positions among coral tines.

Titanophora pikeana

TYPE LOCALITY: Mauritius.
DISTRIBUTION: Houtman Abrolhos, Western Australia, the Great Barrier Reef, Queensland, Lord Howe Island, New South Wales. Warmer waters of the Indo-Pacific.
FURTHER READING: Mshigeni & Papenfuss (1980); Bucher & Norris (1992); Huisman (2018).

Titanophora pikeana (Heron Island, Qld).

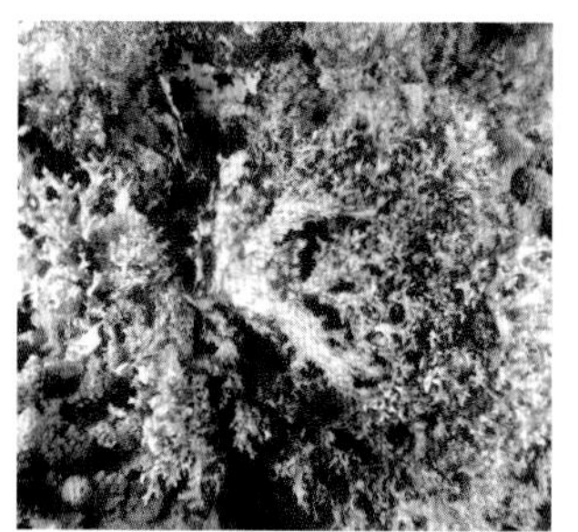

(top) *Titanophora pikeana* – large, mottled plant (Ashmore Reef).

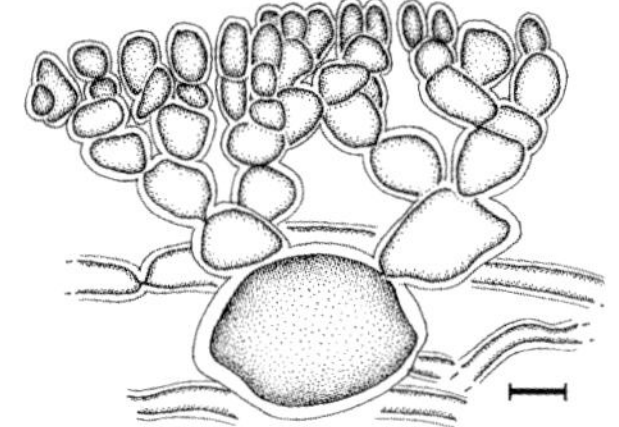

(above) *Titanophora pikeana* – detail of medulla and cortex with vesicular cell (Houtman Abrolhos, WA). Scale = 30 µm.

Incendia

Incendia includes several species with partially encrusting, fan-shaped blades that are often unattached at the margins. The lobed thalli are brittle and moderately to heavily calcified, and typically have bright orange markings. Three species occur in the Kimberley region of north-western Australia, but these are probably more widespread. *Incendia* can be similar in appearance to other genera of the Peyssonneliales, but differs in the combination of secondary pit connections among lower perithallial cells, and multicellular rather than unicellular rhizoids. In the field, species of *Incendia* can often be distinguished by having bright orange markings on raised concentric bands and at the thallus margins.

Incendia undulata

TYPE LOCALITY: Opposite Lenakel market, Lenakel, Tanna, Vanuatu.
DISTRIBUTION: North-western Australia. Vanuatu. Philippines.
FURTHER READING: Dixon & Saunders (2013); Dixon (2018).

Incendia undulata (Rowley Shoals, WA).

Peyssonnelia

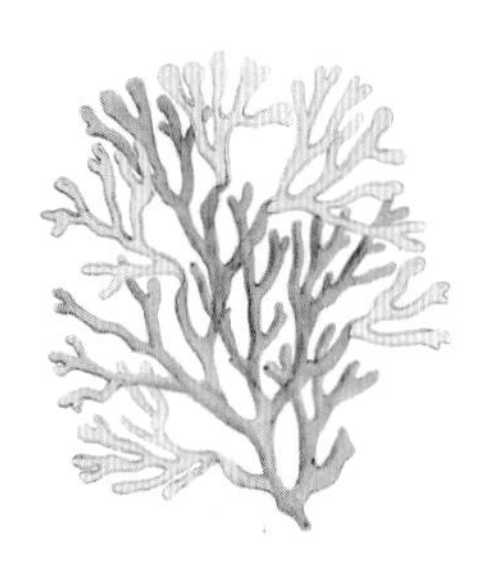

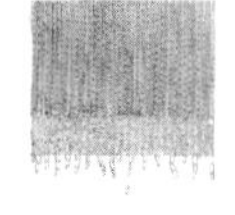

Species of *Peyssonnelia* are common in subtidal habitats, generally attached to rock and often on vertical walls in semi-shaded situations. All are essentially prostrate, although in some species the edges can be detached from the substratum. Thalli are flattened and can be rounded and undivided (e.g. *Peyssonnelia rainboae*) or much-branched (e.g. *Peyssonnelia novae-hollandiae*). The upper surface is usually a deep red to pink colour and has visible concentric growth zones. Growth is via a marginal meristem and forms a filamentous basal layer that cuts off rhizoids towards the substratum and erect or inclined filaments above. Calcification is generally present between the rhizoids and often internally in cells of the erect filaments. A large number of species of *Peyssonnelia* have been described and most require at least an examination of their internal structure to identify with any confidence. Recent studies using DNA sequencing techniques have recognised numerous species that often overlap morphologically, suggesting that accurate discrimination of species in the future will rely on these methods.

Peyssonnelia novae-hollandiae

TYPE LOCALITY: Australia.
DISTRIBUTION: From Flat Rocks, south of Geraldton, Western Australia, around southern mainland Australia to Coffs Harbour, New South Wales, and around Tasmania.

Peyssonnelia rainboae

This species is characterised by its bright-red and often yellow-streaked, lightly to moderately calcified lobes that remain largely free of the substratum.

TYPE LOCALITY: Adele Island, Kimberley, Western Australia.
DISTRIBUTION: Widespread in north-western Australia.

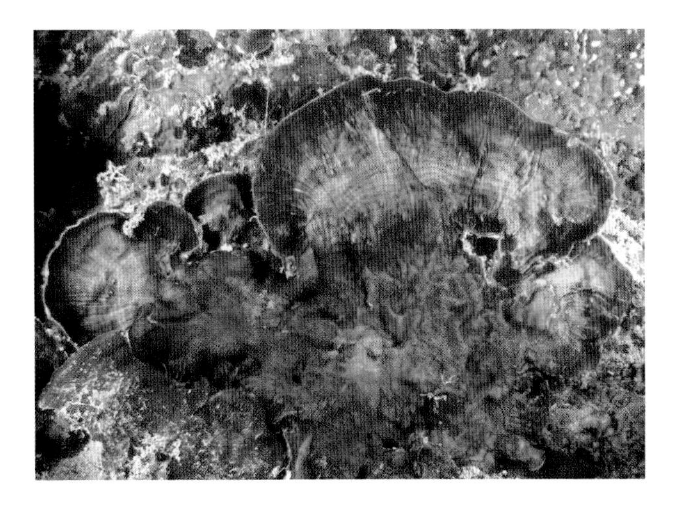

Peyssonnelia tenuiderma

Peyssonnelia tenuiderma has a thin, broadly lobed and weakly calcified thallus.

TYPE LOCALITY: Adele Island, Kimberley, Western Australia.
DISTRIBUTION: Adele Island and Beagle Reef, Western Australia.
FURTHER READING: Womersley (1994); Dixon (2018).

Peyssonnelia novae-hollandiae (Rottnest Island, WA).

(opposite, top) *Peyssonnelia novae-hollandiae* – as illustrated by Harvey (1863, pl. 269, as *Peyssonnelia multifida*).

Peyssonnelia rainboae (Fraser Island, WA).

Peyssonnelia tenuiderma (Beagle Reef, WA).

Plocamium

Plocamium is one of the more easily recognisable genera. Thalli of *Plocamium* all have flattened axes that bear regular alternating series of two or more (depending on the species) lateral branches. The most distal of each group continues to grow and branches in a similar manner, while the lower branches generally remain undivided. Structurally the thallus is uniaxial, with a conspicuous central filament surrounded by a pseudoparenchymatous medulla that grades into a small-celled cortex. Several species of *Plocamium* are found in Australian seas and the genus appears to be at its most diverse in the colder southern waters. The four species included here can be separated by their branch width (narrow in *Plocamium pusillum* and broad in the others), production of hooked branches (*Plocamium hamatum*), and number of branches in each lateral series (two in *Plocamium mertensii* and three in *Plocamium preissianum*). *Plocamium hamatum* is also confined to warmer waters, whereas the other species are more prevalent on the colder south coast.

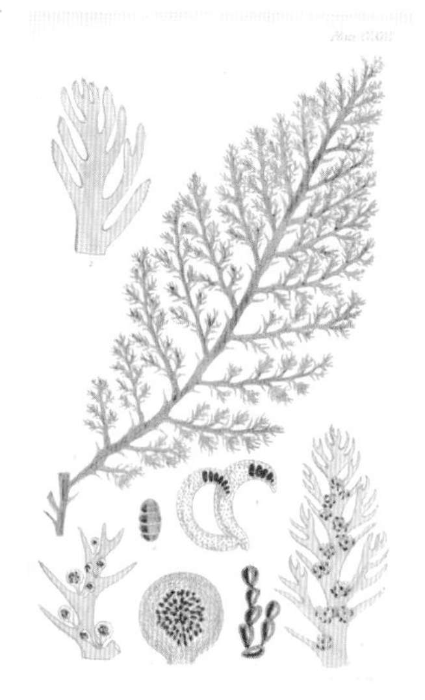

Plocamium pusillum

TYPE LOCALITY: Western Australia.
DISTRIBUTION: Southern Western Australia, Victoria, and New South Wales (Lord Howe Island) and Norfolk Island.

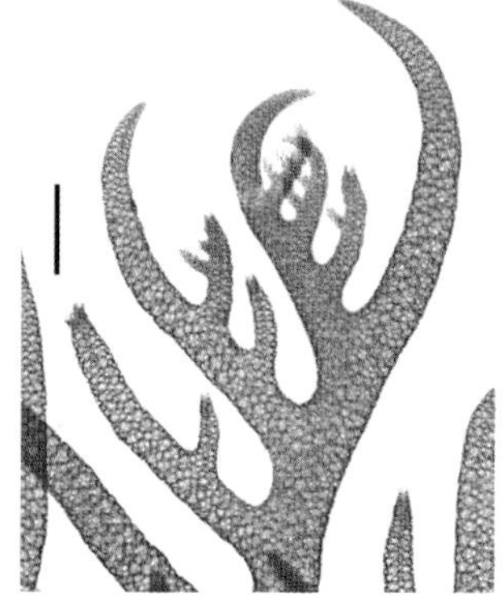

(above) *Plocamium pusillum* – apex of plant (Cape Peron, WA). Scale = 300 µm.

Plocamium pusillum (Cape Peron, WA).

Plocamium hamatum (Heron Island, Qld).

(opposite, top) *Plocamium hamatum* – detail of thallus (Heron Island, Qld). Scale = 1 mm.

Plocamium mertensii (Albany, WA).

(top) *Plocamium mertensii* – as illustrated by Harvey (1862, pl. 223, as *Plocamium procerum*).

Plocamium preissianum (Cape Peron, WA).

Plocamium hamatum

TYPE LOCALITY: Norfolk Island.
DISTRIBUTION: Eastern Australia. New Zealand.

Plocamium mertensii

TYPE LOCALITY: Australia.
DISTRIBUTION: From Nickol Bay, Western Australia, around southern mainland Australia and Tasmania to San Remo, Victoria.

Plocamium preissianum

TYPE LOCALITY: South-west Australia.
DISTRIBUTION: From the Houtman Abrolhos, Western Australia, around southern Australia to Wilsons Promontory, Victoria.
FURTHER READING: Womersley (1994); Saunders & Lehmkuhl (2005); Huisman (2018); Huisman & Saunders (2021).

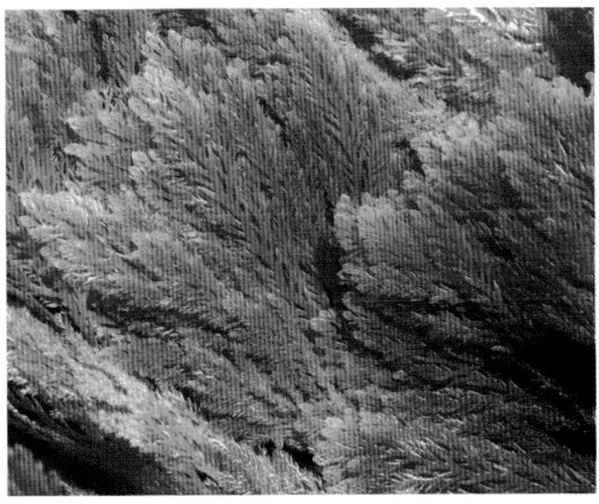

Champia

Species of the genus *Champia* are segmented and often slightly constricted at regular intervals. Branches can be terete or flattened, depending on the species. Structurally all species of *Champia* have a hollow centre (with the exception of the internal membranous diaphragms separating the segments) and an outer cellular layer that is one to two cells thick. The inner surface of the cellular layer has a number of longitudinal filaments that run the length of the axis (in some species these are distributed throughout the internal cavity). These filaments bear spherical gland cells that project into the cavity. The longitudinal filaments meet at the branch apex, where they can be seen as a cluster of smaller cells. Tetrasporangia have tetrahedrally arranged spores and are borne in an intercalary position in the outer cellular layer. Cystocarps are always protuberant and the gonimoblast has terminal carposporangia. Numerous species occur in Australia; the six included here can be recognised by their habitat and by the form of the thallus.

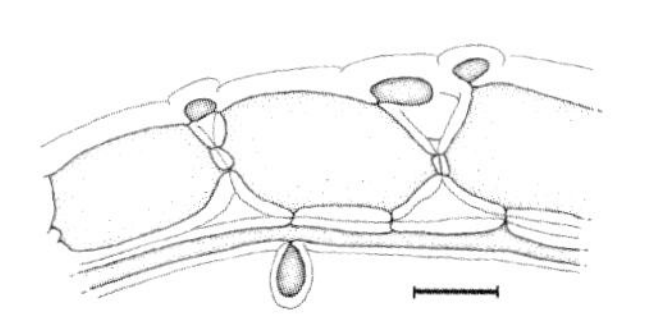

(above) *Champia stipitata* – partial longitudinal section (Rottnest Island, WA). Scale = 30 µm.

Champia bibendum

Champia bibendum has a distinctive flattened thallus that is pinnately branched. Some Australian specimens of this species were previously referred to as *Champia compressa*, a South African species.

TYPE LOCALITY: Oyster Bridge, Coral Bay, Western Australia.
DISTRIBUTION: Known reliably only from the type locality, but probably widespread.

Champia bibendum (Bateman Bay, WA).

Champia patula

A sprawling species with narrow branches.

TYPE LOCALITY: Cape Peron, Western Australia.
DISTRIBUTION: Known only from the type locality.

Champia patula (Cape Peron, WA).

Champia parvula var. *amphibolis* (Cape Peron, WA).

Champia subcompressa (Cassini Island, WA).

Champia stipitata (Rottnest Island, WA).

Champia zostericola (Cape Peron, WA).

Champia parvula var. *amphibolis*

A small species, generally epiphytic on seagrasses.

TYPE LOCALITY: Tiparra Reef, Spencer Gulf, South Australia.
DISTRIBUTION: Known from Tiparra Reef and Cape Peron, Western Australia.

Champia stipitata

This species has a unique solid stipe supporting its terete thallus. It is found exclusively on rock.

TYPE LOCALITY: Roe Reef, Rottnest Island, Western Australia.
DISTRIBUTION: From Cape Peron, Western Australia, north to at least Darwin, Northern Territory.

Champia subcompressa

Champia subcompressa has terete to slightly compressed axes and is iridescent when viewed underwater.

TYPE LOCALITY: Montgomery Reef, Western Australia.
DISTRIBUTION: Montgomery Reef and Cassini Island, Western Australia.

Champia zostericola

Thalli of *Champia zostericola* are generally epiphytic on seagrasses or other algae. They often have curved apices that facilitate attachment.

TYPE LOCALITY: Rottnest Island, Western Australia.
DISTRIBUTION: From the Houtman Abrolhos, Western Australia, around southern Australia to Kiama, New South Wales.
FURTHER READING: Reedman & Womersley (1976); Millar (1990); Huisman (2018); Huisman & Saunders (2020a).

Coelothrix

Coelothrix includes only two species, with *Coelothrix irregularis* found in Australia and common in most tropical regions. Plants form sprawling mats over coral rubble, with both creeping and upright terete axes to about 3 cm high and 0.5 mm in width. Structurally the thallus is multiaxial with a central cavity bordered by a pseudoparenchymatous cortex. The inner wall is lined with longitudinal filaments that bear occasional gland cells. Reproductive specimens are rarely encountered, but tetrasporangia occur in raised patches encircling branch tips. *Coelothrix irregularis* is often iridescent in the living state but pressed specimens are a dark-red colour.

Coelothrix irregularis

TYPE LOCALITY: Key West, Florida.
DISTRIBUTION: From the Houtman Abrolhos, Western Australia, around northern Australia to Lord Howe Island, New South Wales. Tropical regions of the Indo-West Pacific and the western Atlantic.
FURTHER READING: Price & Scott (1992); Huisman (2018).

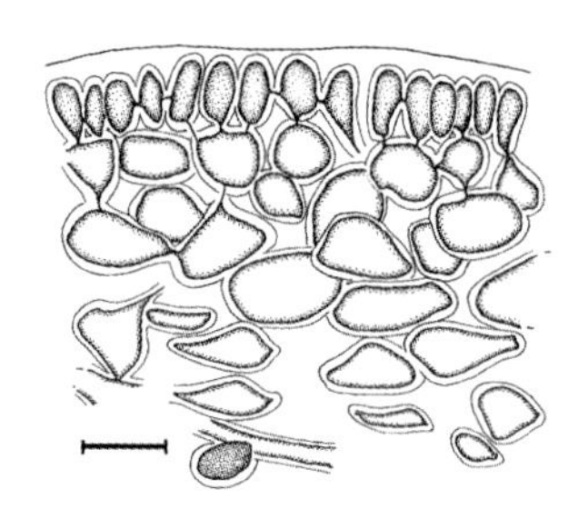

Coelothrix irregularis (Houtman Abrolhos, WA).

(above) *Coelothrix irregularis* – partial section of cortex (Houtman Abrolhos, WA). Scale = 30 µm.

Dictyothamnion

Dictyothamnion includes the single species *Dictyothamnion saltatum*, in Australia known only from New South Wales but also reported from the tropical Pacific. Plants grow as a series of arching segments with multiple attachments to the rock substratum, forming a reticulate network. Structurally the thallus is hollow (although filled with a watery mucilage), with multilayered diaphragms dividing the segments. The inner surface of the cortex is lined by parallel longitudinal filaments that bear secretory cells. Cystocarps are protuberant, and tetrahedrally divided tetrasporangia are scattered in the cortex.

Dictyothamnion saltatum

TYPE LOCALITY: Muttonbird Island, Coffs Harbour, New South Wales.
DISTRIBUTION: Known from the Coffs Harbour region including Split Solitary Island to Korffs Islet and along the New South Wales coast to Sydney Harbour and Montague Island in the south. Samoa.
FURTHER READING: Millar (1990).

Dictyothamnion saltatum (Montague Island, NSW).

Gloioderma

Gloioderma is a widespread genus with several species found in Australian waters. The species range in form from those with distinctly flattened branches (as seen here in *Gloioderma halymenioides*) to several with terete branches. Despite this variation in external appearance, all species have a loosely filamentous cortex and are virtually identical in cross-section. *Gloioderma halymenioides* is relatively common in the south-west of Western Australia. The photographed specimen is distinctly red in colour, while others can have a greenish hue and be iridescent. *Gloioderma australe* has unusual, pad-like structures at the ends of branches that serve to secondarily attach the plant. *Gloioderma iyoense* is a smaller species, with axes that are terete above but become flattened towards the base of the plant.

Gloioderma australe

TYPE LOCALITY: Western Australia.
DISTRIBUTION: Jurien Bay, Western Australia, to Walkerville, Victoria, and the north coast of Tasmania.

Gloioderma halymenioides

TYPE LOCALITY: Fremantle, Western Australia.
DISTRIBUTION: From the Houtman Abrolhos, Western Australia, around southern mainland Australia and Tasmania to Walkerville, Victoria.

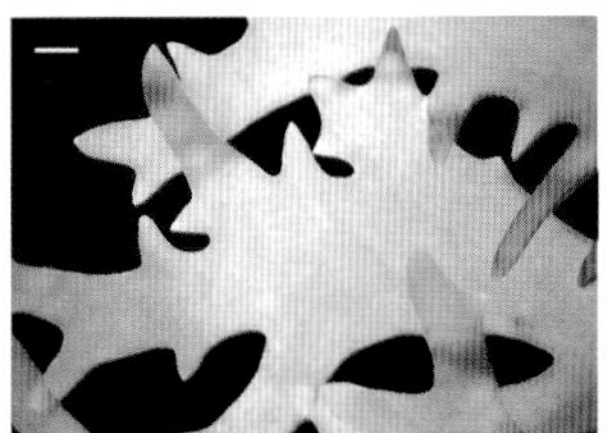

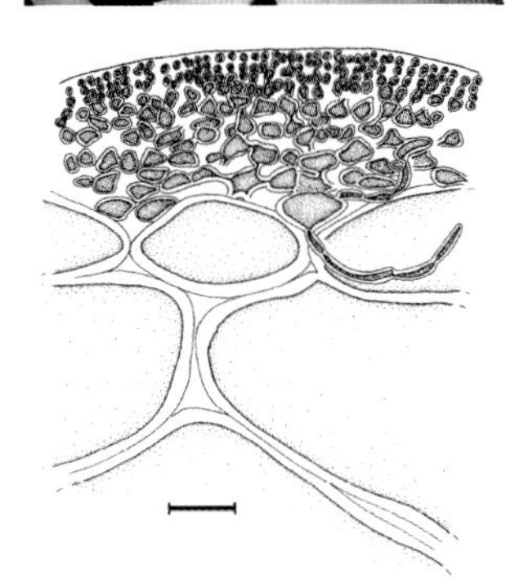

Gloioderma iyoense

TYPE LOCALITY: Otateba, Ehime Prefecture, Japan.
DISTRIBUTION: From the tropics south to the Houtman Abrolhos, Western Australia. Widespread in warmer seas.
FURTHER READING: Norris (1991); Womersley (1996); Huisman (2018).

Gloioderma australe (Kearn Reef, Jurien Bay, WA; photo G.J. Edgar).

(opposite top) *Gloioderma australe* – as illustrated by Harvey (1859b, pl. 194A, as *Horea speciosa*).

Gloioderma halymenioides (Cape Peron, WA).

(top) *Gloioderma halymenioides* – as illustrated by Harvey (1859a, as *Horea halymenioides*).

(above, middle) *Gloioderma halymenioides* – branch detail (Carnac Island, WA). Scale = 1 mm.

(above) *Gloioderma halymenioides* – partial section of medulla and cortex (Rottnest Island, WA). Scale = 30 µm.

Gloioderma iyoense (Houtman Abrolhos, WA).

Leptofauchea

The genus *Leptofauchea* has a pseudoparenchymatous medulla and is therefore similar to *Rhodymenia* in thallus structure, but differs in its thin cortex and the presence of stellate cells lining the walls of the cystocarp cavity. Three species are known from Australian coast: *Leptofauchea coacta*, *Leptofauchea nitophylloides* (not included here), and the recently described *Leptofauchea lucida*. *Leptofauchea lucida* has a distinctive thallus, in which the flattened branches are regularly constricted.

Leptofauchea lucida

TYPE LOCALITY: Bynoe Island, Easter Group, Houtman Abrolhos, Western Australia.

DISTRIBUTION: From Jurien Bay north to the Montebello Islands, Western Australia.

FURTHER READING: Norris & Aken (1985); Fillorama & Saunders (2015); Huisman & Saunders (2020b).

Leptofauchea lucida. (Houtman Abrolhos, WA).

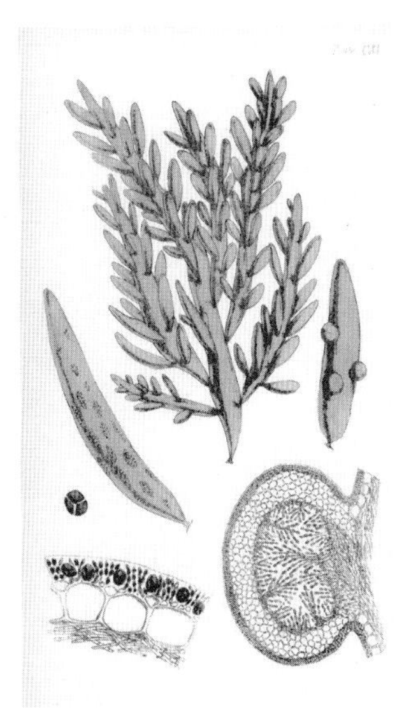

Webervanbossea

Webervanbossea is found only in Australian waters. Thalli have saccate branches that are essentially hollow and filled with mucilage. Structurally the thallus is multiaxial and the inner surface of the outer cellular layer has distinctive filaments that bear gland cells. A further characteristic feature of the genus is that the tetrasporangia are formed in raised, circular patches on the surface of the plant. Three species are recognised, of which *Webervanbossea splachnoides* is included here. Within the genus, *Webervanbossea splachnoides* is characterised by the way the thallus constrictions occur only at the bases of lateral branches. In the other species, thallus constrictions are either additionally present along the branches (*Webervanbossea kaliformis*) or absent altogether (*Webervanbossea tasmanensis*).

Webervanbossea splachnoides

TYPE LOCALITY: Garden Island, near Fremantle, Western Australia.
DISTRIBUTION: Known from the Houtman Abrolhos, Western Australia, around southern mainland Australia to Port Phillip Heads, Victoria, and the north coast of Tasmania.
FURTHER READING: Huisman (1995); Womersley (1996).

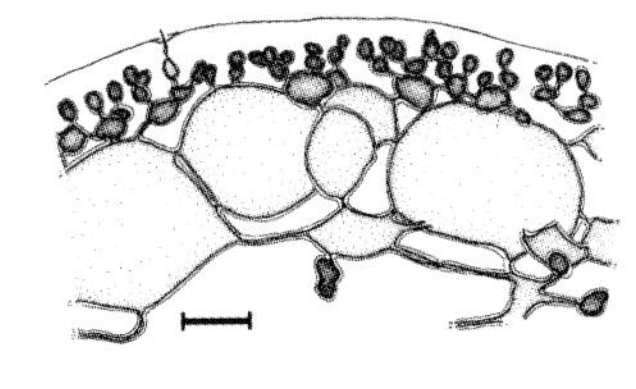

Webervanbossea splachnoides (Houtman Abrolhos, WA).

(top) *Webervanbossea splachnoides* – as illustrated by Harvey (1859a, pl. 111, as *Bindera splachnoides*).

(above) *Webervanbossea splachnoides* – section of cellular layer with vesicular cell (Houtman Abrolhos, WA). Scale = 20 µm.

Asteromenia

Asteromenia is a spectacular plant when viewed underwater, as its iridescent thallus sparkles in colours ranging from white to bronze. Young plants of *Asteromenia* appear as small discs or star-like blades on short, centrally placed stalks. When mature, they often have elongate arms that can re-attach to the substratum. *Asteromenia* is often found growing in clusters on rock surfaces in shaded areas. Structurally the blade has a many-layered medulla of large, colourless cells and a pigmented cortex of smaller cells. Cystocarps are protuberant on the surface of the blade and tetrasporangia generally arise in the cortex of the lower surface. *Asteromenia* was originally thought to include only the single species *Asteromenia peltata* (not shown here), but subsequent studies have recognised several species, including *Asteromenia exanimans* from Western Australia. *Asteromenia poeciloderma* differs in having a mottled iridescence and spiny margins.

Asteromenia exanimans

TYPE LOCALITY: Suomi Island, Easter Group, Houtman Abrolhos, Western Australia.

DISTRIBUTION: From Jurien Bay to the Dampier Archipelago, Western Australia.

Asteromenia poeciloderma

TYPE LOCALITY: Mermaid Reef, Rowley Shoals, Western Australia.

DISTRIBUTION: Known only from Mermaid and Clerke Reefs, Rowley Shoals, Western Australia.

FURTHER READING: Huisman & Millar (1996); Saunders et al. (2006); Huisman & Saunders (2018; 2022).

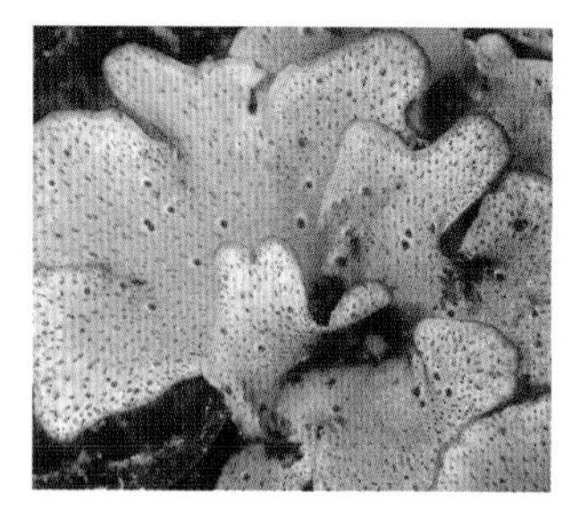

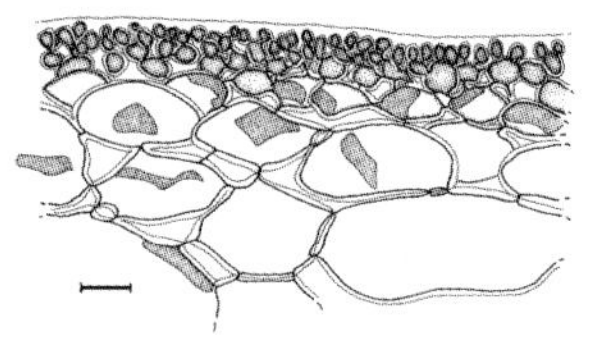

Asteromenia exanimans (Houtman Abrolhos, WA; photo A. Gunning)

(left) *Asteromenia exanimans* – partial section of thallus showing medulla and cortex (Houtman Abrolhos, WA). Scale = 20 µm.

(far left) *Asteromenia exanimans* (Houtman Abrolhos, WA; photo A. Gunning)

Asteromenia poeciloderma (Rowley Shoals, WA).

Hymenocladia

Hymenocladia includes six species worldwide, most of which can be found in southern and south-western Australia. Plants are generally much branched, with terete or distinctly flattened axes. Structurally the thallus is multiaxial, and mature axes have a pseudoparenchymatous medulla grading to a smaller celled cortex. Filaments and smaller cells are often intermixed with the large medullary cells. *Hymenocladia* is unusual in the Rhodymeniaceae in producing tetrahedrally divided, intercalary tetrasporangia. Cystocarps are protuberant. *Hymenocladia dactyloides* and *Hymenocladia usnea* are a widespread species often found growing epiphytically on the seagrass *Amphibolis*.

Hymenocladia conspersa

TYPE LOCALITY: Garden Island, Western Australia.
DISTRIBUTION: From Port Denison, Western Australia, around southern Australia to Walkerville, Victoria, and Kent Island in Bass Strait.

Hymenocladia conspersa (Cape Peron, WA).

(top) *Hymenocladia conspersa* – as illustrated by Harvey (1862, pl. 237, as *Calliblepharis conspersa*).

Hymenocladia dactyloides

TYPE LOCALITY: Western Australia.
DISTRIBUTION: Jurien Bay to Esperance, Western Australia.

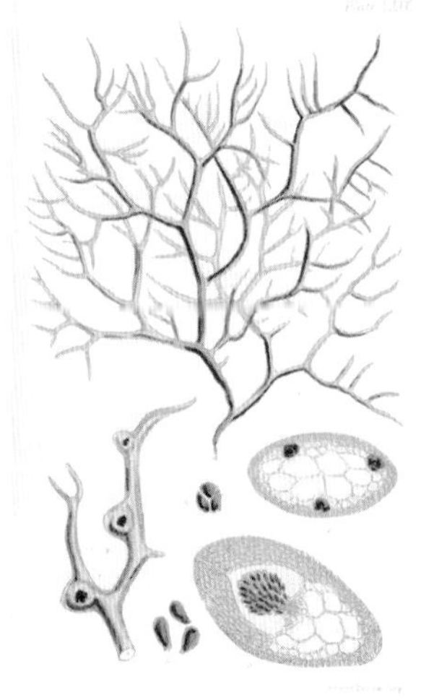

Hymenocladia usnea

TYPE LOCALITY: Kent Island, Bass Strait.
DISTRIBUTION: From Port Denison, Western Australia, around southern Australia to Walkerville, Victoria, and Kent Island in Bass Strait.
FURTHER READING: Womersley (1996); Filloramo & Saunders (2016).

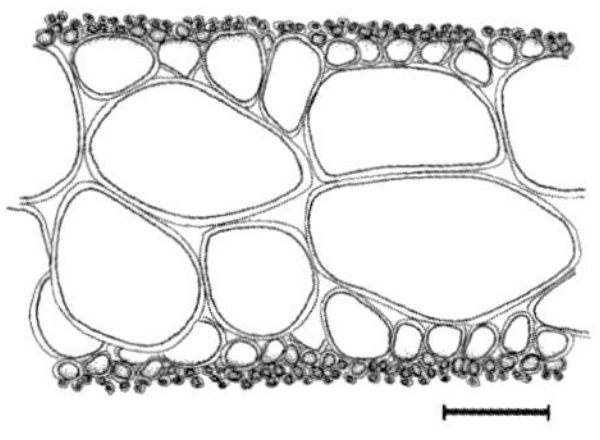

Hymenocladia dactyloides – epiphytic on *Amphibolis antarctica*, with *Amphibolis griffithii* in the background (Cape Peron, WA).

(top) *Hymenocladia dactyloides* – as illustrated by Harvey (1859a: pl. 80, as *Gracilaria dactyloides*).

Hymenocladia usnea (Seabird, WA).

(above) *Hymenocladia usnea* – partial section of thallus (Jurien Bay, WA). Scale = 100 µm.

Perbella

Not the largest of seaweeds, but an attention grabber. *Perbella minuta* generally has an unusual iridescence on the surface, which can form a striking mottled pattern or zig-zag lines. Several other red and brown algae can also be iridescent, but, as with *Perbella*, this is only seen when they are freshly collected or observed underwater; it is always lost when plants are dried. Plants occur as clusters of paddle-like, flattened fronds arising from a small holdfast. They can grow to about 10 cm in height and have a firm, leathery texture. Each frond is generally unbranched, but occasionally new fronds can arise from the margins. The shape of the fronds and iridescent surface are distinctive features of this species. This species was formerly known as *Erythrymenia minuta*.

Perbella minuta

TYPE LOCALITY: Port Phillip Heads, Victoria.
DISTRIBUTION: From the Perth region, Western Australia, around southern Australia to Portsea, Victoria, and around Tasmania.
FURTHER READING: Huisman et al. (2006, as *Erythrymenia minuta*); Filloramo & Saunders (2016); Scott (2017).

Perbella minuta (Rottnest Island, WA).

Ceratodictyon

Ceratodictyon previously comprised only the single species *Ceratodictyon spongiosum*, but it now incorporates species previously included in the genus *Gelidiopsis*, despite their very different appearances. *Ceratodictyon spongiosum* is the name given to the algal component of a commonly found (in the tropics at least) symbiosis between it and a sponge known as *Haliclona cymiformis*. The algal tissue is rarely visible and can only be seen if dissected out from under the sponge coating. It is composed of many thick axes that are secondarily attached to one another under the sponge tissue. Reproductive structures of the alga are borne in specialised structures that protrude from the tips of the sponge, thereby allowing for dispersal of the spores. *Ceratodictyon scoparium* is not associated with a sponge and forms low turfs to a few centimetres tall, with the upper branches flattened and palmate.

Ceratodictyon scoparium

TYPE LOCALITY: Réunion.
DISTRIBUTION: Widespread in the Indo-Pacific, in Western Australia south to the Perth region.

Ceratodictyon spongiosum

TYPE LOCALITY: Aru Island, Indonesia.
DISTRIBUTION: From the Houtman Abrolhos, Western Australia, around northern Australia to the southern Great Barrier Reef, Queensland. Widespread in the tropical Indo-Pacific.
FURTHER READING: Price & Kraft (1991); Huisman (2018).

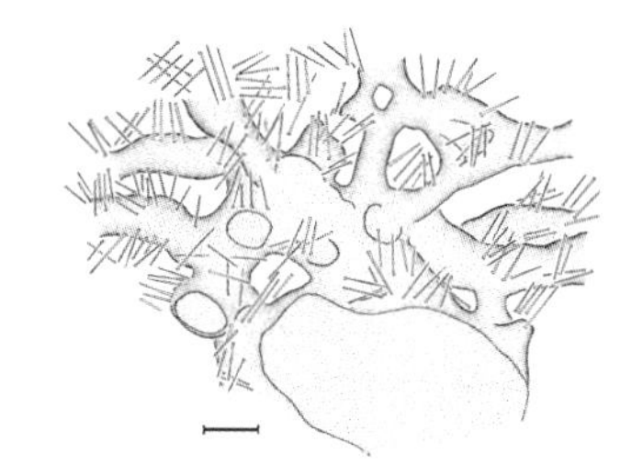

Ceratodictyon scoparium (Heron Island, Qld).

Ceratodictyon spongiosum – dark green branches growing among coral (Lizard Island, Qld).

(right) *Ceratodictyon spongiosum* – partial section of thallus showing the much-divided algal tissue and associated sponge spicules (Houtman Abrolhos, WA). Scale = 200 μm.

Lomentaria

Lomentaria is represented by three species in southern Australia. Thalli are much branched, with terete or compressed saccate branches that are basally constricted. Structurally the thallus is multiaxial, with a medulla of filaments bearing vesicular cells and a solid outer cortex. An unusual feature of the genus is the production of tetrasporangia in pits, a feature it shares with the closely related *Semnocarpa*. The two genera can be separated by the protuberant cystocarps in *Lomentaria*, whereas those of *Semnocarpa* are immersed. *Lomentaria pyramidalis* is often found as an epiphyte on seagrasses and can grow to 45 cm in height; the plants pictured from Cape Peron are much smaller, which is often the case with species at the edges of their range in Western Australia.

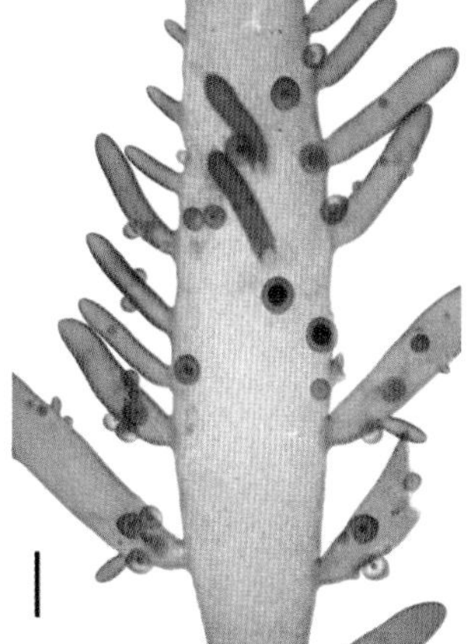

Lomentaria pyramidalis

TYPE LOCALITY: Port MacDonnell, South Australia, Australia.
DISTRIBUTION: From Cape Peron, Western Australia, to Flinders, Victoria.
FURTHER READING: Womersley (1996).

(top) *Lomentaria pyramidalis* – detail of thallus with protuberant cystocarps (Cape Peron, WA). Scale = 1 mm.

(above) *Lomentaria pyramidalis* – detail of thallus with tetrasporangia in pits (Cape Peron, WA). Scale = 1 mm.

(above) *Lomentaria pyramidalis* – epiphytic on the seagrass *Posidonia australis* (Cape Peron, WA).

Semnocarpa

A small genus with only two species, *Semnocarpa* includes the comparatively large *Semnocarpa corynephora* (not shown) and the small epiphyte *Semnocarpa minuta*. Thalli are saccate and unbranched or irregularly branched; lateral branches are constricted at their bases. Structurally the thallus has a central cavity and a peripheral cellular layer composed of a large-celled medulla and smaller-celled cortex. Longitudinal filaments, bearing lateral vesicular cells, line the inner surface of the medullary layer. Like all members of the Lomentariaceae, *Semnocarpa* has tetrahedrally divided tetrasporangia aggregated in small depressions in the cellular layer. A distinctive feature of *Semnocarpa* is the immersed position of the cystocarps – in most other genera of the Rhodymeniales, the cystocarps are protuberant. *Semnocarpa minuta* is a common epiphyte on the stems of the seagrass *Amphibolis* and has also been found on calcified algae such as *Halimeda versatilis* and *Amphiroa anceps*.

Semnocarpa minuta

TYPE LOCALITY: Green Island, adjacent to Rottnest Island, Western Australia.
DISTRIBUTION: From the Houtman Abrolhos south to Cape Peron, Western Australia.
FURTHER READING: Huisman et al. (1993).

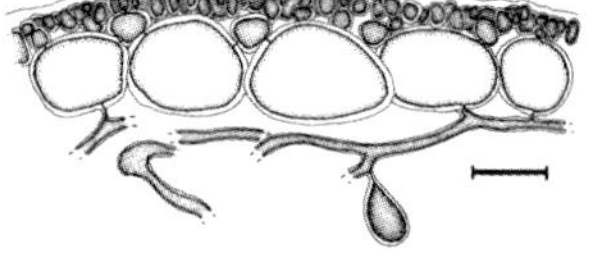

Semnocarpa minuta – growing on *Amphibolis antarctica* (Cape Peron, WA).

(above) *Semnocarpa minuta* – section of cellular layer with vesicular cell (Rottnest Island, WA). Scale = 30 µm.

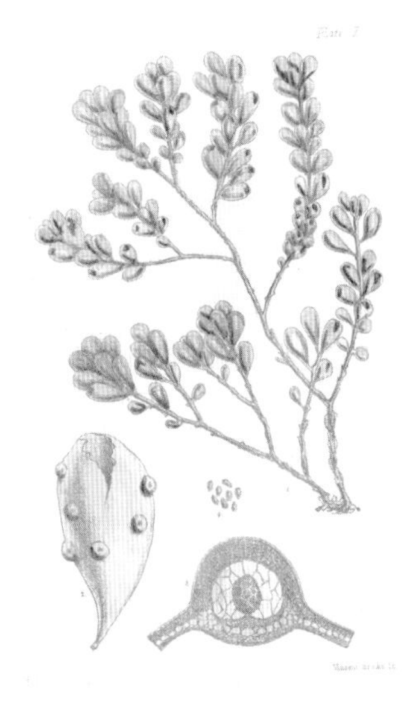

Botryocladia

The name *Botryocladia* literally means 'with grape-like branches' and it is not difficult to see how it was derived. Species of *Botryocladia* all have spherical to ovoid vesicles borne on a solid stem. The vesicles have hollow centres that are filled with mucilage. The outer, cellular layer is composed of an inner medulla of large, colourless cells and an outer cortex of smaller, pigmented cells. In some species, such as *Botryocladia sonderi*, a secondary layer of smaller cells is formed on the inner surface of the medullary cells. Spherical or pear-shaped vesicular cells are borne on specialised cells and project into the cavity. Tetrasporangia are scattered in the cortex and have cruciately arranged spores. Cystocarps have a thick pericarp and can be protuberant or somewhat immersed in the thallus. Several species are found in Australian seas, mostly – as is typical of the genus – in warmer waters. *Botryocladia sonderi* is unusual in being found in the colder areas of southern Australia. *Botryocladia leptopoda* can be recognised by its elongate axes with smaller vesicles.

Botryocladia leptopoda

TYPE LOCALITY: Moreton Bay, Queensland.

DISTRIBUTION: From Carnac Island, Western Australia, around northern Australia to Lord Howe Island, New South Wales. Widespread in warmer waters of the Indo-West Pacific.

Botryocladia sonderi

TYPE LOCALITY: Western Australia.

DISTRIBUTION: From Port Denison, Western Australia, around southern mainland Australia to North Waratah Bay, Victoria, and the north coast of Tasmania.

FURTHER READING: Cribb (1996); Womersley (1996).

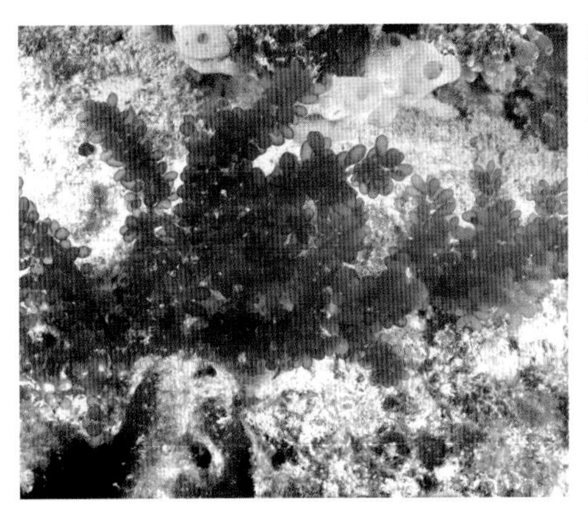

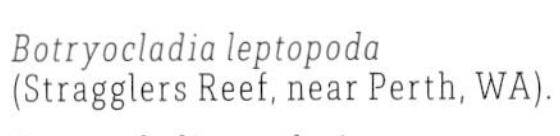

Botryocladia leptopoda (Stragglers Reef, near Perth, WA).

Botryocladia sonderi (Cape Peron, WA).

(top) *Botryocladia sonderi* – as illustrated by Harvey (1858, pl. 10, as *Chrysymenia obovata*).

(right) *Botryocladia sonderi* – section of cellular layer with vesicular cells (Jurien Bay, WA). Scale = 30 µm.

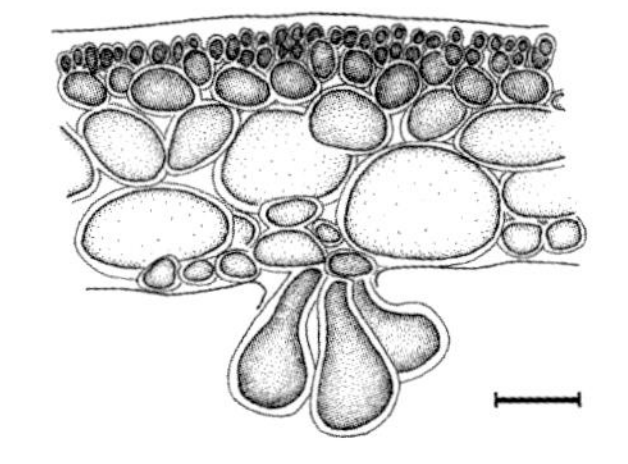

Campylosaccion

A monotypic genus thus far known only from northern Western Australia, *Campylosaccion* has a vesiculate thallus and is somewhat similar to certain species of *Botryocladia*, but differs in its apical structure, wherein the medulla forms a weft of stellate cells that stretch as the tissue matures, initially remaining interconnected by thin processes, but eventually detaching and leaving only specialised inner medullary cells that bear vesicular cells. *Campylosaccion* also has tetrahedrally divided tetrasporangia, whereas those of *Botryocladia* and *Webervanbossea*, another similar genus, are cruciately divided. *Campylosaccion decumbens* is a distinctive species, with an often decumbent habit and secondary attachment structures formed at the apices of vesicles.

Campylosaccion decumbens

TYPE LOCALITY: Quondong Point, Broome, Western Australia.
DISTRIBUTION: Known from Broome, Barrow Island and the Dampier Archipelago, northern Western Australia.
FURTHER READING: Huisman (2018).

Campylosaccion decumbens (Broome, WA).

Chamaebotrys

Chamaebotrys is known by four species, of which the widespread *Chamaebotrys boergesenii* is found in Australia. Plants have terete, hollow axes that are regularly segmented and have internal septa. The axes typically re-attach to the substratum and also to adjacent branches, forming an intricate network in which it is difficult to separate individual axes. Internally the branches have a hollow, mucilage-filled centre and an outer, cellular layer. The cellular layer has two distinct regions: an inner medulla of large, colourless cells and an outer cortex of smaller, pigmented cells. Spherical vesicular cells are borne on the inner surface of the medullary cells. Tetrasporangia have cruciately arranged spores and are borne in nemathecia. Cystocarps are protuberant.

Chamaebotrys boergesenii

TYPE LOCALITY: Sailus Besar, Isles Paternoster, Indonesia.
DISTRIBUTION: Known from Rottnest Island, Western Australia, the Great Barrier Reef, Queensland, and Lord Howe Island, New South Wales. Warmer waters of the Indo-Pacific.
FURTHER READING: Huisman (1996); Schils et al. (2003).

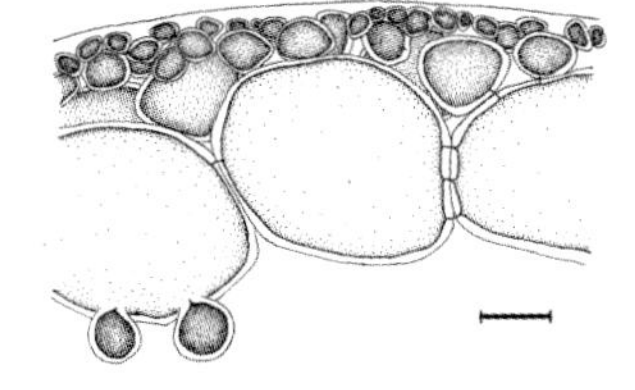

Chamaebotrys boergesenii (Rottnest Island, WA).

(above) *Chamaebotrys boergesenii* – partial section of cellular layer with vesicular cells (Rottnest Island, WA). Scale = 30 µm.

Chrysymenia

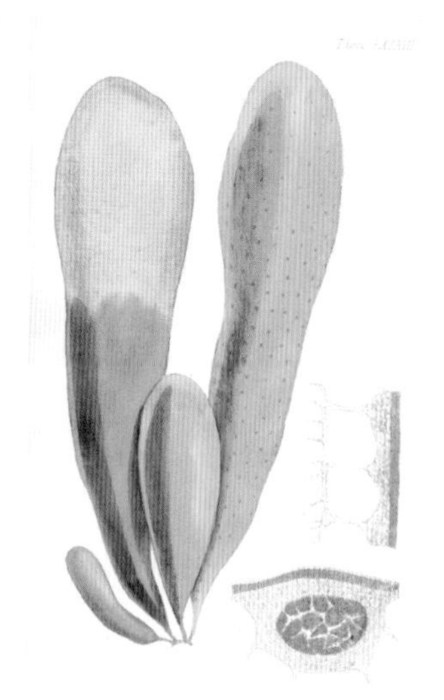

Chrysymenia includes a number of species, mostly found in the northern hemisphere. The genus is characterised by a hollow, branched or unbranched thallus that lacks internal diaphragms. Structurally the thallus is multiaxial, although the meristematic cells at the branch apices are generally difficult to discern and appear only as a vague cluster of smaller cells. Mature branches have a largely cell-free centre that is filled with mucilage and occasionally traversed by rhizoidal filaments. The outer, cellular layer is composed of an inner medulla of large, hyaline cells that grade into the smaller celled, pigmented cortex. Vesicular cells are borne on the inner surface of the medulla. *Chrysymenia brownii* was previously included in the genus *Gloiosaccion*, but that genus has now been subsumed into *Chrysymenia*. It is usually found growing on rocks or shells embedded in sand, often with several individuals growing together. Intact plants can often be found washed up on the beach. *Chrysymenia kaernbachii* is apparently widely distributed in tropical waters of the Indo-Pacific.

Chrysymenia brownii

TYPE LOCALITY: Georgetown, Tasmania.

DISTRIBUTION: From the Houtman Abrolhos, Western Australia, around southern mainland Australia to Walkerville, Victoria, and around Tasmania.

Chrysymenia kaernbachii

TYPE LOCALITY: New Guinea.

DISTRIBUTION: Known from the Houtman Abrolhos to the Dampier Peninsula, Western Australia, and the southern Great Barrier Reef, Queensland. Warmer seas of the Indo-West Pacific.

FURTHER READING: Cribb (1983); Womersley (1996); Schmidt et al. (2016).

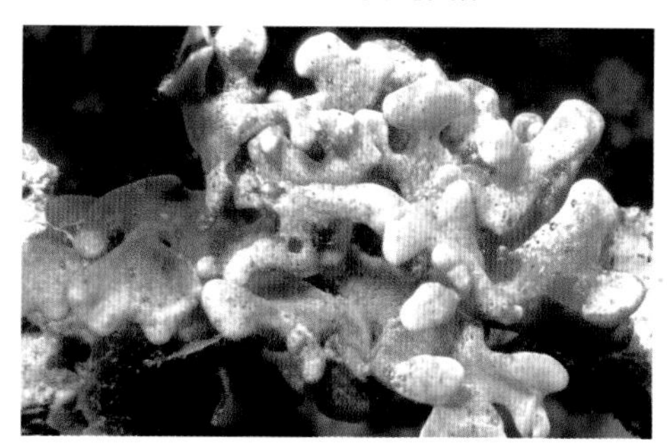

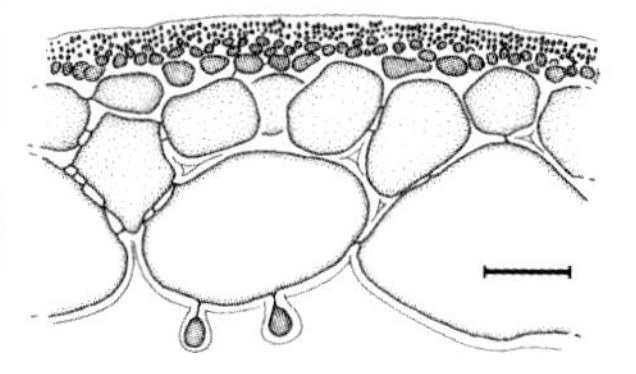

Chrysymenia brownii (Augusta, WA).

(top) *Chrysymenia brownii* – as illustrated by Harvey (1859a, pl. 83, as *Gloiosaccion brownii*).

(above) *Chrysymenia brownii* – partial section of medulla and cortex (Cottesloe, WA). Scale = 100 µm.

Chrysymenia kaernbachii (Houtman Abrolhos, WA).

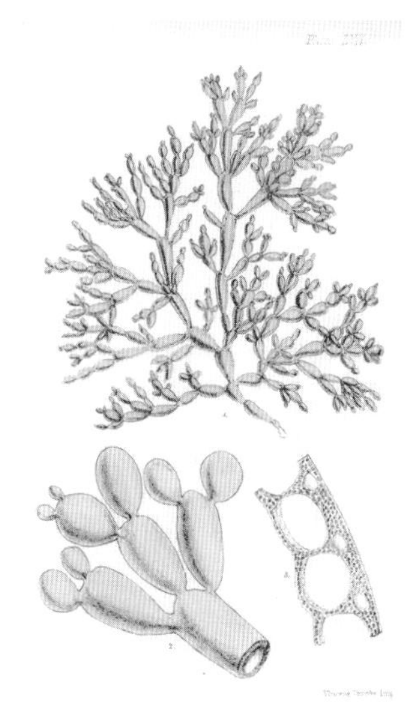

Coelarthrum

Plants of *Coelarthrum* are recognisable by their terete, dichotomously to polychotomously branched axes that are regularly segmented. The segments can be spherical in shape or, as is often the case in older parts of the plant near the base, elongate and sausage-like. Internally the axes have a central, mucilage-filled cavity bordered by an outer, cellular layer. The outer layer has an inner medulla of large colourless cells and an outer cortex of smaller, pigmented cells. As in many genera of the Rhodymeniaceae with internal cavities, spherical vesicular cells are borne on the inner surface of the medullary cells. Tetrasporangia are scattered throughout the cortex and have cruciately arranged spores. The axes of *Coelarthrum cliftonii* are generally light pink to red and soft to the touch. Branches often re-attach to adjacent branches. Within the genus, this species can be recognised by the broad joints between segments and largely immersed cystocarps. *Coelarthrum opuntia* is typically more robust, dark purple to red in colour and firm to the touch. The joints between segments are stalk-like and narrow.

Coelarthrum cliftonii

TYPE LOCALITY: Western Australia.

DISTRIBUTION: From the Montebello Islands, north-western Australia, south and east to Troubridge and Kangaroo Islands, South Australia. New Zealand. Canary Islands. Natal, southern Africa. Mauritius. West Indies. Hawaiian Islands. Papua New Guinea, Indonesia.

Coelarthrum opuntia

TYPE LOCALITY: Indian Ocean.

DISTRIBUTION: From Darwin, Northern Territory, around Western Australia to Wilsons Promontory, Victoria, and northern Tasmania. Rarely in Queensland. Indian Ocean. Japan.

FURTHER READING: Huisman (1996); Womersley (1996).

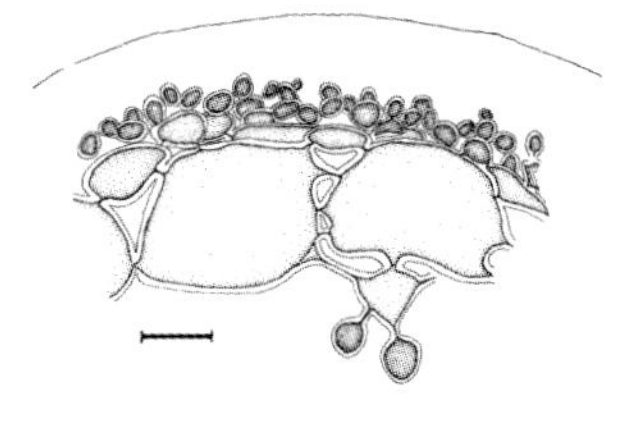

Coelarthrum cliftonii (Houtman Abrolhos, WA).

(top) *Coelarthrum cliftonii* – as illustrated by Harvey (1858, pl. 57, as *Chylocladia cliftonii*).

(above) *Coelarthrum cliftonii* – section of cellular layer (Rottnest Island, WA). Scale = 20 µm.

Coelarthrum opuntia (James Price Point, WA).

Halopeltis

Several of the species currently included in *Halopeltis* were previously placed in *Rhodymenia*, but were segregated based primarily on the results of DNA sequence analyses. Plants are mostly flattened and regularly or irregularly subdichotomously branched. Structurally the thalli have a medulla of large cells, with the spaces between them filled with smaller cells. This feature can serve to separate *Halopeltis* from *Rhodymenia*, which otherwise can be very similar in appearance, but is inconsistent in species from locations other than Australia. Tetrasporangia are formed in sori near branch tips; these are initially circular to oval in shape, but as they mature they become oblong. Cystocarps are protuberant, spherical or occasionally hemispherical. In addition to the typical growth form, the genus includes *Halopeltis austrina* (not shown), a small parasite that grows on the surface of other species of *Halopeltis*.

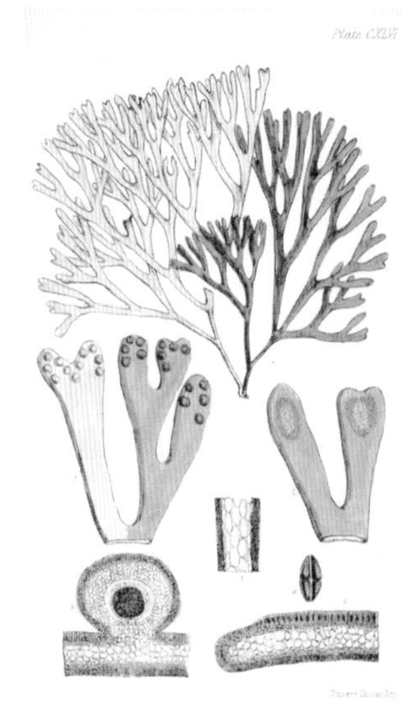

(top) *Halopeltis australis* – as illustrated by Harvey (1860, pl. 146, as *Rhodymenia australis*).

Halopeltis australis

TYPE LOCALITY: Western Australia.
DISTRIBUTION: Horrocks Beach, Western Australia, around southern Australia to Gabo Island, Victoria (possibly further north), and around Tasmania.
FURTHER READING: Saunders & McDonald (2010); Schneider et al. (2012); Scott (2017).

Halopeltis australis (Cape Peron, WA).

Rhodymenia

Species of *Rhodymenia* have flattened, cartilaginous axes that are generally dichotomously divided. Structurally the thallus is multiaxial and has a totally pseudoparenchymatous construction with a large-celled medulla grading into a smaller-celled cortex. Cystocarps are protuberant and tetrasporangia are cruciately divided and borne in largely unmodified sori at the branch apices. As reproductive specimens are rarely found, identification of *Rhodymenia* is generally based on vegetative features. *Rhodymenia novaehollandica* is common in southern Australia. In habit, it can be virtually identical to *Halopeltis australis*, but can be separated by the lack of smaller intercalary cells in the medulla and the cortex associated with tetrasporangial sori remaining unmodified. *Rhodymenia leptophylla* can be distinguished by its terete stolons and lower branches that grade into the flattened, dichotomously divided upper branches. *Rhodymenia prolificans* has slender branches broadening upwardly and marginal and often also surface proliferations.

Rhodymenia leptophylla

TYPE LOCALITY: Bay of Islands, New Zealand.
DISTRIBUTION: On most coasts of Australia and New Zealand. Widespread in the western and central Pacific.

Rhodymenia leptophylla (Heron Island, Qld).

Rhodymenia novaehollandica

TYPE LOCALITY: Queenscliff Jetty, Port Phillip Heads, Victoria.
DISTRIBUTION: Probably widespread in southern Australia.

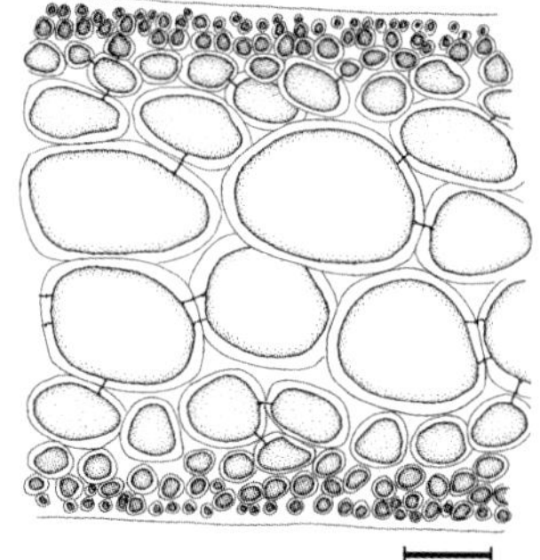

Rhodymenia prolificans

TYPE LOCALITY: Georgetown, Tasmania.
DISTRIBUTION: Known from Cape Peron, Western Australia, and Portland and Gabo Island, Victoria, and around Tasmania.
FURTHER READING: Womersley (1996); Huisman et al. (2006); Saunders & McDonald (2010); Nelson (2013).

Rhodymenia novaehollandica (Cape Peron, WA).

(top) *Rhodymenia novaehollandica* – section of thallus (Rottnest Island, WA). Scale = 30 µm.

Rhodymenia prolificans (Cape Peron, WA).

Aglaothamnion

Species of *Aglaothamnion* have filamentous, tufted thalli that are generally uncorticated, thereby allowing easy examination of the thallus structure. The thalli are uniaxial, with each axial cell producing one lateral branch that divides in a similar fashion to the main axes. This type of branching is common in the Callithamniaceae and can be found in a number of genera. *Aglaothamnion* is difficult to separate from the closely related genus *Callithamnion*. The form of the reproductive structures can be of assistance, as the carpospores of *Callithamnion* are generally arranged in spherical lobes, while those of *Aglaothamnion* are in divided lobes. These differences do not define the genera, however, which are separated by the presence of uninucleate cells in *Aglaothamnion* and multinucleate cells in *Callithamnion*. Tetrasporangia are tetrahedrally divided.

Aglaothamnion cordatum

TYPE LOCALITY: Off Cruz Bay, between St Thomas and St John, Virgin Islands.

DISTRIBUTION: Widely distributed in tropical to warm-temperate waters. In Australia, known from the Houtman Abrolhos, Western Australia, north and east to the southern Great Barrier Reef, Queensland.

FURTHER READING: Price & Scott (1992, as *Callithamnion cordatum*); Huisman (2018).

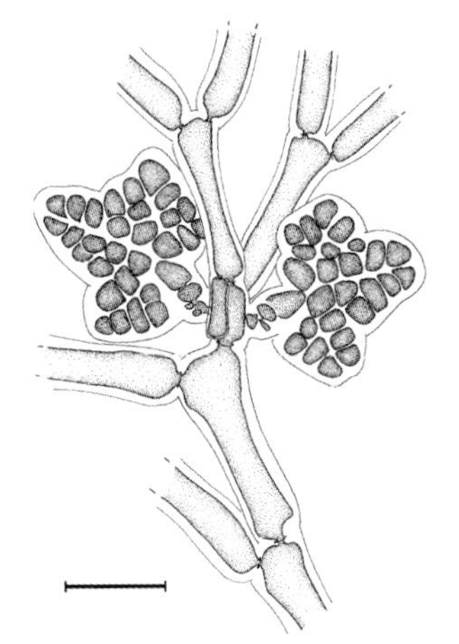

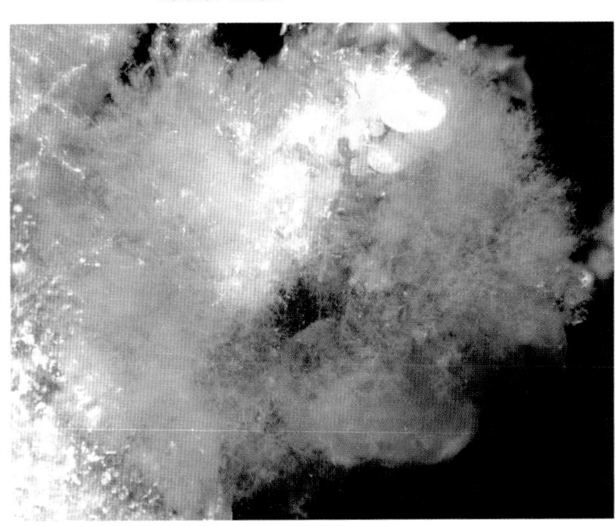

Aglaothamnion cordatum (Houtman Abrolhos, WA).

(above) *Aglaothamnion cordatum* – detail of mature carposporophyte with divided lobes of carposporangia (Houtman Abrolhos, WA). Scale = 50 µm.

Carpothamnion

The robust thallus of *Carpothamnion* belies its affinities with the Callithamniaceae, a family composed mostly of delicate, filamentous plants. Younger axes of *Carpothamnion* are obviously filamentous, but as they mature they become thickly corticated by secondary rhizoidal filaments that arise from the basal cells of lateral branches. These corticating filaments eventually produce a distinctive layer of small cells that forms a coherent outer layer, and the main axis is obscured. The structure of the thallus is best observed in the younger branches, where the uniaxial nature can be more clearly seen. A single lateral branch arises per axial cell. Unfortunately, these lateral and younger branches are often lost from drift specimens. *Carpothamnion gunnianum* is the only species of the genus found in Australia, where it is a widely distributed but uncommon species in colder waters.

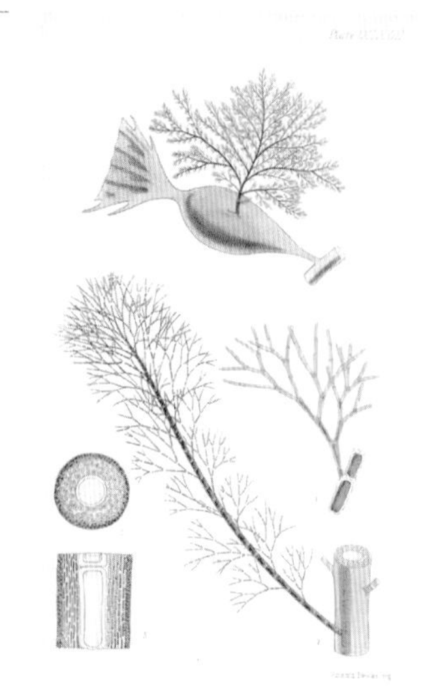

Carpothamnion gunnianum

TYPE LOCALITY: Port Arthur, Tasmania.
DISTRIBUTION: From the Houtman Abrolhos, Western Australia, around southern mainland Australia to Phillip Island, Victoria, and Tasmania.
FURTHER READING: Wollaston (1992, as *Thamnocarpus gunnianus*); Womersley & Wollaston (1998b).

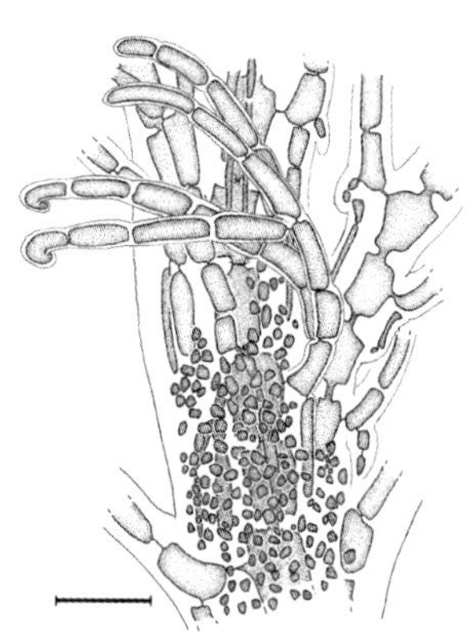

Carpothamnion gunnianum (Mosman, WA).

(top) *Carpothamnion gunnianum* – as illustrated by Harvey (1863, pl. 273, as *Callithamnion penicillatum*).

(above) *Carpothamnion gunnianum* – detail of thallus showing developing cortical layer (Houtman Abrolhos, WA). Scale = 50 µm.

Crouania

With its uniaxial filamentous construction, *Crouania* is similar in structure to *Euptilocladia*, differing in the production of three whorl branches per axial cell instead of four. The genus is widely distributed in Australia, with several larger species found in the colder waters of southern Australia. Some species of *Crouania* are mucilaginous and accumulate sand and other detritus, giving them the appearance of being slightly calcified. *Crouania dampieriana* was described from tropical Western Australia, where it can be quite common.

Crouania dampieriana

TYPE LOCALITY: Barrow Island, Western Australia.
DISTRIBUTION: From the Barrow Island north to Cassini Island, Western Australia.
FURTHER READING: Wollaston (1968, for southern Australian species); Huisman (2018).

Crouania dampieriana (Barrow Island, WA).

(above) *Crouania dampieriana* – epiphytic plant on *Codium dwarkense* (James Price Point, WA). Scale = 200 µm.

Euptilocladia

Euptilocladia is a genus of three species, two endemic to Australia, of which *Euptilocladia spongiosa* is included here. Thalli are flattened and alternately branched, with a mixture of short branches of limited growth and longer lateral branches. Structurally the thallus is uniaxial, but the central filament is obscured by rhizoidal filaments that form a dense cortical layer and give the thallus a spongy texture. Each axial cell bears four whorl branches. The only non-Australian species is *Euptilocladia magruderi* (not shown), which occurs in the Hawaiian Islands.

Euptilocladia spongiosa

TYPE LOCALITY: Robe, South Australia.
DISTRIBUTION: From the Houtman Abrolhos, Western Australia, around southern mainland Australia to Waratah Bay, Victoria.
FURTHER READING: Wollaston (1968); Wollaston & Womersley (1998a).

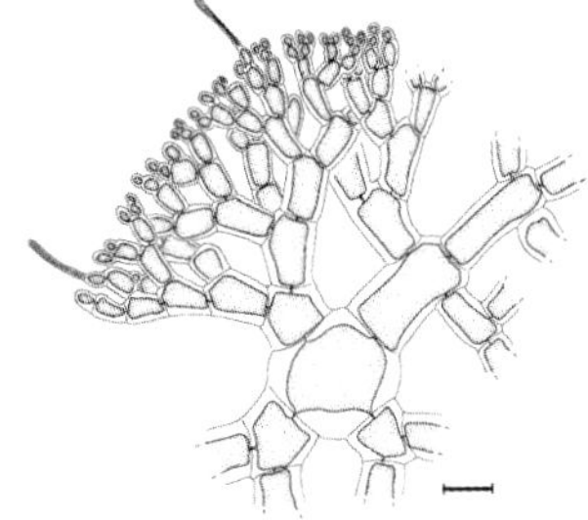

Euptilocladia spongiosa (Two Rocks, WA).

(above) *Euptilocladia spongiosa* – partial section of thallus (Two Rocks, WA). Scale = 30 µm.

Euptilota

Euptilota is represented by *Euptilota articulata*, a relatively common and distinctive species. Plants are flattened and alternately distichously branched. Structurally the thallus is uniaxial, with each axial cell bearing a single lateral branch. Major axes become secondarily corticated by rhizoidal filaments that arise from the basal cells of lateral branches.

Euptilota articulata

TYPE LOCALITY: Australia (probably Fremantle, Western Australia).
DISTRIBUTION: From the Houtman Abrolhos, Western Australia, around southern mainland Australia and eastern Tasmania to Coffs Harbour and Lord Howe Island, New South Wales. Japan. India.
FURTHER READING: Millar (1990); Womersley (1998); Scott (2017).

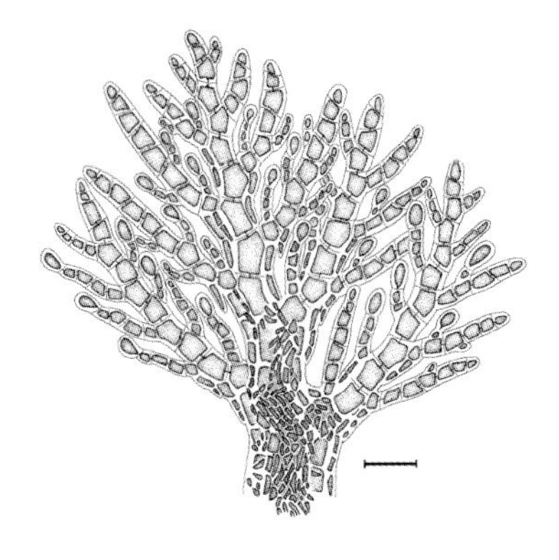

Euptilota articulata (Cape Peron, WA).

(above) *Euptilota articulata* – detail of thallus (Carnac Island, WA). Scale = 200 µm.

(right) *Euptilota articulata* – detail of apical region (Rottnest Island, WA). Scale = 100 µm.

Hirsutithallia

Hirsutithallia includes six species found in the colder waters of southern Australia. Thalli are filamentous with each axial cell bearing a single lateral branch. The primary axes become corticated by downwardly growing filaments that arise from the lower cells of lateral branches. A characteristic feature of the genus is the production of short anticlinal filaments from the cells of the cortical filaments, giving the axis a furry appearance. However, in some specimens, particularly in Western Australia, these are rare or lacking altogether, making identification uncertain. *Hirsutithallia laricina* is usually epiphytic on *Posidonia* but can also grow on other algae and seagrasses. In the photo here, it is epiphytic on the seagrass *Amphibolis antarctica*.

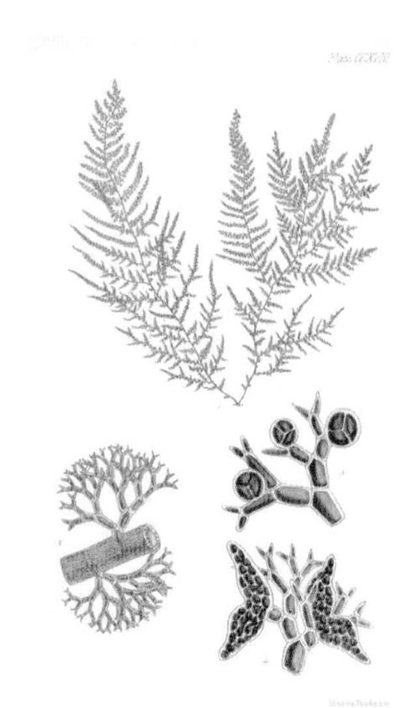

Hirsutithallia laricina

TYPE LOCALITY: Rottnest Island, Western Australia.
DISTRIBUTION: From Yanchep, Western Australia, around southern mainland Australia to Port Phillip Heads, Victoria.
FURTHER READING: Womersley & Wollaston (1998b).

Hirsuthallia laricina (Cape Peron, WA).

(top) *Hirsuthallia laricina* – as illustrated by Harvey (1862, pl. 218, as *Callithamnion laricinum*).

(above) *Hirsuthallia laricina* – detail of thallus (Cape Peron, WA). Scale = 1 mm.

Seirospora

Seirospora includes small, filamentous plants in which each axial cell bears a single lateral branch. This type of branching is common in the Callithamniaceae, and sterile specimens of *Seirospora* are impossible to distinguish from *Callithamnion* and *Aglaothamnion* (and others). Carposporophytes are essential for positive identification – in *Seirospora* the carposporangia are arranged in short chains (compare with *Aglaothamnion*, see p. 165). Tetrasporangia are tetrahedrally divided. *Seirospora orientalis* grows on a variety of subtidal algae and invertebrates.

Seirospora orientalis

TYPE LOCALITY: One Tree Island, Great Barrier Reef, Queensland.
DISTRIBUTION: Known from the Houtman Abrolhos north to the Montebello Islands, Western Australia, and the Capricorn Group of the Great Barrier Reef, Queensland. Yemen. Japan.
FURTHER READING: Kraft (1988b); Huisman (2018).

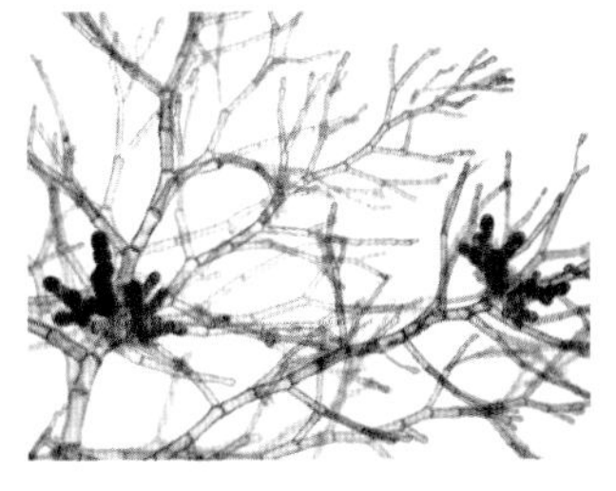

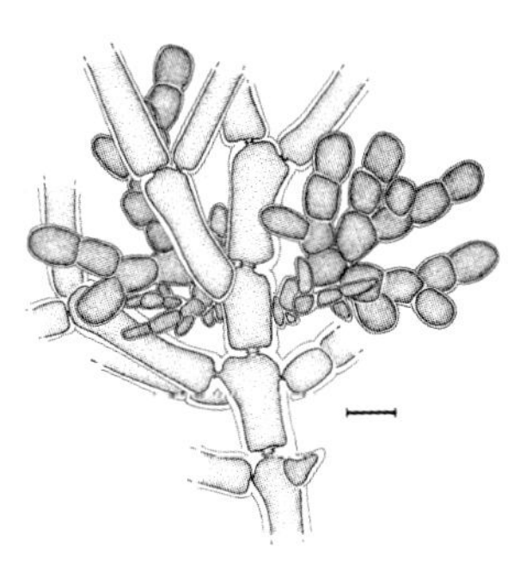

Seirospora orientalis (Heron Island, Qld).

(top) *Seirospora orientalis* (Sykes Reef, Qld).

(above) *Seirospora orientalis* – detail of mature carposporophyte (Houtman Abrolhos, WA). Scale = 30 µm.

Acrothamnion

Acrothamnion is an epiphyte common on a variety of algae, where it forms creeping axes that bear tufts of erect filamentous branches. All branches are uncorticated, with each axial cell producing whorls of four short branches. In *Acrothamnion preissii*, the branches in each whorl consist of two opposite, comparatively long branches with two shorter branches lying between them. All branches in the whorl are pinnately branched and most bear terminal vesicular cells. This last-mentioned feature is unique to *Acrothamnion*.

Acrothamnion preissii

TYPE LOCALITY: Rottnest Island, Western Australia.
DISTRIBUTION: From the Shark Bay region, Western Australia, around southern mainland Australia to Wilsons Promontory, Victoria.
FURTHER READING: Wollaston (1968); Wollaston & Womersley (1998b).

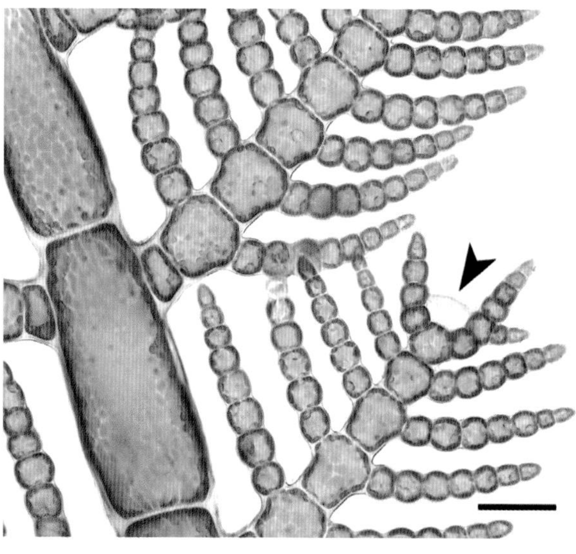

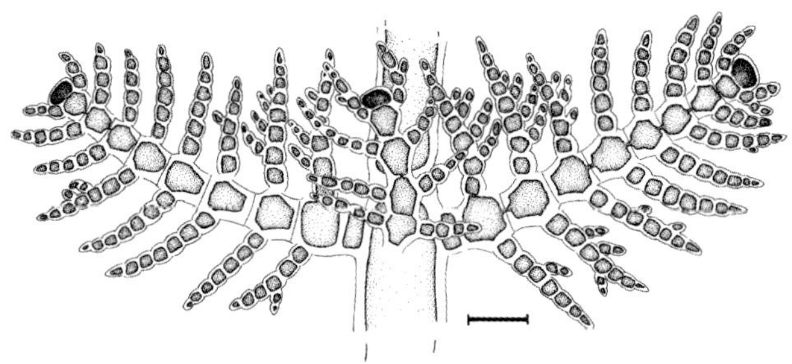

Acrothamnion preissii – detail of thallus with terminal vesicular cells (arrow) (Cape Peron, WA). Scale – 25 µm.

(left) *Acrothamnion preissii* – detail of thallus (Houtman Abrolhos, WA). Scale = 30 µm.

Amoenothamnion

The two Australian species of *Amoenothamnion* are mostly restricted to the south coast, but one, *Amoenothamnion planktonicum,* also occurs on the west coast, north to Dongara. Thalli are small and filamentous, generally dichotomously divided with each axial cell bearing a whorl of four short branches. These 'whorl branches' have a distinctive structure – the basal cell is relatively large and bears a single central branch that terminates in an acute tip, as well as several shorter or single-celled lateral branches. *Amoenothamnion planktonicum* commonly occurs as an epiphyte on larger algae and seagrasses.

Amoenothamnion planktonicum

TYPE LOCALITY: Sports Beach, Bridgewater Bay (near Portland), Victoria.

DISTRIBUTION: From Dongara, Western Australia, around southern mainland Australia and south-eastern Tasmania to Twofold Bay, New South Wales.

FURTHER READING: Wollaston (1968); Womersley & Wollaston (1998a).

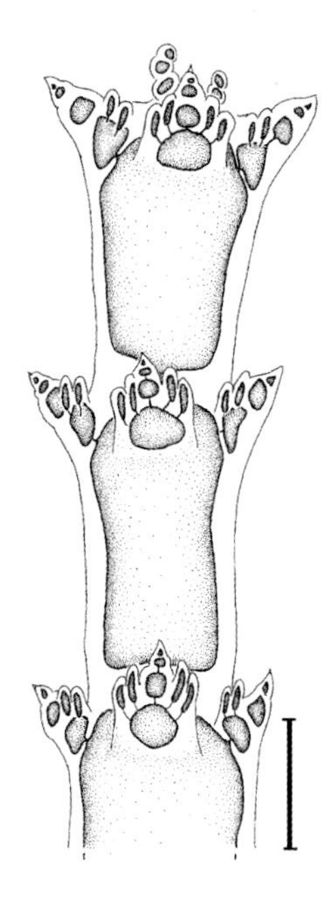

Amoenothamnion planktonicum – detail of axial cells and whorl branches (Rottnest Island, WA). Scale = 50 µm.

Antithamnion

Thalli of *Antithamnion* are generally small and filamentous, with creeping and erect axes that are uncorticated. Distinguishing features of the genus include the formation of short lateral branches in opposite pairs and vesicular cells borne on special two- to four-celled branches. In *Antithamnion armatum*, the lateral branches are curved towards the base and are secundly branched from the upper face. Terminal cells of lateral branches taper to an acute spine. *Antithamnion hanovioides* is the most common species of *Antithamnion* on southern Australian coasts, growing as an epiphyte on seagrasses and larger algae.

Antithamnion armatum

TYPE LOCALITY: Australia.
DISTRIBUTION: From Shark Bay, Western Australia, around southern Australia to Nora Creina, South Australia.

Antithamnion hanovioides

TYPE LOCALITY: Gulf St Vincent, South Australia.
DISTRIBUTION: From the Houtman Abrolhos, Western Australia, to Barranjoey Head, New South Wales, and around Tasmania.
FURTHER READING: Wollaston (1968); Wollaston & Womersley (1998b).

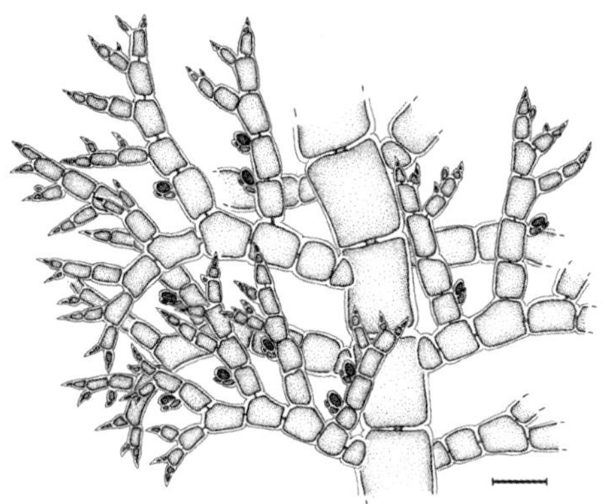

Antithamnion armatum (Cape Peron, WA).

(left) *Antithamnion armatum* – detail of thallus (Houtman Abrolhos, WA). Scale = 100 µm.

Antithamnion hanovioides – growing on *Amphibolis antarctica* (Cape Peron, WA).

Antithamnionella

Antithamnionella is a genus of small, filamentous plants with creeping and erect axes. Somewhat similar to *Antithamnion*, thalli of *Antithamnionella* can be distinguished by the production of lens-shaped vesicular cells from cells of whorl branches, as opposed to the more elliptical vesicular cells on specialised short branches in *Antithamnion*. In addition, the number of whorl branches per axial cell is variable within the genus, with those of *Antithamnionella graeffei* generally produced in threes.

Antithamnionella graeffei

TYPE LOCALITY: Tonga.
DISTRIBUTION: In Australia known from the Great Barrier Reef, Queensland, and the Houtman Abrolhos north to Cassini Island, Western Australia. Widespread in tropical and warm-temperate seas.
FURTHER READING: Athanasiadis (1996); Huisman (2018).

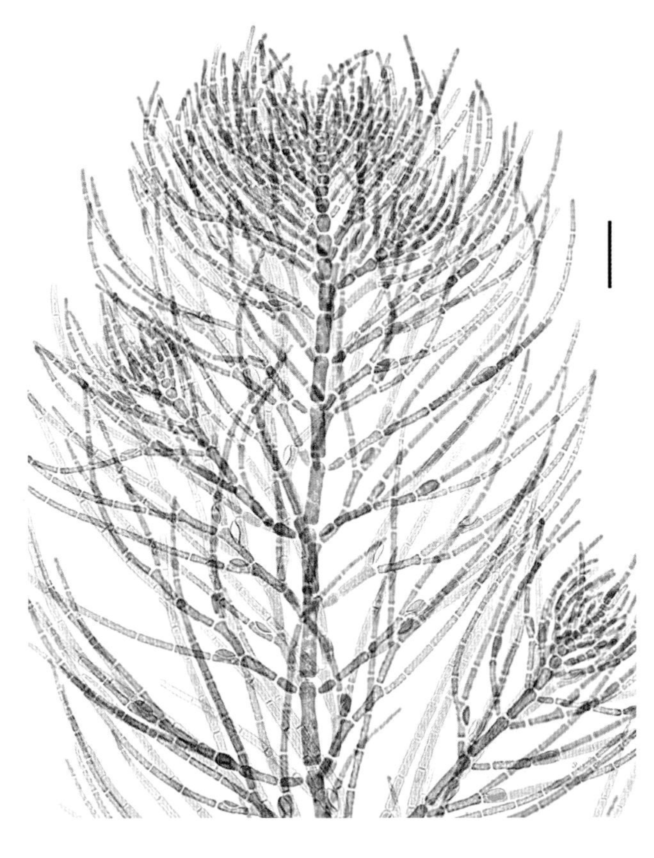

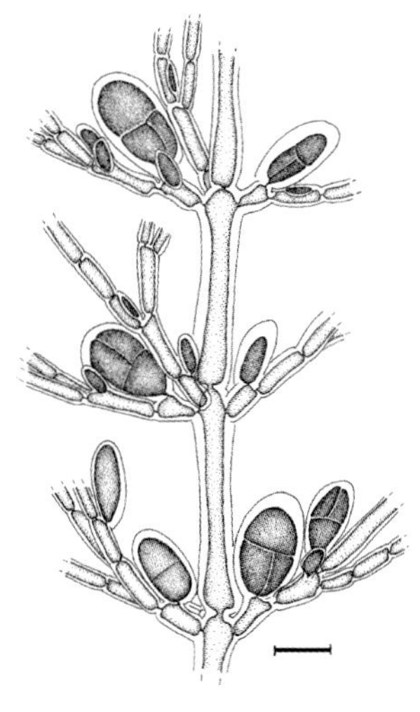

Antithamnionella graeffii (Cassini Island, WA). Scale = 50 µm.

(above) *Antithamnionella graeffei* – portion of thallus with vesicular cells and tetrasporangia (Houtman Abrolhos, WA). Scale = 30 µm.

Balliella

Most species of the genus *Balliella* are small and inconspicuous, with *Balliella hirsuta* one of the few exceptions. Thalli are filamentous, with the major axes becoming secondarily corticated by filaments arising from the basal cells of lateral branches. Each axial cell bears two lateral branches. The genus is characterised by the presence of unusual vesicular cells that are spherical and attached to the bearing cell by a thin cytoplasmic thread. In most species of the genus, these arise on the lower (or abaxial) surface of the cells of lateral branches. In *Balliella hirsuta*, however, the vesicular cells are borne on the upper (or adaxial) surface. *Balliella crouanioides* occurs in northern Western Australia and is often the only alga present in areas of low light; it often has distinctive orange tips. *Balliella grandis* is a larger species found only on the Great Barrier Reef.

Balliella crouanioides

TYPE LOCALITY: Mage Island, Japan.
DISTRIBUTION: Known from Adele Island to Long Reef, northern Western Australia. Kenya. Mozambique. South Africa. Taiwan. Southern Japan. Korea.

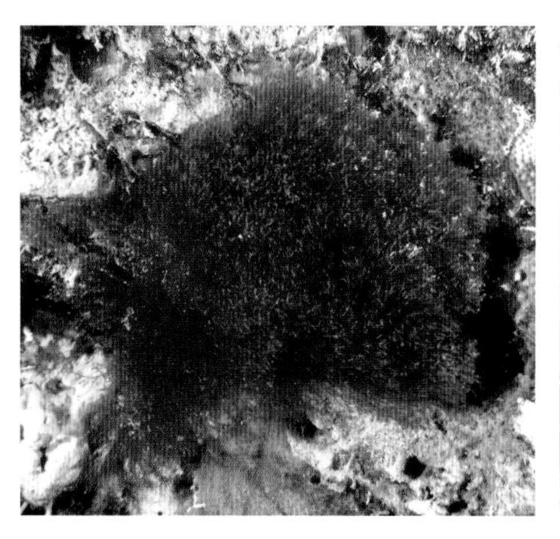

Balliella grandis

TYPE LOCALITY: Wistari Reef, near Heron Island, Queensland.
DISTRIBUTION: Known only from the southern Great Barrier Reef, Queensland.

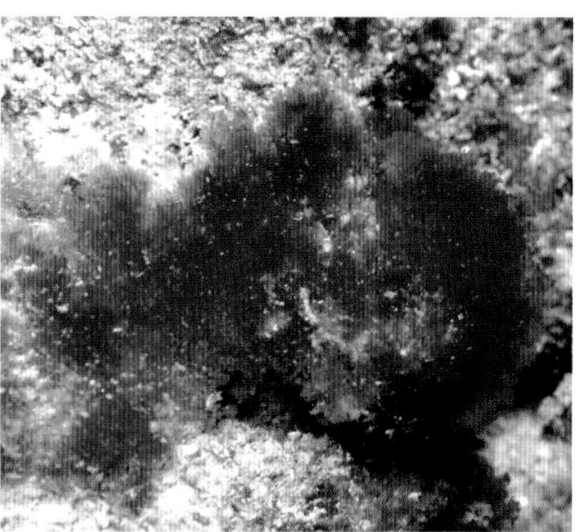

Balliella crouanioides (Cassini Island, WA).

Balliella grandis (Broomfield Reef, Qld).

Balliella hirsuta

TYPE LOCALITY: Green Island, adjacent to Rottnest Island, Western Australia.
DISTRIBUTION: Known only from Rottnest Island, Western Australia.
FURTHER READING: Huisman (1988); Huisman & Kraft (1984); Huisman (2018).

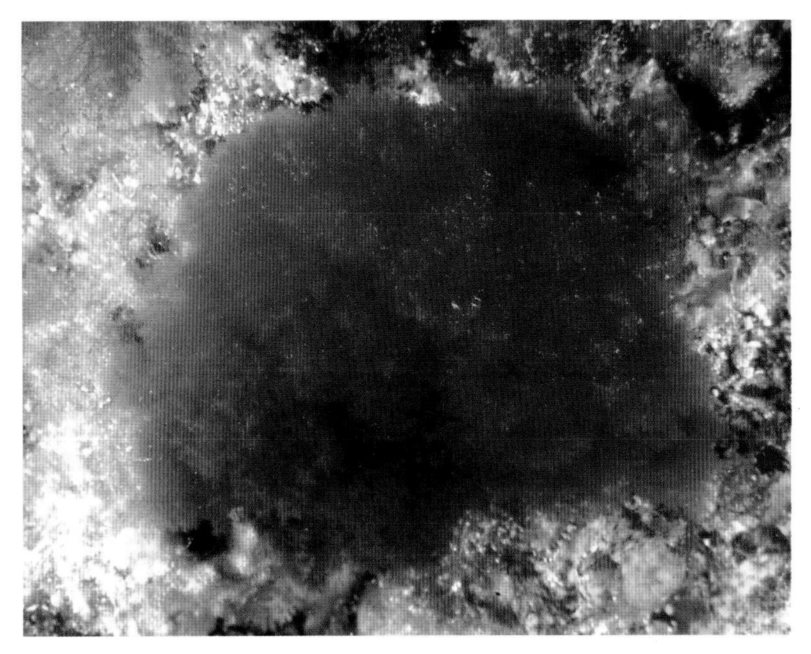

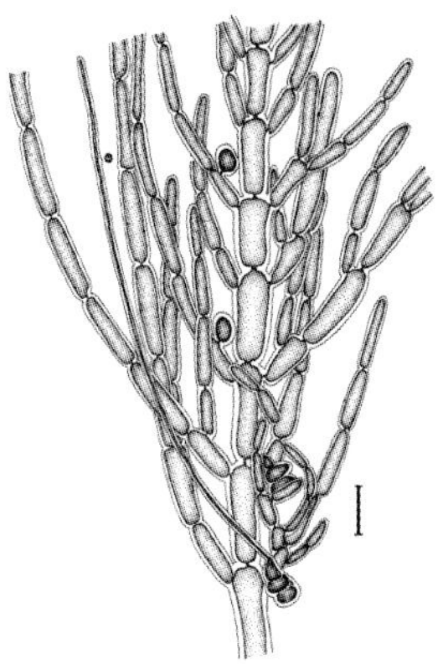

Balliella hirsuta
(Rottnest Island, WA).

(above) *Balliella hirsuta* –
detail of thallus with carpogonial
branches (Rottnest Island, WA).
Scale = 20 µm.

Centroceras

Several species of *Centroceras* have been recorded from Australia, of which the widely distributed and common *Centroceras gasparrinii* is included here. Thalli are filamentous and uniaxial, and grow to around 4 cm in height. Axes are pseudodichotomously branched and apices are generally slightly to distinctly curled. All axes are corticated, initially around the joints between axial cells (nodal cortication), then progressively to cover the entire filament. Distinctive features of the species are the regular arrangement of cortical cells in longitudinal rows and the presence of short spines on the nodal cortication. In the past, this and other species would have been included in a broadly defined *Centroceras clavulatum* complex, but recent studies have recognised several more precisely demarcated species.

Centroceras gasparrinii

TYPE LOCALITY: Palermo, Sicily, Italy.
DISTRIBUTION: Throughout Australia. Widely distributed in temperate and tropical seas.
FURTHER READING: Huisman (2018).

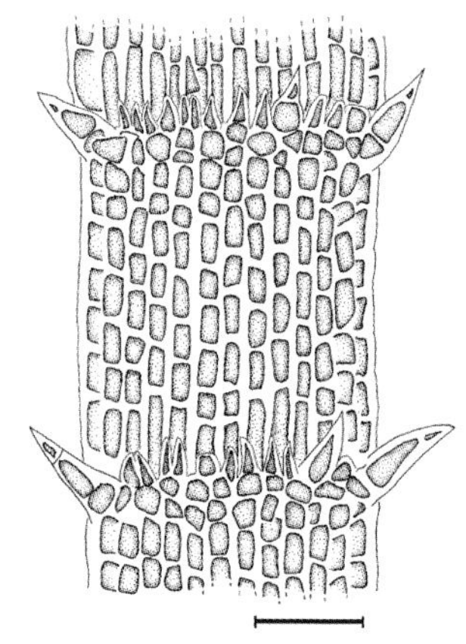

Centroceras gasparrinii (Barrow Island, WA).

(above) *Centroceras gasparrinii* – detail of axis and cortication (Houtman Abrolhos, WA). Scale = 50 µm.

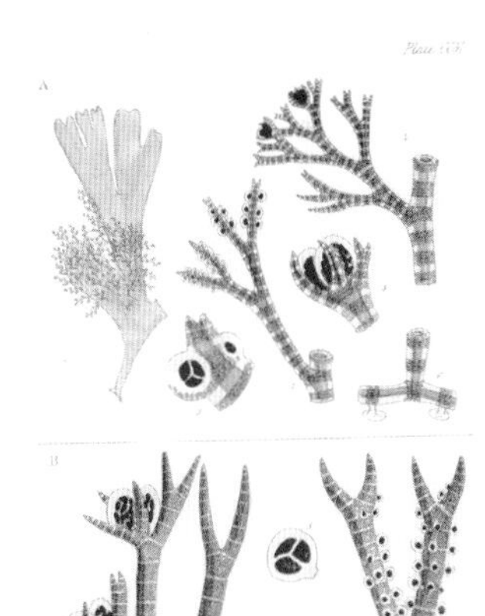

Ceramium

The genus *Ceramium* is represented in Australia by a large number of species. Most have small, tufted thalli that grow on rock or epiphytically on larger algae and seagrasses. Growth can be exclusively erect or can include prostrate axes that creep along the substratum. Thalli are pseudodichotomously, irregularly, or alternately branched and apices are often slightly to distinctly curved. All species are filamentous and uniaxial, with axes that are corticated to some degree. Cortication is initially around the joints between axial cells (nodal cortication) but in some species spreads to cover the entire axial filament, as in *Ceramium virgatum*. The pattern and degree of cortication is used to separate the various species, but some care must be taken as the extent of cortication in mature axes is often different from that closer to the apices. The examples included here are common along the Australian coast but many other species are likely to be encountered.

Ceramium filicula

Ceramium filicula grows almost exclusively on larger brown and (rarely) red algae. The species has a distinctive holdfast of densely clumped rhizoids. Thalli are regularly alternately branched and generally prostrate.

TYPE LOCALITY: Port Noarlunga, South Australia.
DISTRIBUTION: From the Houtman Abrolhos, Western Australia, around southern Australia to Coffs Harbour, New South Wales. Seychelles.

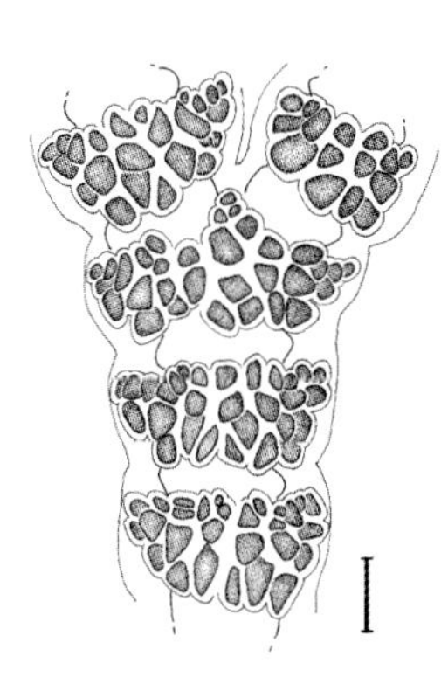

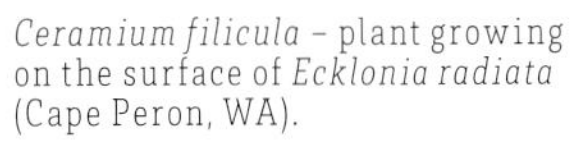

Ceramium filicula – plant growing on the surface of *Ecklonia radiata* (Cape Peron, WA).

(above) *Ceramium filicula* – detail of nodal cortication (Houtman Abrolhos, WA). Scale = 30 µm.

(top) *Ceramium filicula* (A) and *Ceramium isogonum* (B) – as illustrated by Harvey (1862, pl. 206, *C. filicula* as *C. miniatum*).

Ceramium isogonum

A relatively large species, often with conspicuously inrolled apices, *Ceramium isogonum* can also be recognised by the slightly rounded edges of the nodal cortication.

TYPE LOCALITY: Garden Island, Western Australia.
DISTRIBUTION: Widespread in the Indo-Pacific and on most Australian coasts.

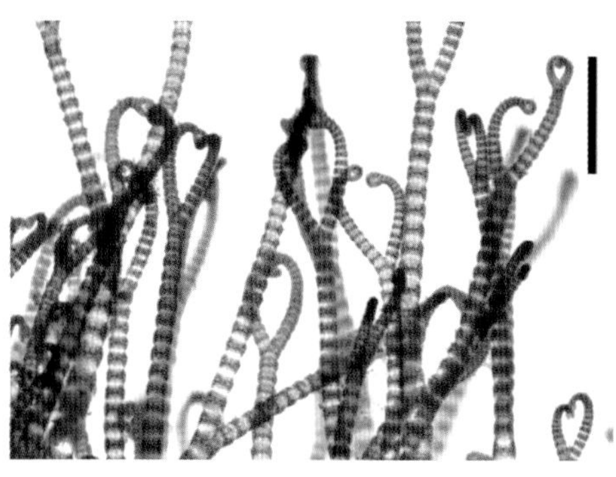

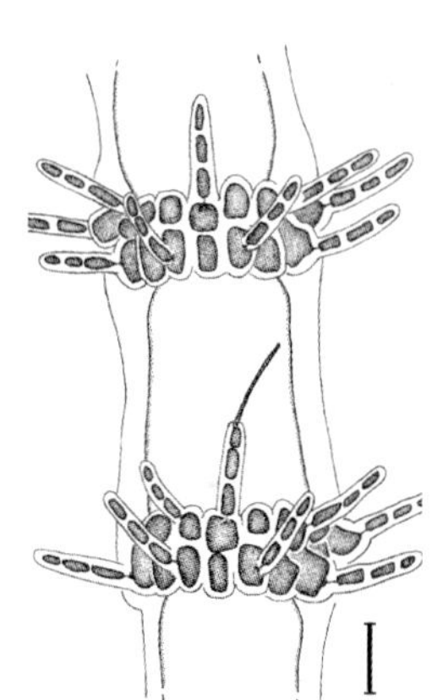

Ceramium isogonum (Long Reef, WA).

(middle) *Ceramium isogonum* – detail of thallus (Cape Peron, WA). Scale = 1 mm.

(above) *Ceramium shepherdii* – detail of nodal cortication with short filaments (Houtman Abrolhos, WA). Scale = 30 µm.

Ceramium phillipsiae

A distinctive species that grows as an epiphyte on cartilaginous red algae. It has a basal 'stalk' formed by the naked lowermost cell.

TYPE LOCALITY: South Scott Reef, Western Australia.
DISTRIBUTION: Rowley Shoals, Scott Reef and Ningaloo, Western Australia.

Ceramium phillipsiae (Ningaloo, WA). Scale = 500 µm.

Ceramium shepherdii

Ceramium shepherdii is always epiphytic on the seagrasses *Amphibolis* and *Posidonia*, or on the algae associated with them. It has characteristic short, blunt filaments that arise in a double whorl from the nodal cortical cells.

TYPE LOCALITY: Four kilometres south of Redcliff Point, northern Spencer Gulf, South Australia.
DISTRIBUTION: Barrow Island, Western Australia, around southern Australia to Port Stephens and Lord Howe Island, New South Wales.

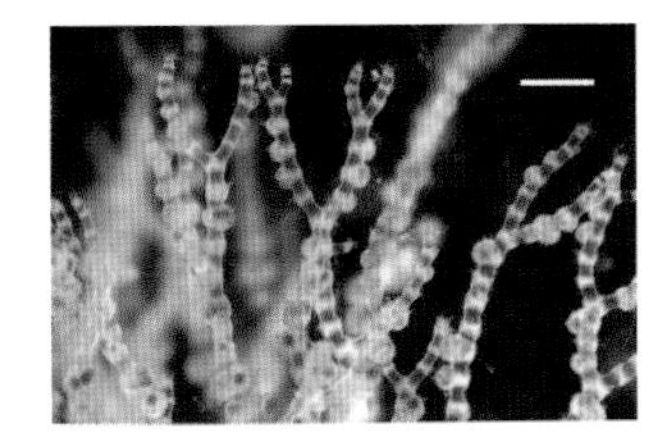

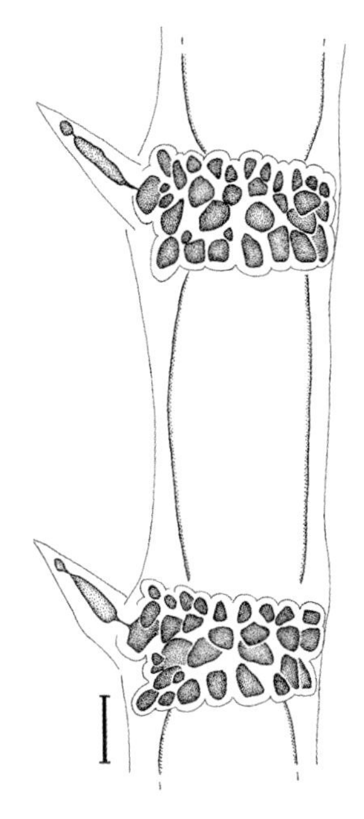

Ceramium puberulum

A common epiphyte on the seagrasses *Posidonia* (p. 429) and *Amphibolis* (p. 420), *Ceramium puberulum* has distinctive short spines arising from the nodal cortication.

TYPE LOCALITY: Western Australia.
DISTRIBUTION: Shark Bay, Western Australia, around southern Australia to Jervis Bay, New South Wales, and northern Tasmania.

Ceramium cliftonianum

A common species along southern Australian coasts.

TYPE LOCALITY: Western Australia.
DISTRIBUTION: Houtman Abrolhos, Western Australia, around southern Australia to Port Stephens, New South Wales, and around Tasmania.

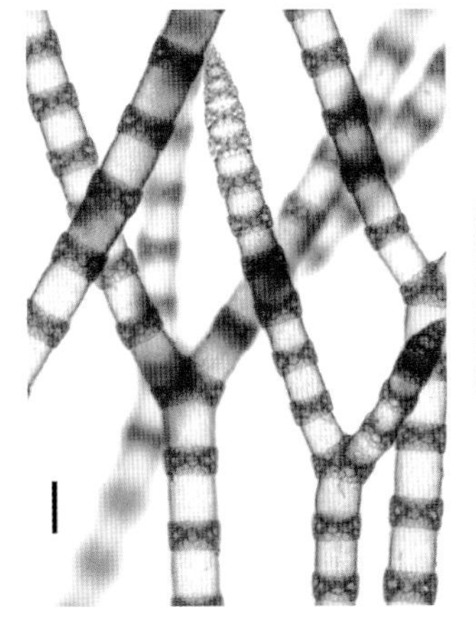

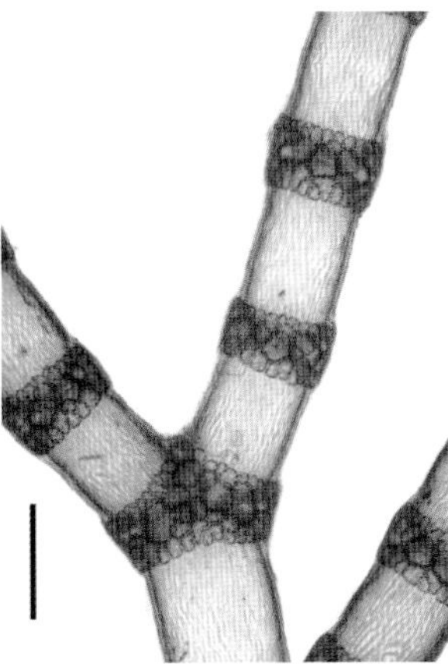

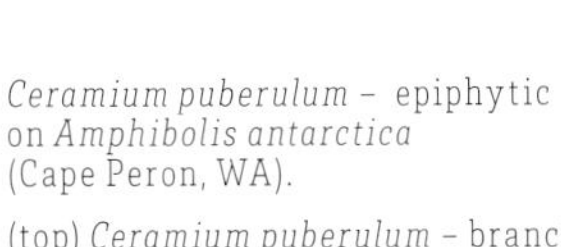

Ceramium puberulum – epiphytic on *Amphibolis antarctica* (Cape Peron, WA).

(top) *Ceramium puberulum* – branch detail (Cape Peron, WA). Scale = 500 µm.

(above) *Ceramium puberulum* – detail of nodal cortication with two-celled spines (Houtman Abrolhos, WA). Scale = 30 µm.

(right) *Ceramium cliftonianum* – upper branches (Cape Peron, WA). Scale = 100 µm.

(far right) *Ceramium cliftonianum* – detail of nodal cortication (Cape Peron, WA). Scale = 100 µm.

Ceramium virgatum

A readily recognised species due to its complete cortical layer, with no obvious banding.

TYPE LOCALITY: South Harbour, Helgoland, North Sea.

DISTRIBUTION: Fremantle, Western Australia to Wilsons Promontory, Victoria, and around Tasmania. Widespread in most seas.

FURTHER READING: Womersley (1978); Womersley (1998, *Ceramium virgatum* as *Ceramium rubrum*), Huisman (2010), Maggs et al. (2002).

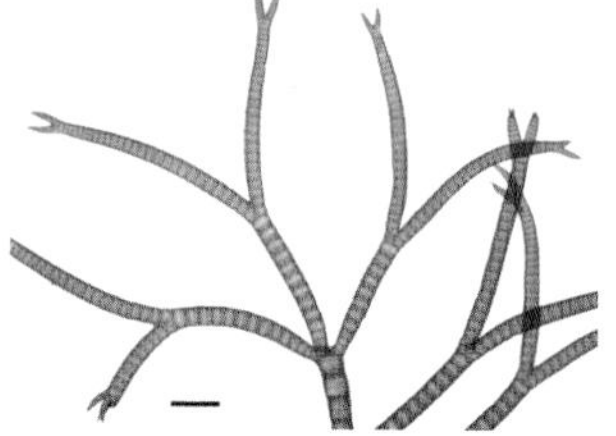

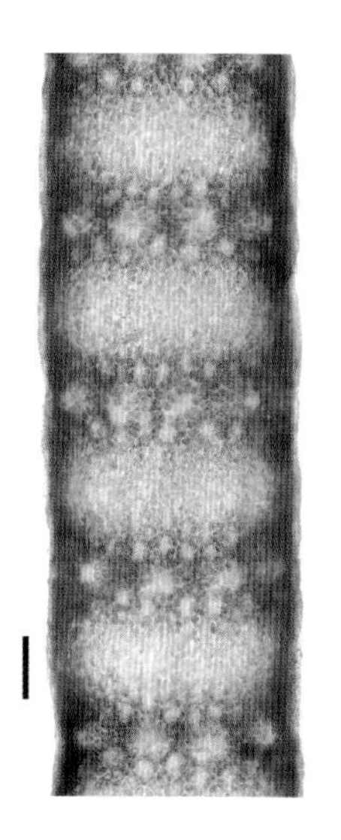

Ceramium virgatum (Cape Peron, WA).

(top) *Ceramium virgatum* – upper branches (Cape Peron, WA). Scale = 1 mm.

(above) *Ceramium virgatum* – detail of filaments (Cape Peron, WA). Scale = 100 µm.

Corallophila

Species of *Corallophila* have filamentous thalli in which the axes are entirely corticated by secondary filaments arising from the nodes. These filaments are predominantly basipetally directed and a pair of filaments is formed from each periaxial cell. In this feature, *Corallophila* contrasts with the closely related genera *Ceramium* (in which the growth of corticating filaments – where present – is mainly acropetal; p. 179) and *Centroceras* (in which a single corticating filament is produced per periaxial cell; p. 178). *Corallophila huysmansii* has prostrate axes that subtend sparingly branched, upright axes to about 1 cm high. The cortical cells are generally transversely elongate and are arranged in vague longitudinal rows.

Corallophila huysmansii

TYPE LOCALITY: Lucipara Island, Indonesia.
DISTRIBUTION: From the Houtman Abrolhos, Western Australia, around northern Australia to Lord Howe Island, New South Wales. Warmer seas of the Indo-Pacific.
FURTHER READING: Price & Scott (1992, as *Ceramium huysmansii*); Norris (1993); Huisman (2018).

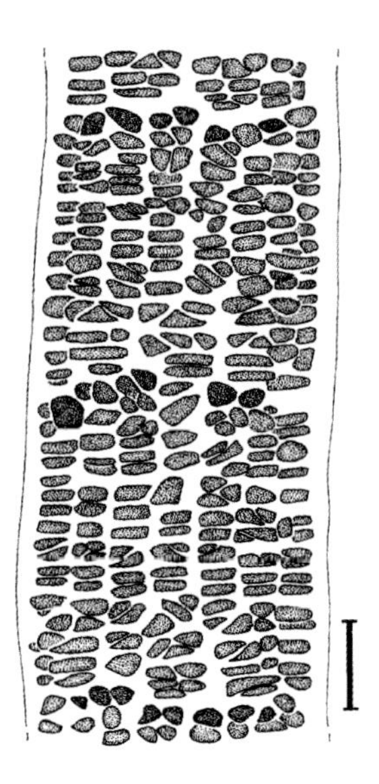

Corallophila huysmansii – detail of cortication (Houtman Abrolhos, WA). Scale = 30 µm.

Gayliella

Gayliella includes several small, but often prolific, species. Most species of the genus were previously included in *Ceramium* (p. 179), with which they share a filamentous construction with distinct cortication arising in bands at the nodes between axial cells. In *Gayliella*, the lower cells of the nodal bands are elongated transversely so that they appear to partially wrap around the primary filament, whereas in *Ceramium* the cells are more equidimensional. *Gayliella transversalis* is dichotomously to subalternately branched and is an extremely common epiphyte on a variety of seagrasses and other algae.

Gayliella transversalis

TYPE LOCALITY: Spanish Rock, Bermuda.
DISTRIBUTION: Widely distributed in warmer seas.
FURTHER READING: Womersley (1978, as *Ceramium*), (1998, as *Ceramium*); Huisman (2018).

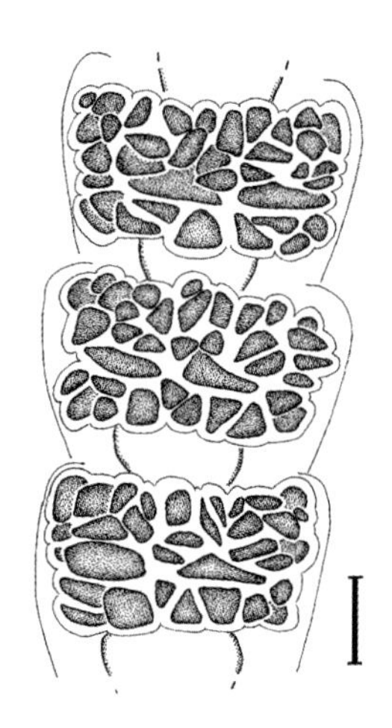

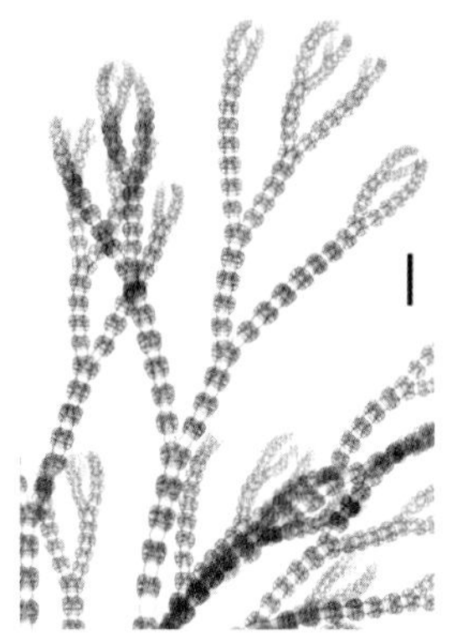

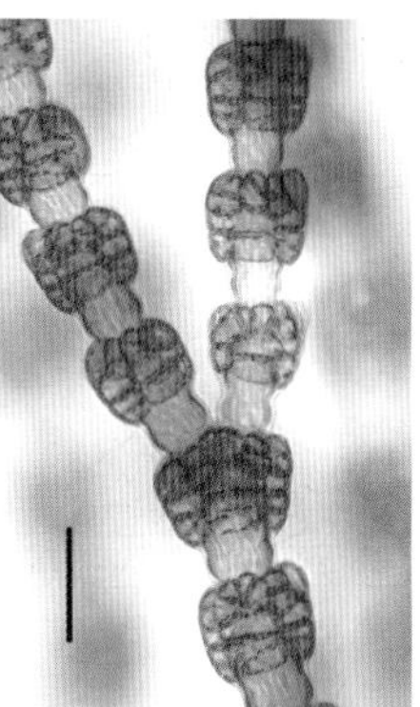

Gayliella transversalis – growing on the seagrass *Heterozostera* (Rottnest Island, WA).

(above) *Gayliella transversalis* – detail of nodal cortication (Houtman Abrolhos, WA). Scale = 30 µm.

(far left) *Gayliella transversalis* – closer view of thallus (Cape Peron, WA). Scale - 100 µm.

(left) *Gayliella transversalis* – closer view of nodes (Cape Peron, WA). Scale = 50 µm.

Spyridia

Spyridia is characterised by its unusual vegetative structure, in which two types of cortical patterns occur. On the main axes, the central filament is corticated by alternating bands of short and long periaxial cells that form a complete cover. In contrast, the lateral branches are uncorticated except for a short band of cells at each node, reminiscent of the pattern found in some species of *Ceramium* (see p. 179). Several species of *Spyridia* are found in Australia, but by far the most common and widespread is *Spyridia filamentosa*, a species that occurs in a wide variety of habitats.

Spyridia filamentosa

TYPE LOCALITY: Adriatic Sea.
DISTRIBUTION: Australia-wide. Widespread in most seas.
FURTHER READING: Womersley & Cartledge (1975); Womersley (1998); Huisman (2018).

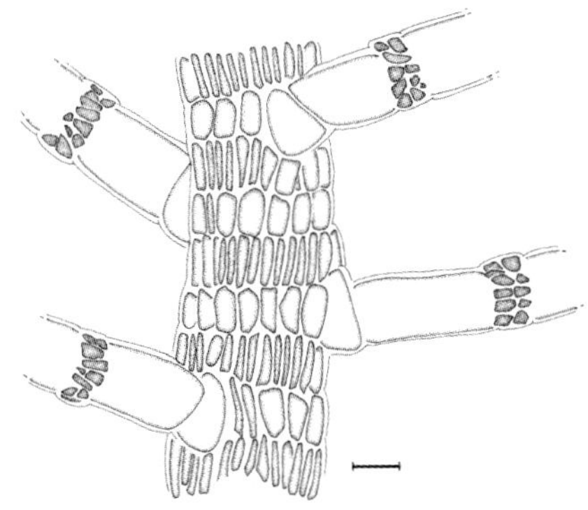

Spyridia filamentosa (Cape Peron, WA).

(right) *Spyridia filamentosa* – detail of thallus (Penguin Island, WA). Scale = 30 µm.

Anotrichium

The genus *Anotrichium* includes filamentous species with relatively large cells and with only one lateral branch per axial cell. A distinctive feature of the genus is that tetrasporangia and spermatangia are borne on one celled pedicels that can arise singly or in whorls, depending on the species. A number of species occur in Australia. *Anotrichium tenue* is a distinctive species commonly found in most regions worldwide. It can be recognised by its extensive prostrate axes and the production of upright branches from the proximal end of axial cells. In addition, distinctive whorls of colorless trichoblasts are borne at the apices of upright branches.

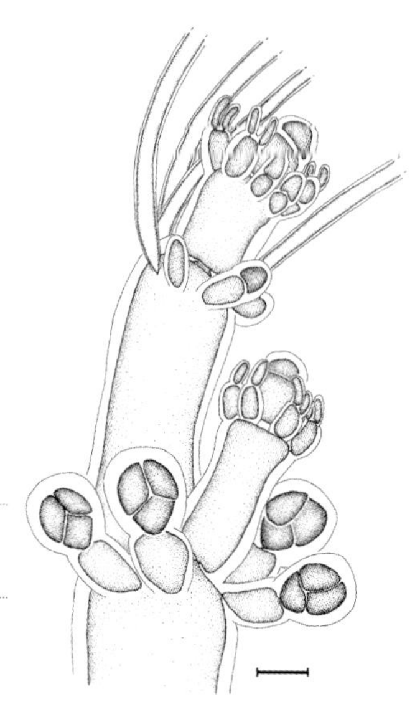

Anotrichium tenue

TYPE LOCALITY: Venice, Adriatic Sea.
DISTRIBUTION: Throughout Australia. Widespread in tropical and warmer seas.
FURTHER READING: Baldock (1976), (1998).

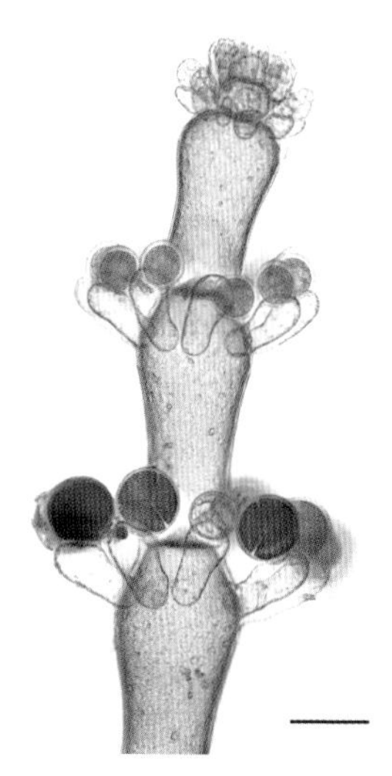

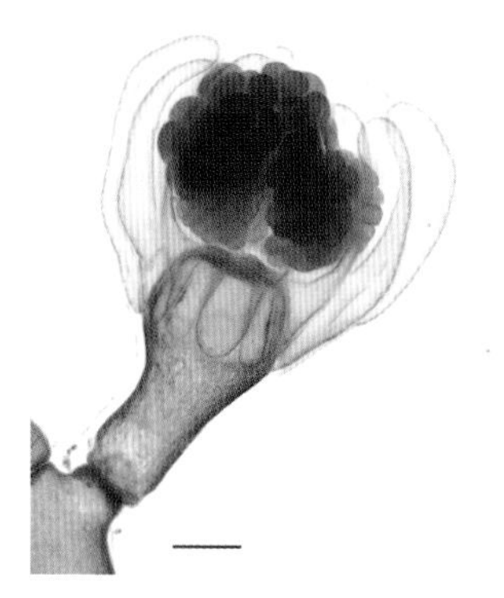

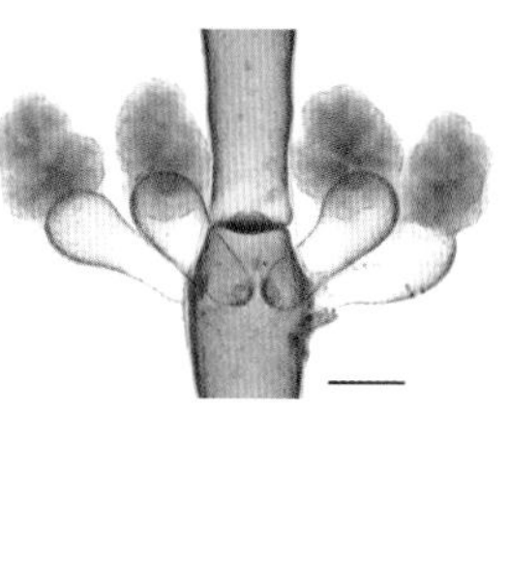

Anotrichium tenue – growing on the seagrass *Amphibolis antarctica* (Cape Peron, WA)

(above) *Anotrichium tenue* – detail of tetrasporangial plants (Cape Peron, WA). Scale = 100 µm.

(far left) *Anotrichium tenue* – detail of carposporophyte (Cape Peron, WA). Scale = 100 µm.

(left) *Anotrichium tenue* – detail of male plants (Cape Peron, WA). Scale = 100 µm.

(top) *Anotrichium tenue* – apex of tetrasporangial thallus (Houtman Abrolhos, WA). Scale = 30 µm.

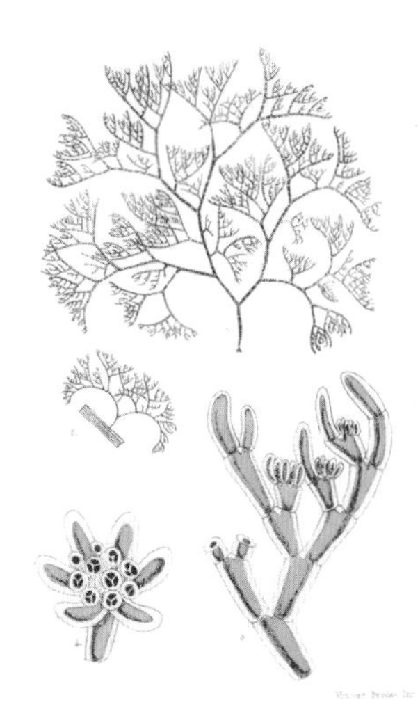

Bornetia

Bornetia is represented in Australia by two species, of which the common *Bornetia binderiana* is included here. Plants are filamentous with large, multinucleate cells, somewhat resembling species of the genus *Griffithsia* but differing in the lack of synchronous branches. Thalli of *Bornetia binderiana* are essentially flabellate, resulting from all branches arising in one plane, although adjacent branches often overlap. The darker regions in the detail photograph are tetrasporangia, which in *Bornetia* arise in clusters and are surrounded by involucral filaments.

Bornetia binderiana

TYPE LOCALITY: Western Australia.
DISTRIBUTION: Champion Bay, Western Australia, around southern mainland Australia to Port Phillip, Victoria.
FURTHER READING: Baldock & Womersley (1968), (1998).

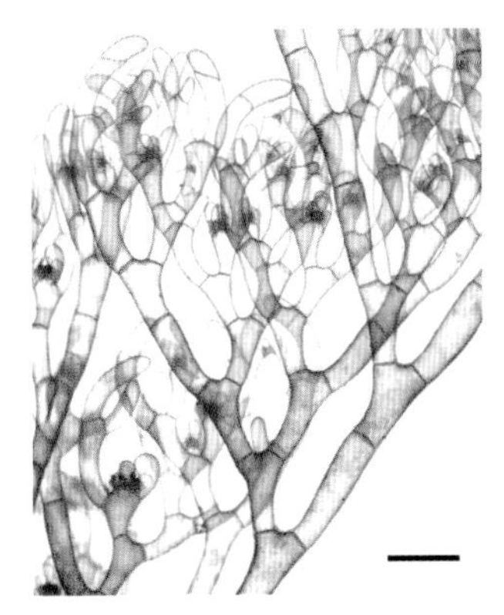

Bornetia binderiana (Cape Peron, WA).

(top) *Bornetia binderiana* – as illustrated by Harvey (1858, pl. 52, as *Griffithsia binderiana*).

(above) *Bornetia binderiana* – detail of thallus (Cape Peron, WA). Scale = 1 mm.

Griffithsia

A well-known genus with several species found in Australia, *Griffithsia* is readily recognised by the large size of its vegetative cells. Those of *Griffithsia monilis* – as seen in the photograph – can be up to 2.5 mm in diameter, making them some of the largest cells known. Thalli of all species of the genus are filamentous, with uncorticated axes that are often dichotomously divided. Vegetative cells can be globose (as in *Griffithsia monilis* and *Griffithsia ovalis*) or cylindrical, depending on the species. Reproductive structures are found on the distal region of cells, often associated with the constriction between cells. *Griffithsia monilis* is generally found growing on rock, but is also occasionally epiphytic. *Griffithsia ovalis* is always epiphytic on seagrasses or other algae. The common *Griffithsia teges* grows in coarse mats on a variety of firm substrata, generally on rough-water coasts.

Griffithsia monilis

TYPE LOCALITY: Swan River, Fremantle, Western Australia.

DISTRIBUTION: Rottnest Island, Western Australia, around southern mainland Australia to Sydney, New South Wales, and isolated occurrences in Tasmania and southern Queensland.

Griffithsia ovalis

TYPE LOCALITY: King George Sound, Western Australia.

DISTRIBUTION: Houtman Abrolhos, Western Australia, to Penneshaw, Kangaroo Island, South Australia.

Griffithsia monilis – growing on *Ecklonia radiata* (Cape Peron, WA).

Griffithsia ovalis – epiphytic on *Posidonia* (Bremer Bay, WA).

(top) *Griffithsia ovalis* – as illustrated by Harvey (1862, pl. 203).

Griffithsia teges

TYPE LOCALITY: Swan River, Fremantle, Western Australia.
DISTRIBUTION: Rottnest Island, Western Australia, around southern mainland Australia to Sydney, New South Wales, and isolated occurrences in Tasmania and southern Queensland.
FURTHER READING: Baldock (1976), (1998); Scott (2017).

Griffithsia teges (Cape Peron, WA).

Guiryella

Guiryella is a genus with only one species, *Guiryella repens*, which is a small, tufted plant that grows epiphytically on a variety of brown and red algae. Thalli are filamentous and uncorticated, with prostrate axes that creep across the host and issue branched upright axes. A single lateral branch is produced per axial cell and these arise in a slight rotation. *Guiryella* is one of a small group of genera in the Wrangeliaceae that produce bulbous propagules that serve to vegetatively reproduce the plant (see also *Mazoyerella*, p. 192, and *Tanakaella*, p. 194). These vary in form and position, depending on the genus – those of *Guiryella* are two-celled and arise in a terminal position. Sexual reproduction is also known.

Guiryella repens

TYPE LOCALITY: Mostyn's Lump, Houtman Abrolhos, Western Australia.
DISTRIBUTION: Houtman Abrolhos, Cape Peron and Rottnest Island, Western Australia, and Port Noarlunga, South Australia.
FURTHER READING: Huisman & Kraft (1992); Huisman & Gordon-Mills (1994).

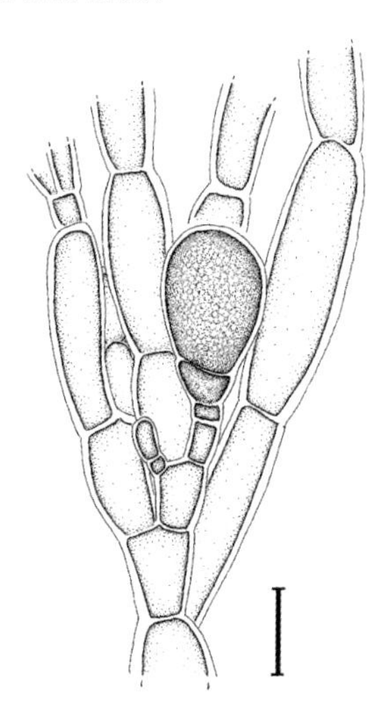

Guiryella repens – growing on *Dictyota* (Houtman Abrolhos, WA).

(above) *Guiryella repens* – portion of thallus with a propagule (Houtman Abrolhos, WA). Scale = 100 µm.

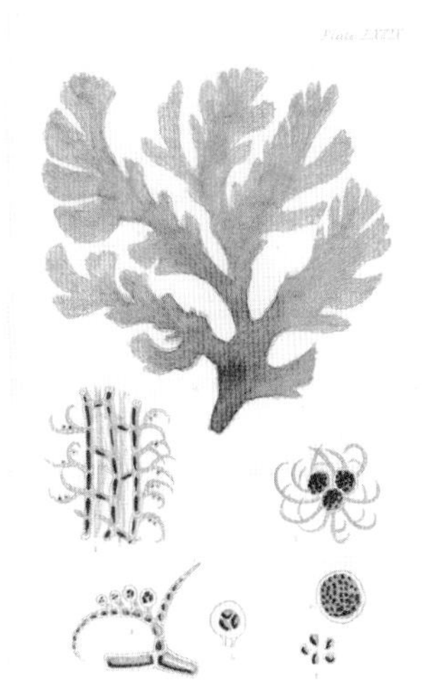

Haloplegma

Haloplegma has the filamentous construction typical of the Wrangeliaceae, but is unusual in that the lateral branches secondarily fuse and form a flattened, spongiose thallus. Two species have been recorded from Australia. *Haloplegma duperreyi* generally occurs in tropical waters but is also occasionally found in colder regions. *Haloplegma preissii* is more prevalent in temperate regions and is common on reefs in the south-west region. It is characterised by its dense corticating layer that obscures the primary branches. Surface filaments in *Haloplegma duperreyi* are short and straight, with blunt or tapering apical cells, whereas those of *Haloplegma preissii* are elongate and curved.

Haloplegma duperreyi

TYPE LOCALITY: Martinique, West Indies.
DISTRIBUTION: In Australia known from the tropics on the west and east coasts. Widespread in tropical seas.

Haloplegma preissii

TYPE LOCALITY: Western Australia.
DISTRIBUTION: From Broome, Western Australia, around southern mainland Australia to Walkerville, Victoria, and Tasmania. Possibly widespread in tropical waters.
FURTHER READING: Fuhrer et al. (1981); Womersley & Wollaston (1998c); Huisman (2018).

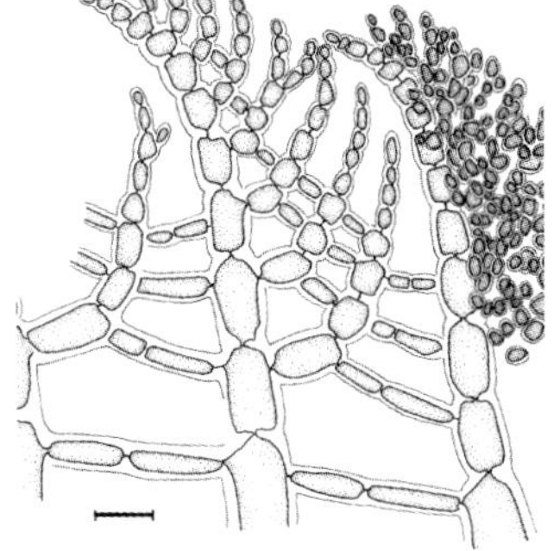

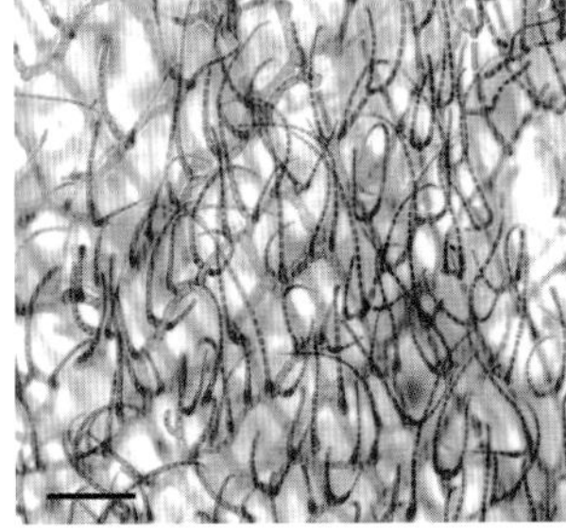

Haloplegma duperreyi (Heron Island, Qld).

Haloplegma preissii (James Price Point, WA).

(top) *Haloplegma preissii* – as illustrated by Harvey (1859a, pl. 79).

(far right) *Haloplegma preissii* – detail of thallus surface (Cape Peron, WA). Scale = 250 µm.

(right) *Haloplegma preissii* – detail of thallus: at left, an optical section showing the anastomosing lateral branches; at right, a view of the surface of the thallus (Jurien Bay, WA). Scale = 30 µm.

Mazoyerella

Mazoyerella is one of a small group of genera (see also *Tanakaella*, p. 194) in the Wrangeliaceae that reproduce vegetatively by means of bulbous propagules. In most, a typical sexual life history also occurs, although it is rarely observed in field-collected material, and in some genera is known only from laboratory culture. The propagules in *Mazoyerella* are composed of a single, starch-filled cell, but in other genera they can be up to three-celled. *Mazoyerella australis* is a small, filamentous plant that is often epiphytic on seagrasses and other algae.

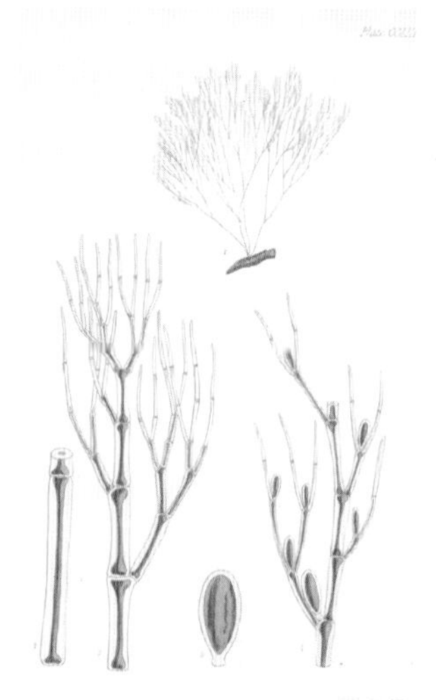

Mazoyerella australis

TYPE LOCALITY: Rottnest Island, Western Australia.
DISTRIBUTION: Rottnest Island, Western Australia, to Western Port, Victoria.
FURTHER READING: Gordon-Mills & Womersley (1974); Huisman & Gordon-Mills (1994); Huisman & Womersley (1998).

Mazoyerella australis (Rottnest Island, WA).

(top) *Mazoyerella australis* – as illustrated by Harvey (1863, pl. 253, as *Corynospora australis*).

Spongoclonium

Thalli of *Spongoclonium* are filamentous, with alternately branched axes that remain uncorticated or can become secondarily corticated by rhizoids arising from the basal cells of lateral branches, depending on the species. The cortication becomes very dense and eventually obscures the branching pattern, which is therefore best observed in apical regions. *Spongoclonium conspicuum* grows to around 30 cm in height and thalli have a spongy texture. The plant in the photograph appears to be calcified, but this is due to the accumulation of sand and other detritus in the corticating filaments. *Spongoclonium crispulum* is a smaller uncorticated species that grows on rock or as an epiphyte.

Spongoclonium crispulum

TYPE LOCALITY: Rottnest Island, Western Australia.
DISTRIBUTION: Perth region of Western Australia.

Spongoclonium conspicuum

TYPE LOCALITY: Cape Liptrap, Victoria.
DISTRIBUTION: From Cape Peron, Western Australia, around southern mainland Australia to Waratah Bay, Victoria, and the north coast of Tasmania.
FURTHER READING: Wollaston (1990), Womersley & Wollaston (1998d); Huisman (2018).

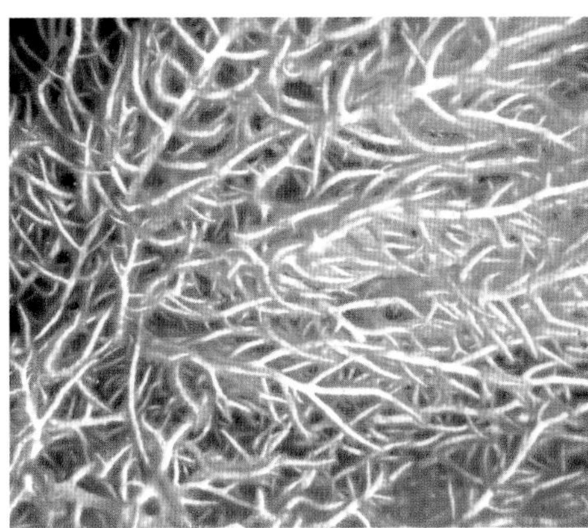

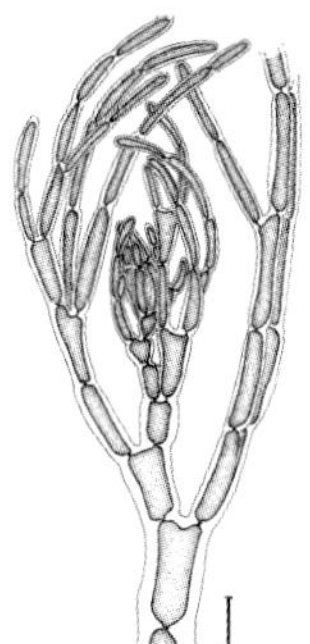

Spongoclonium crispulum (Cape Peron, WA).

Spongoclonium conspicuum (Augusta, WA).

(left) *Spongoclonium conspicuum* – apex of thallus (Augusta, WA). Scale = 30 µm.

Tanakaella

Plants of *Tanakaella* form small filamentous tufts to 5 mm high, arising from prostrate basal filaments. The genus includes three species, with only *Tanakaella itonoi* known in Australia. Thalli grow exclusively as epiphytes on blade-like members of the Kallymeniaceae, including *Leiomenia cribrosa* as seen in the photo. *Tanakaella* reproduces vegetatively by unusual bulbous propagules, which are borne laterally on upper cells and are similar to those found in the related genera *Guiryella* (see p. 190) and *Mazoyerella* (see p. 192). Sexual reproduction also occurs and involves a three-phased life history (gametophyte, carposporophyte and tetrasporophyte), similar to that of most red algae.

Tanakaella itonoi

TYPE LOCALITY: Elliston, South Australia.
DISTRIBUTION: Houtman Abrolhos and Rottnest Island, Western Australia, and Egg Island, Isles of St Francis, to Troubridge Island, South Australia.
FURTHER READING: Huisman & Gordon-Mills (1994); Huisman & Womersley (1998).

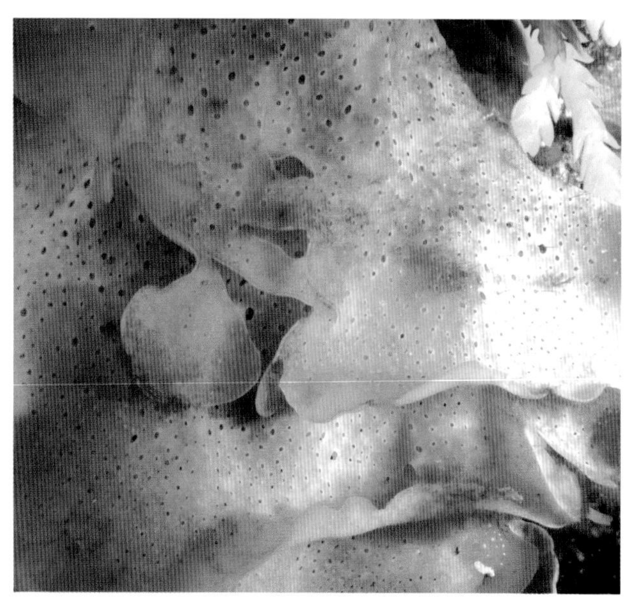

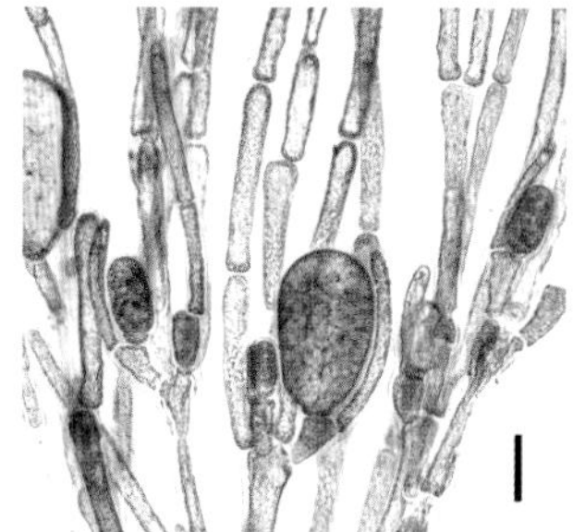

Tanakaella itonoi – growing on the surface of *Leiomenia cribrosa* (Rottnest Island, WA).

(above) *Tanakaella itonoi* – detail of thallus with bulbous propagules (Rottnest Island, WA). Scale = 50 µm.

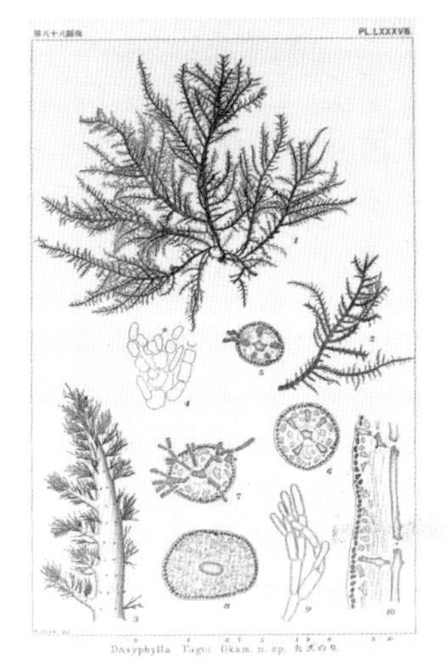

Wrangelia

Species of *Wrangelia* are widespread on most Australian coasts. Thalli are filamentous and wholly erect or have prostrate and erect axes, with branch apices typically curved. The number of whorl branches on each cell of the primary axis varies from three to five, depending on the species, and the development of whorl branches can be unequal, resulting in various combinations of short and long branches. Cortication can be present or absent. Spermatangia and tetrasporangia are borne on lower cells of whorl branches and mostly have a protective filamentous involucre, and carposporophytes are formed at the apices of branches. Separation of species of *Wrangelia* is based primarily on vegetative features such as habit, presence and form of cortication, and number of whorl branches per axial cell, with five in most species but only four in some. *Wrangelia tagoi* is unusual in having a secondary layer of cortication composed of irregularly arranged cells.

Wrangelia decumbens

TYPE LOCALITY: South Scott Reef, Western Australia.
DISTRIBUTION: Known only from the type locality.

Wrangelia elegantissima

TYPE LOCALITY: Makapu'u Beach, Oahu, Hawaiian Islands.
DISTRIBUTION: Ningaloo Reef north to Broome, Western Australia. Hawaiian Islands.

Wrangelia decumbens (Scott Reef, WA).

Wrangelia elegantissima (James Price Point, WA).

(top) *Wrangelia tagoi* – as illustrated by Okamura (1912, pl. 88, as *Dasyphila tagoi*).

Wrangelia plumosa

TYPE LOCALITY: Georgetown, Tasmania.
DISTRIBUTION: Shark Bay, Western Australia, to Coffs Harbour, New South Wales, and around Tasmania.

Wrangelia tagoi

TYPE LOCALITY: Wagu (Prov. Shima), Japan.
DISTRIBUTION: South Scott Reef, Western Australia. Japan. China.
FURTHER READING: Gordon (1972); Norris (1994); Womersley (1998); Huisman (2018).

Wrangelia plumosa – growing on *Amphibolis antarctica* (Cape Peron, WA).

Wrangelia tagoi (Scott Reef, WA).

Acrosorium

Like many members of the Delesseriaceae, *Acrosorium* has a delicate, membranous thallus. It can be distinguished from other genera by its pattern of growth, which involves many marginal growing points, and the presence of microscopic veins in the thallus that are visible only with the aid of a microscope. A central midrib, as occurs in several closely related genera, is not found in *Acrosorium*. Several species occur in Australia, of which *Acrosorium ciliolatum* has distinctive hooked apices that latch on to other seaweeds, such as the green *Struvea plumosa* in the photo.

Acrosorium ciliolatum

TYPE LOCALITY: King George Sound, Western Australia.
DISTRIBUTION: From the Perth region, Western Australia, around southern Australia to Queensland. Widespread in most oceans.
FURTHER READING: Womersley (2003); Wynne (2014).

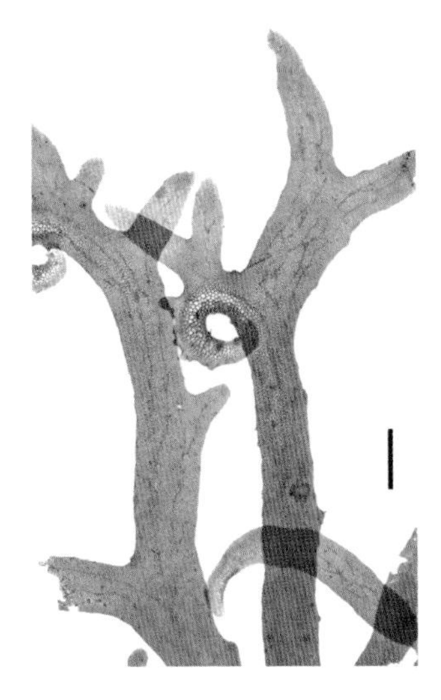

Acrosorium ciliolatum – attached to the stipes of *Struvea plumosa* just below water level (Cape Peron, WA).

(above) *Acrosorium ciliolatum* – detail of thallus showing microscopic veins and hooked apices (Cape Peron, WA). Scale = 1 mm.

Caloglossa

Species of *Caloglossa* commonly occur on high intertidal rocks and mangrove pneumatophores, often associated with other red algae such as *Bostrychia*. They form low mats composed of arching, pale-brown to reddish-brown, leaf-like blades loosely attached to the substratum at intervals by clusters of rhizoids. *Caloglossa leprieurii* is the most commonly encountered species. Despite its often shabby appearance, particularly when growing in muddy environments, *Caloglossa* lives up to its name (which translates to 'beautiful tongue') when viewed under magnification.

Caloglossa leprieurii

TYPE LOCALITY: Cayenne, French Guiana.
DISTRIBUTION: Widespread in warm and temperate seas.
FURTHER READING: King & Puttock (1994); Kamiya et al. (2003); Womersley (2003).

Caloglossa leprieurii – detail of thallus. The dark patches are reproductive structures (tetrasporangia) (Cape Preston, WA). Scale = 500 µm.

(above) *Caloglossa leprieurii* – growing on mangrove pneumatophores (Shark Bay, WA).

Claudea

Claudea is represented in Australia by the single species *Claudea elegans*, regarded by Irish botanist William Henry Harvey (1858) as 'the most beautiful of all algae' and accorded prime position in the first volume of his *Phycologia Australica*. Thalli have net-like fronds that develop unilaterally on the primary axes. The nets are composed of series of parallel leaflets that cut off perpendicular leaflets that attach to opposite leaflets, the process replicated at smaller scales to produce the fine nets. *Claudea elegans* is more common in south-eastern Australia and only rarely encountered on the west coast.

Claudea elegans

TYPE LOCALITY: Australia.
DISTRIBUTION: Houtman Abrolhos, Western Australia, to Walkerville, Victoria, and northern Tasmania. India. Pakistan. Natal. Brazil.
FURTHER READING: Womersley (2003); Wynne (2014).

Claudea elegans (Bunker Bay, WA; photo G.J. Edgar).

(top) *Claudea elegans* – as illustrated by Harvey (1858, pl. 1).

Hemineura

Hemineura includes the single species *Hemineura frondosa*. Thalli are membranous and light pink in colour, and have a pinnate branching pattern. A midrib is present and can be seen with the naked eye in most parts of the plant, but it becomes indistinct near the apices and basal regions of branches. This feature is characteristic of *Hemineura*. Tetrasporangia and spermatangia are borne in raised patches on either side of the midrib, while cystocarps are hemispherical and occur on the midrib. *Hemineura frondosa* is widely distributed in southern Australia, reaching its western limit at the Houtman Abrolhos. Plants from the west coast are generally smaller and more delicate than those from the colder south coast.

Hemineura frondosa

TYPE LOCALITY: Tasmania.
DISTRIBUTION: From the Houtman Abrolhos, Western Australia, around southern mainland Australia to Gabo Island, Victoria, and around Tasmania.
FURTHER READING: Lin et al. (2001); Womersley (2003); Scott (2017).

Hemineura frondosa (Houtman Abrolhos, WA).

(top) *Hemineura frondosa* – as illustrated by Harvey (1860, pl. 179, as *Delesseria frondosa*).

Heterodoxia

Heterodoxia includes the single species *Heterodoxia denticulata*. Plants have a branched, membranous thallus with a broad central midrib. The margin of the blade has numerous short spines. The blade is mostly one cell thick with the exception of the midrib, where the thickened section is composed of numerous regularly aligned cells. Cystocarps are scattered and sunken within the blades, with a hemispherical pericarp. Tetrasporangia are formed in patches scattered on the blades or on small leaflets arising from the midribs.

Heterodoxia denticulata

TYPE LOCALITY: Rottnest Island, Western Australia.
DISTRIBUTION: Houtman Abrolhos, Western Australia, to Nora Creina, South Australia.
FURTHER READING: Fuhrer et al. (1981); Wynne (1988), (2014); Womersley (2003).

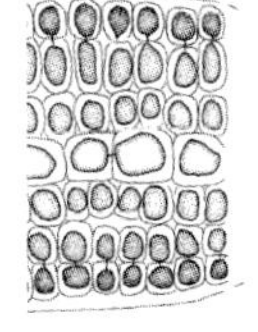

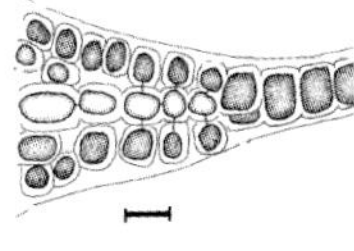

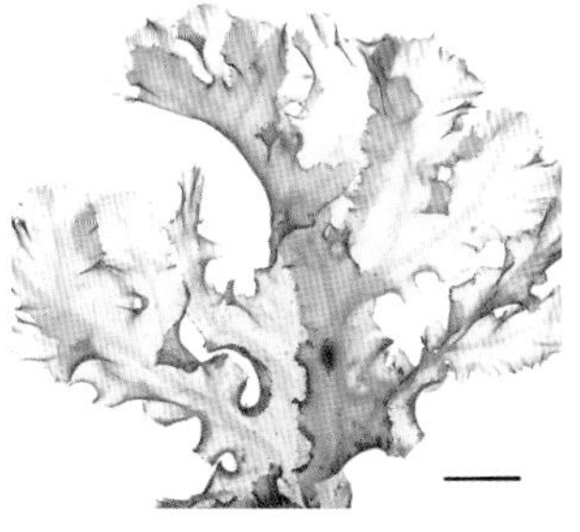

Heterodoxia denticulata (Cape Peron, WA).

(top) *Heterodoxia denticulata* – as illustrated by Harvey (1863, pl. 244, as *Delesseria denticulata*).

(far right) *Heterodoxia denticulata* – closer view of thallus (Cape Peron, WA). Scale = 5 mm.

(right) *Heterodoxia denticulata* – partial section of thallus showing thickened midrib grading into a one-cell-thick blade (Rottnest Island, WA). Scale = 50 µm.

Hypoglossum

Hypoglossum is a widespread genus with about 30 species. Plants have flat, branched blades, with a midrib and one-celled thick lateral wings, but no lateral veins. Growth is from a single transversely dividing apical cell, with all apical cells of second- and third-order cell rows reaching the margins of the blade. Cystocarps occur on the midrib and tetrasporangia are usually arranged in patches on both sides of the midrib. All species of *Hypoglossum* are leafy and delicate, but their size varies considerably, with some quite large and up to 20 cm tall. Most, however, like *Hypoglossum barbatum*, form small thalli with prostrate axes that creep over the substratum.

Hypoglossum barbatum

TYPE LOCALITY: Hyuga, Miyazaki Prefecture, Japan.
DISTRIBUTION: Possibly widespread in tropical seas.
FURTHER READING: Wynne (1988), (2014); Schneider (2000); Womersley (2003).

Hypoglossum barbatum (Coral Bay, WA).

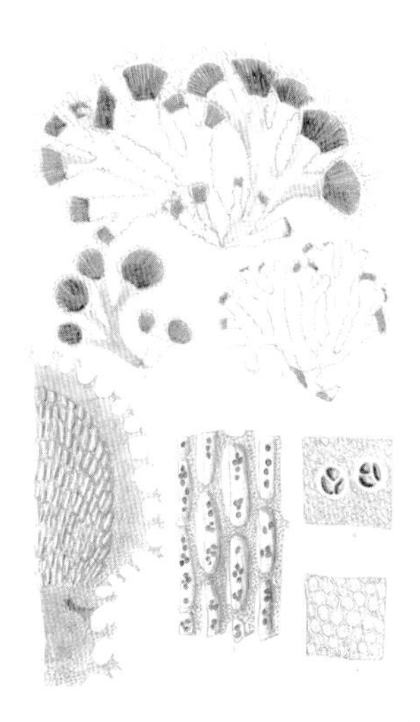

Martensia

One of the more unusual and attractive of the red algae, *Martensia* has a thin, membranous central region bordered (to varying degrees) by a lace-like mesh. On occasions, the two regions can be produced in an alternating sequence, as in *Martensia denticulata*, one of the species included here. The plant of *Martensia denticulata* illustrated has an unusually coarse meshwork in lower portions, and the large ball-like structures are cystocarps. *Martensia denticulata* grows on rock at a range of depths, with specimens from the intertidal region tending to be stunted and with a correspondingly smaller mesh. This species was previously thought to be a variation of *Martensia fragilis*, but is now regarded as a separate species, distinguished by the teeth-like growths from the margins. *Martensia australis* is characterised by its large fronds with a coarse mesh. It has been widely reported but is rarely collected from the type locality. *Martensia millarii* is known from tropical regions of Western Australia; it has only the single mesh region and smooth margins.

Martensia australis

TYPE LOCALITY: King George Sound, Western Australia.

DISTRIBUTION: Apparently widespread in the Indo-Pacific.

Martensia millarii

TYPE LOCALITY: Champagney Island, Western Australia.

DISTRIBUTION: From Barrow Island north to Long Reef, Western Australia.

Martensia australis – as illustrated by Harvey (1858, pl. 8).

Martensia millarii (James Price Point, WA).

(top) *Martensia denticulata* – as illustrated by Harvey (1860, pl. 127).

Martensia denticulata

TYPE LOCALITIES: Garden and Rottnest Islands, Western Australia.
DISTRIBUTION: Known reliably only from the Perth region, Western Australia.
FURTHER READING: Wynne (2014); Lin et al. (2009), (2013); Huisman & Lin (2018b).

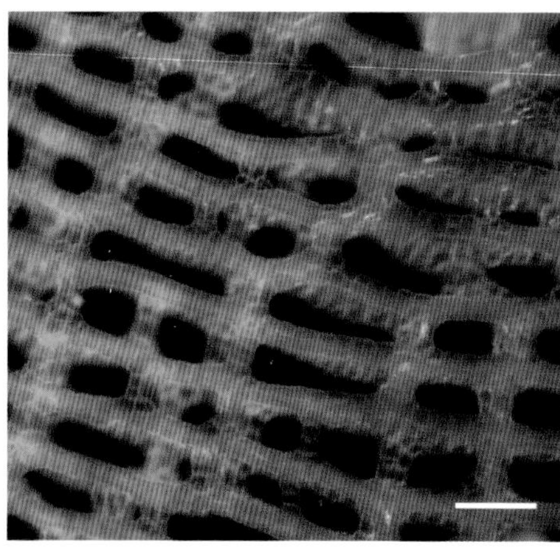

Martensia denticulata (Cape Peron, WA).

(left) *Martensia denticulata* – detail of mesh (Rottnest Island, WA). Scale = 250 µm.

Nitophyllum

Nitophyllum, which translates to 'shiny leaf', includes approximately 30 species worldwide, of which three occur in southern Australia. Plants are foliose and delicate, with blades one cell thick in upper parts but with extra layers lower down. The genus lacks any form of midrib or veins and growth is by divisions of cells at the margins. *Nitophylum pulchellum* is a rare Western Australian species that grows to about 5 cm tall. Thalli are subdichotomously divided and have distinctive convoluted margins. The only reproductive specimens that have been observed are tetrasporophytes, with tetrasporangia occurring in scattered patches.

Nitophyllum pulchellum

TYPE LOCALITY: King George Sound, Western Australia.
DISTRIBUTION: From Rottnest Island south to King George Sound, Western Australia.
FURTHER READING: Womersley (2003); Wynne (2014).

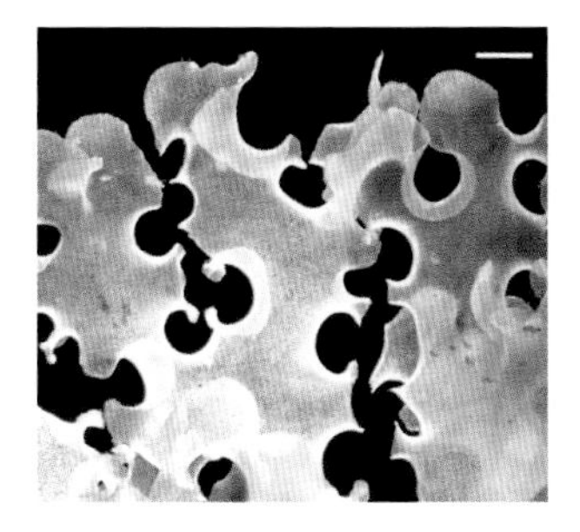

Nitophyllum pulchellum (Cape Peron, WA).

(above) *Nitophyllum pulchellum* – branch detail (Cape Peron, WA). Scale = 1 mm.

Patulophycus

Thalli of *Patulophycus eclipes*, the only species of the genus, are small and paddle-shaped, with serrated margins. When viewed underwater, plants have a bluish-pink iridescence, which stands out from the surroundings and makes them obvious. Structurally the blades are several cells thick and growth is by means of a single transversely dividing apical cell. A midrib is present, but lateral veins are absent. *Patulophycus* is similar in form to several other genera of Delesseriaceae, but is distinguished by the production of carpogonial branches from only the dorsal surface of the blade. Tetrasporangia occur in patches between the midrib and the margins of upper branches.

Patulophycus eclipes

TYPE LOCALITY: Bowen Island, Jervis Bay, New South Wales.
DISTRIBUTION: Known from Lord Howe Island and Twofold Bay, New South Wales, north to Miami, southern Queensland. South Africa.
FURTHER READING: Millar & Wynne (1992); Wynne (2014).

Patulophycus eclipes (Montague Island, NSW).

Platyclinia

The family Delesseriaceae includes some of the more attractive red algae, most with a delicate, leafy or blade-like thallus. *Platyclinia* includes six species restricted to the southern hemisphere, three from colder waters of South America and subantarctic islands, and three from southern Australia. Plants are foliose, irregularly branched, and veins are absent. Structurally the thallus is one cell thick at the margins but becomes thicker centrally. Growth is marginal, without a distinct apical cell. *Platyclinia ramosa* can be distinguished from the other Australian species by its smaller size and smooth margins. Tetrasporangia are borne in scattered, oval to irregularly shaped patches.

Platyclinia ramosa

TYPE LOCALITY: Elliston, South Australia.
DISTRIBUTION: Only known from Cape Peron, Western Australia, and the Eyre Peninsula, South Australia.
FURTHER READING: Womersley (2003).

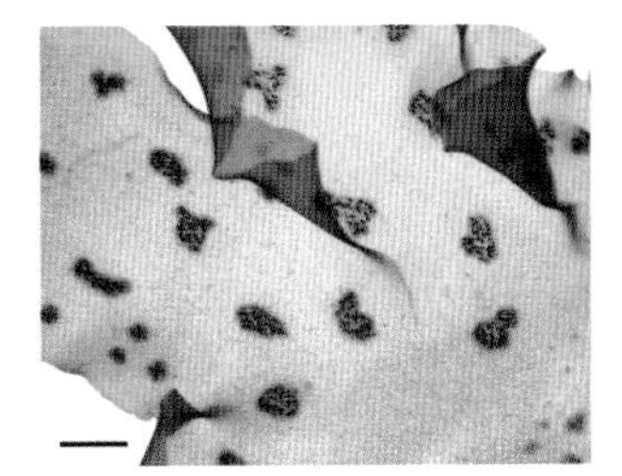

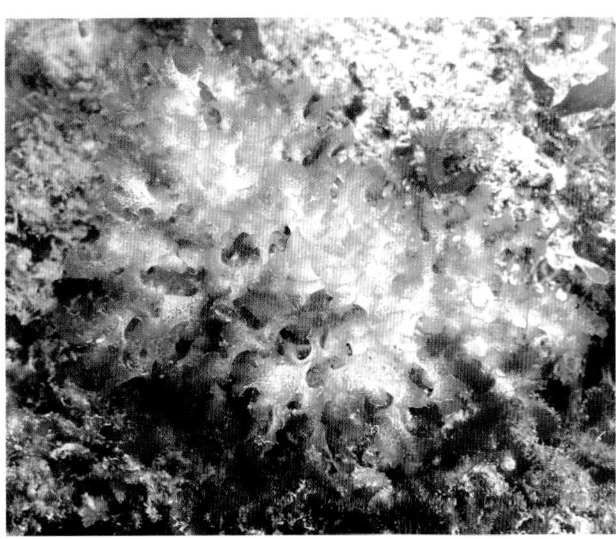

Platyclinia ramosa (Cape Peron, WA).

(above) *Platyclinia ramosa* – detail of thallus with tetrasporangia (Cape Peron, WA). Scale = 1 mm.

Taenioma

Thalli of *Taenioma* have prostrate axes that produce a small turf of flattened upright branches. A distinctive feature of the genus is the presence of colourless hairs at the tips of the branches. Structurally the thallus is uniaxial, with each axial cell ringed by four pericentral cells. In the flattened branches, the lateral pericentral cells each produce a pair of flanking cells. *Taenioma perpusillum* is a common epiphyte but is inconspicuous because of its small size (to a couple of millimetres in height).

Taenioma perpusillum

TYPE LOCALITY: San Agustin, Oaxaca, Mexico.
DISTRIBUTION: From Cape Peron, Western Australia, around northern Australia to Lord Howe Island, New South Wales. Widespread in warmer seas.
FURTHER READING: Price & Scott (1992); Womersley (2003); Wynne (2014).

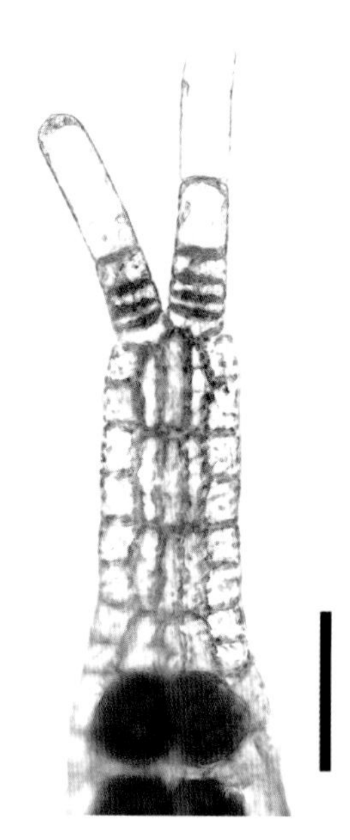

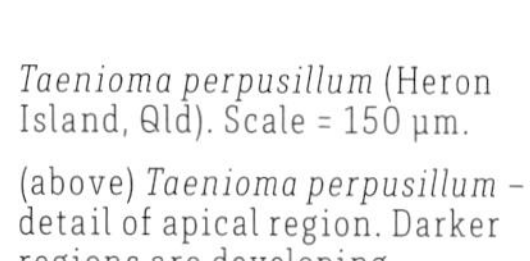

Taenioma perpusillum (Heron Island, Qld). Scale = 150 µm.

(above) *Taenioma perpusillum* – detail of apical region. Darker regions are developing tetrasporangia (Heron Island, Qld). Scale = 50 µm.

(left) *Taenioma perpusillum* – detail of apical region with terminal hairs (Rottnest Island, WA). Scale = 30 µm.

Vanvoorstia

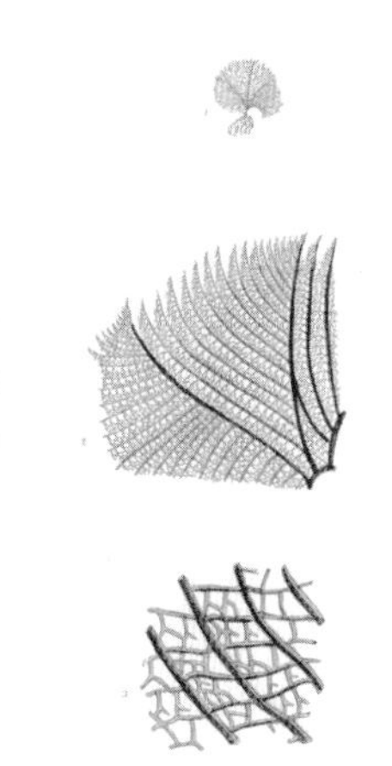

Vanvoorstia includes four species and has the unusual distinction of including a presumed extinct species, the eastern Australian *Vanvoorstia bennettiana*, which once occurred in Sydney Harbour but has not been seen for well over 100 years. Other species of the genus, however, are widely distributed in the tropical Indo-West Pacific. Plants have net-like blades, with the individual branches interlinked to form a mesh. The two species included here differ in the degree of cortication of their axes, with those of *Vanvoorstia coccinea* heavily corticated and robust, whereas *Vanvoorstia spectabilis* lacks cortication in all but the primary axes. The latter is also primarily found in the intertidal, while *Vanvoorstia coccinea* is subtidal. Cystocarps occur on midribs of upper blade portions that are free from other blades, and tetrasporangia are borne in younger blades of the network.

Vanvoorstia coccinea

TYPE LOCALITY: Sri Lanka.
DISTRIBUTION: Widespread in the tropical Indian and western Pacific Oceans.

Vanvoorstia spectabilis

TYPE LOCALITY: Weligama, south coast of Sri Lanka.
DISTRIBUTION: Known from the tropical Indian and western Pacific Oceans, the easternmost locality being the Hawaiian Islands.
FURTHER READING: Yoshida & Mikami (1994); De Clerck et al. (1999); Millar (2003); Wynne (2014); Huisman (2018).

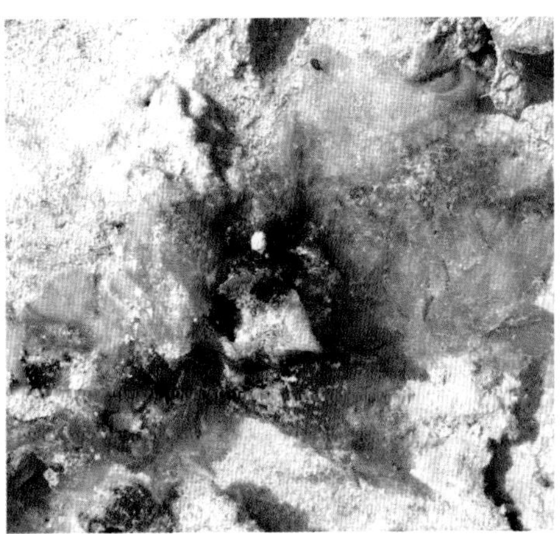

Vanvoorstia coccinea (Broomfield Reef, Qld).

Vanvoorstia spectabilis (Champagney Islands, WA).

(top) *Vanvoorstia bennettiana* – as illustrated by Harvey (1859a, pl. 61, as *Claudea bennettiana*).

Zellera

Zellera is a monotypic genus with the single species *Zellera tawallina*, a net-like species similar in form to *Vanvoorstia* (see p. 209) but differing in the branching pattern and arrangement of blade filaments at the apices. In *Vanvoorstia*, the lateral branches arise at various angles, resulting in a net-like blade with irregularly shaped spaces. In contrast, lateral branches in *Zellera* are perpendicular to the bearing branch and the spaces are rectangular. *Zellera tawallina* is widespread although seemingly never abundant. Plants are feather-like and often have orange tips.

Zellera tawallina

TYPE LOCALITY: Tawaliketjil Island, Indonesia.
DISTRIBUTION: Widespread in tropical regions of the north-eastern Indian and western Pacific Oceans.
FURTHER READING: Wynne (2014); Huisman (2018).

Zellera tawallina (Brue Reef, WA).

Cottoniella

A genus of several species, but apparently rarely encountered. Plants are generally small and tufted, with both prostrate and erect axes. Thalli are uniaxial, with each axial cell surrounded by four pericentral cells, the two lateral pericentral cells each dividing further to produce a pair of 'flanking' cells. A distinctive feature of the genus is the production of unbranched filaments from one surface of the thallus. These arise either singly or in pairs from each segment. Three species are known from Australia, of which *Cottoniella filamentosa* is included here.

Cottoniella filamentosa

TYPE LOCALITY: Biscayne Key, Florida, USA.
DISTRIBUTION: From the Houtman Abrolhos north to the Dampier Archipelago, Western Australia. Apparently widely, but sporadically, distributed in warmer seas.
FURTHER READING: Cormaci et al. (1978); Huisman (2018).

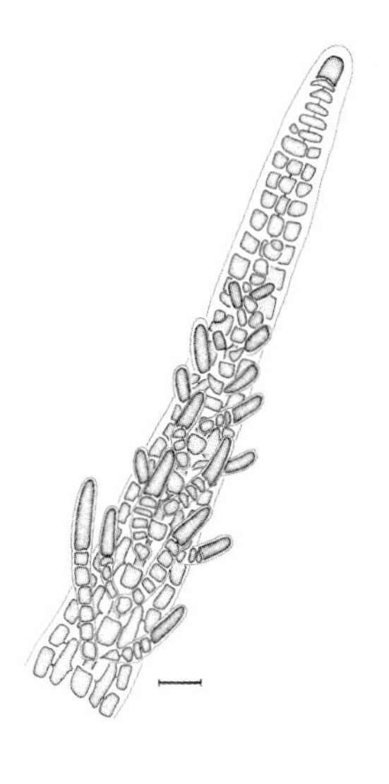

Cottoniella filamentosa (Houtman Abrolhos, WA).

(above) *Cottoniella filamentosa* – apex of thallus showing the origin of paired filaments (Houtman Abrolhos, WA). Scale = 20 µm.

Platysiphonia

Species of *Platysiphonia* display a variety of thallus forms, from delicate, uncorticated plants (e.g. *Platysiphonia delicata*) to the more robust, corticated species (such as *Platysiphonia hypneoides*). All have compressed young branches – in the larger species these become secondarily corticated and terete. Structurally the thallus is uniaxial, with each axial cell producing four pericentral cells. Of these, the lateral pericentral cells each cut off two flanking cells to form the flattened branch. Some species of *Platysiphonia* are similar in form to *Sarcomenia delesserioides* (see opposite page), but differ in that the flanking cells remain undivided (except for cortical derivatives). Those of *Sarcomenia* divide further to produce broad, flat branches. In *Platysiphonia hypneoides*, corticating cells are initiated about 10 cells from the apex and rapidly cover the axis, obscuring the growth pattern. When living, the species is an iridescent grey, but reverts to bright pink when exposed to air.

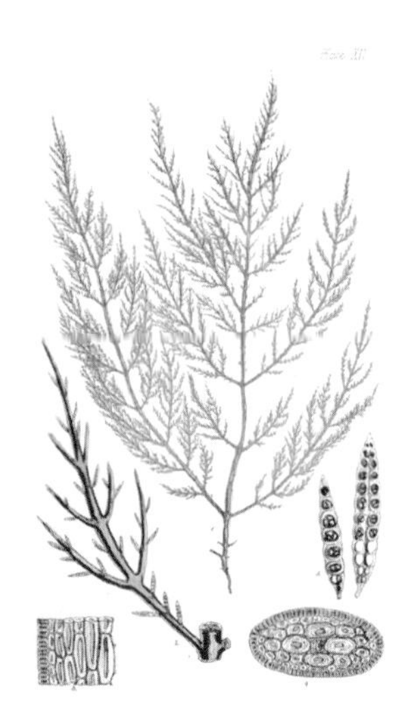

Platysiphonia delicata

TYPE LOCALITY: Sanlúcar de Barrameda, Cádiz, Spain.
DISTRIBUTION: Widespread in most seas.

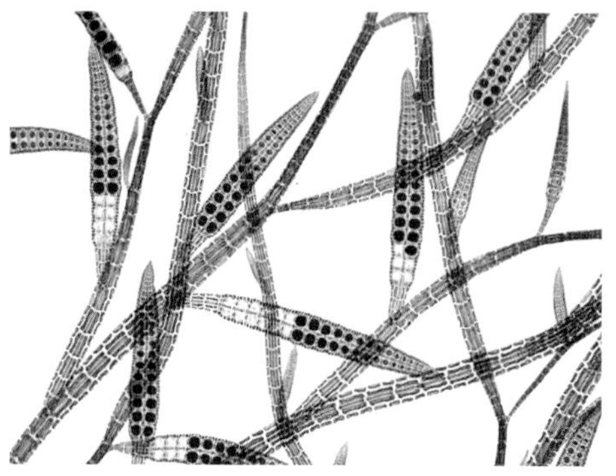

Platysiphonia hypneoides

TYPE LOCALITIES: Fremantle and Garden Island, Western Australia.
DISTRIBUTION: Perth region, Western Australia.
FURTHER READING: Womersley & Shepley (1959); Womersley (2003).

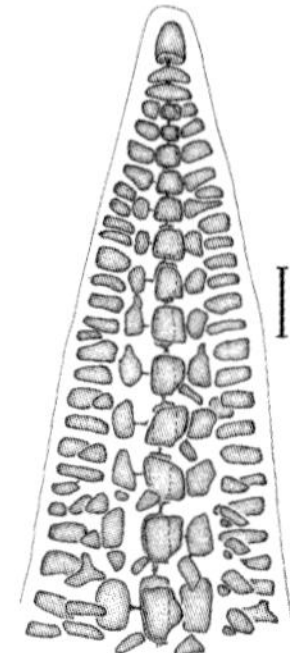

Platysiphonia delicata (Cape Peron, WA).

(far left) *Platysiphonia delicata* – detail of plants with tetrasporangia (Cape Peron, WA).

Platysiphonia hypneoides (Rottnest Island, WA).

(left) *Platysiphonia hypneoides* – detail of apical region (Rottnest Island, WA). Scale = 20 µm.

(top) *Platysiphonia hypneoides* – as illustrated by Harvey (1858, pl. 12, as *Sarcomenia hypneoides*).

Sarcomenia

The genus *Sarcomenia* includes only the single species *Sarcomenia delesserioides*. Thalli have distinctly flattened and membranous axes with a noticeable midrib. The lateral branches arise from a point midway between the midrib and the margin of the main axis. *Sarcomenia delesserioides* generally grows on rock or other hard surfaces in areas of high water movement. The blue iridescence seen in the photograph is typical of this species and disguises its affinities with the red algae. When plants are removed from the water, they become a vivid pink. Plants can grow to approximately 50 cm in height with branches 1–3 cm broad.

Sarcomenia delesserioides

TYPE LOCALITY: Western Australia.
DISTRIBUTION: From the Houtman Abrolhos, Western Australia, around southern Australia to Western Port, Victoria.
FURTHER READING: Womersley & Shepley (1959); Womersley (2003); Wynne (2014).

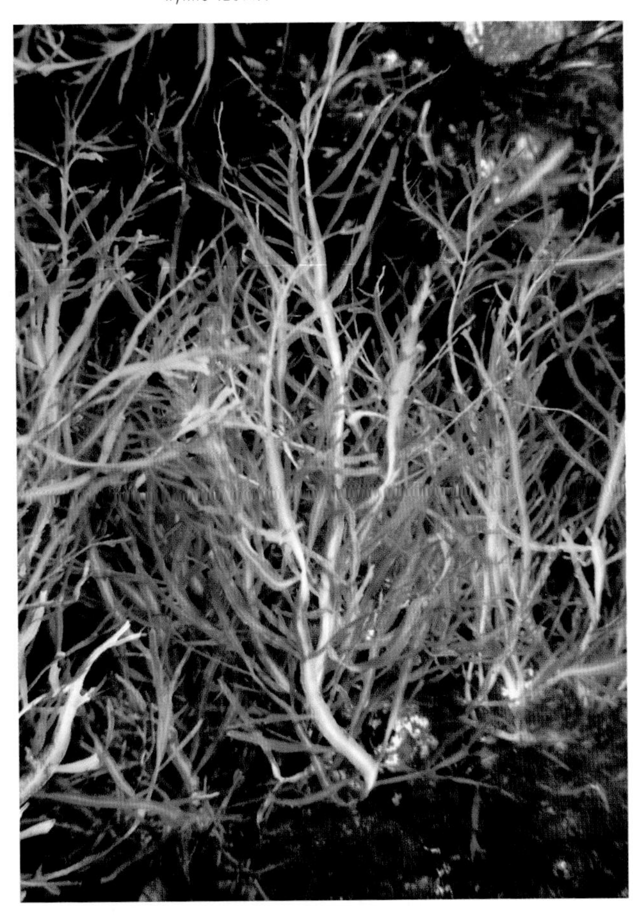

Sarcomenia delesserioides (Houtman Abrolhos, WA).

(top) *Sarcomenia delesserioides* – as illustrated by Harvey (1860, pl. 121).

Acanthophora

Thalli of *Acanthophora* are irregularly or alternately branched, with firm, terete axes that bear regularly arranged spines either restricted to special side branches (*Acanthophora spicifera*) or over most of the branches (*Acanthophora dendroides* and *Acanthophora ramulosa*). *Acanthophora pacifica* has compressed branches that can be a bluish colour when living. Colourless filaments, known as trichoblasts, are often present at the tips of branches, but these can be secondarily lost as the branch matures. Structurally the axes are pseudoparenchymatous with a visible central axis surrounded by five pericentral cells. Reproductive structures are either sessile on short lateral branches, as is the case with the cystocarps, or borne in specialised dwarf branches known as stichidia (the tetrasporangia and spermatangia). *Acanthophora spicifera* is eaten in Tahiti, but is a pest species in the Hawaiian Islands.

Acanthophora dendroides

TYPE LOCALITY: Rottnest Island, Western Australia.
DISTRIBUTION: From the Perth region, Western Australia, around northern Australia to Port Stephens and Lord Howe Island, New South Wales.

Acanthophora pacifica

TYPE LOCALITY: Arue point, Tahiti.
DISTRIBUTION: In Australia known only from Heron Island in the Great Barrier Reef. Elsewhere known from several tropical Pacific islands.

Acanthophora ramulosa

TYPE LOCALITY: Angola.
DISTRIBUTION: Warmer waters of the Indian Ocean and western Atlantic.

Acanthophora spicifera

TYPE LOCALITY: St Croix, Virgin Islands.
DISTRIBUTION: From the Houtman Abrolhos, Western Australia, around northern Australia to Lord Howe Island, New South Wales. Widely distributed in tropical seas.
FURTHER READING: Cribb (1983); Kraft (1979); Jong et al. (1999); Huisman (2018).

Acanthophora dendroides (Cape Peron, WA).

Acanthophora pacifica (Heron Island, Qld).

(left) *Acanthophora dendroides* – partial section of thallus (Jurien Bay, WA). Scale = 100 µm.

Acanthophora ramulosa (Long Reef, WA).

Acanthophora spicifera (Scott Reef, WA).

Amansia

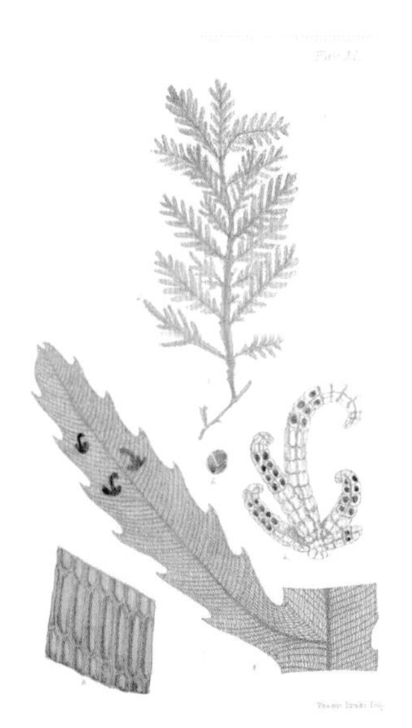

Thalli of *Amansia* have flattened branches of a consistent width and with a well-defined midrib. The edges of the branches are often serrated. Structurally the thallus is uniaxial with a ring of five pericentral cells surrounding the central filament. Divisions of the pericentral cells result in the two-cells-thick lateral wings that are present on either side of the midrib. Branches are dorsiventral in that there are developmental differences between the two faces of the blade. There have been differing treatments of the species herein included in *Amansia*, with some authors placing species in a segregate genus *Melanamansia* based on the presence of two pseudopericentral cells that arise secondarily from the dorsal pericentral cells, resulting in a central axis ringed by seven cells. *Amansia rhodantha* and *Amansia glomerata* have a similar rosette appearance, differing in the presence of the aforementioned pseudopericentral cells in the latter. *Amansia serrata* has more elongate branches with serrated margins.

Amansia glomerata

TYPE LOCALITY: Hawaiian Islands.
DISTRIBUTION: Widespread in warmer waters of the Indo-Pacific.

Amansia rhodantha

TYPE LOCALITY: Cape Malheureux, Mauritius.
DISTRIBUTION: From Jurien Bay (rarely) and Ningaloo Reef, Western Australia, around northern Australia to Lord Howe Island, New South Wales. Widespread in warmer seas of the Indian and Pacific Oceans.

Amansia serrata

TYPE LOCALITY: Rottnest Island, Western Australia.
DISTRIBUTION: Flat Rocks, Western Australia, to Robe, South Australia.
FURTHER READING: Norris (1988), (1995); Womersley (2003); Sherwood et al. (2011); Huisman (2018).

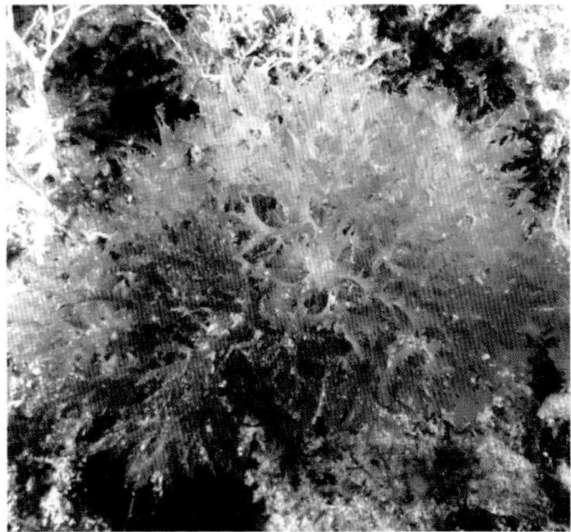

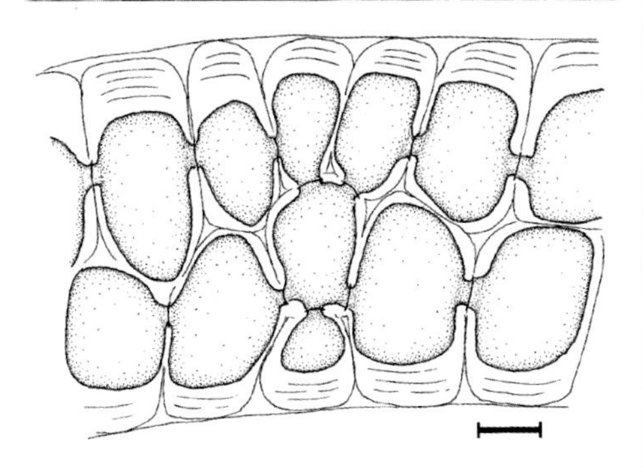

Amansia glomerata (Heron Island, Qld).

(opposite, top) *Amansia serrata* – as illustrated by Harvey (1858, pl. 51, as *Amansia kuetzingioides*).

Amansia rhodantha (Ningaloo Reef, WA).

(above) *Amansia rhodantha* – partial section of thallus showing central axis ringed by five cells (Jurien Bay, WA). Scale = 20 µm.

Amansia serrata (Cape Peron, WA).

(above right) *Amansia serrata* – closer view of thallus (Cape Peron, WA). Scale = 1 mm.

Bostrychia

Bostrychia is a common genus in the intertidal zone of tropical and temperate seas, where it is often associated with mangroves. *Bostrychia tenella* also grows on rock, and in tropical Australia can form thick bands in association with barnacles in the upper intertidal region. It is often a dull purple colour and thalli are alternately branched and structurally uniaxial with two tiers of pericentral cells per axial cell. Tetrasporangia occur in stichidia on lateral branches and are tetrahedrally divided.

Bostrychia tenella

TYPE LOCALITY: St Croix, Virgin Islands.
DISTRIBUTION: From the North West Cape region, Western Australia, around northern Australia to Lord Howe Island, New South Wales. Widespread in tropical regions.
FURTHER READING: King & Puttock (1989); Womersley (2003); Coppejans et al. (2009); Huisman (2018).

Bostrychia tenella – growing on rock (Barrow Island, WA).

(left) *Bostrychia tenella* (Kathleen Island, WA).

Chondria

Species of *Chondria* display a variety of habits, from large, robust thalli with terete axes to minute species with flattened branches. Branching is generally radial or subdistichous and in most species the branches are tapered towards the bases. Thalli are uniaxial and apices are capped by non-pigmented filaments known as trichoblasts. The axial filament is discernible throughout the thallus and each axial cell is surrounded by a ring of five large pericentral cells. These pericentral cells produce a smaller celled cortex. Several species have been recorded from Australia, and those from colder water are included in Gordon-Mills and Womersley (1987). The species included here are mainly tropical in distribution.

Chondria armata

Chondria armata has distinctive clusters of short spiny branches at the tips.

TYPE LOCALITY: Wagap, New Caledonia.

DISTRIBUTION: From the North West Cape region, Western Australia, around northern Australia to Arrawarra and Lord Howe Island, New South Wales. Warmer seas of the Indo-West Pacific.

Chondria curdieana

A common species on rough-water coasts of southern Australia, where it is often epiphytic on seagrasses.

TYPE LOCALITY: Southern Australia.

DISTRIBUTION: From the Houtman Abrolhos, Western Australia, to Waratah Bay, Victoria, and northern Tasmania.

Chondria armata (North West Island, Qld).

Chondria curdieana (Cape Peron, WA).

(right) *Chondria curdieana* – branch detail (Cape Peron, WA). Scale = 1 mm.

Chondria dangeardii

One of several flattened species, *Chondria dangeardii* is often an iridescent blue underwater.

TYPE LOCALITY: Cape Verde, Senegal.
DISTRIBUTION: From Cape Peron, Western Australia, around northern Australia to Lord Howe Island, New South Wales. Warmer seas of the Indo-West Pacific.

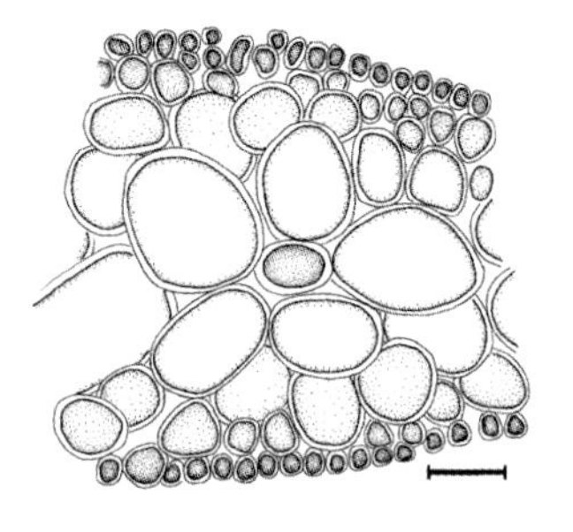

Chondria simpliciuscula

Chondria simpliciuscula is a small species with terete branches that creeps across the surface of rocks and other algae.

TYPE LOCALITY: Aldabra Island, Seychelles.
DISTRIBUTION: Warmer seas of the Indo-Pacific.

Chondria dangeardii (Houtman Abrolhos, WA).

(top) *Chondria dangeardii* – closer view of thallus (Cape Peron, WA). Scale = 2 mm.

(above) *Chondria dangeardii* – partial section of thallus showing five pericentral cells (Bundegi Reef). Scale = 30 µm.

Chondria simpliciuscula (North West Island, Qld).

Chondria succulenta

A radially branched species, often found growing in the intertidal.

TYPE LOCALITY: King George Sound, Western Australia.
DISTRIBUTION: Cape Peron, Western Australia, along southern Australia to New South Wales.

Chondria thielei

A larger, upright species, often with a distinctive secund branching pattern.

TYPE LOCALITY: Between Malus and Gidley Islands, Dampier Archipelago, Western Australia.
DISTRIBUTION: Known from Cape Peron, Barrow Island and the Dampier Archipelago, Western Australia.
FURTHER READING: Gordon-Mills & Womersley (1987); Millar (1990); Price & Scott (1992); Womersley (2003); Huisman (2018).

Chondria succulenta (Cape Peron, WA).

Chondria thielei (Cape Peron, WA).

Chondrophycus

The genus *Chondrophycus* was, until recently, regarded as a subgenus of *Laurencia* and the two taxa share many characteristics. Both include cartilaginous species in which branches have an apical depression and are structurally uniaxial. *Chondrophycus* differs from *Laurencia* in producing two (as opposed to four) pericentral cells per axial cell, and in lacking secondary pit connections between epidermal cells. *Chondrophycus brandenii* is often found in shallow subtidal habitats on moderately rough coasts.

Chondrophycus brandenii

TYPE LOCALITY: Elliston Bay, South Australia.
DISTRIBUTION: From Port Gregory, Western Australia, around southern mainland Australia to Lake Macquarie, New South Wales.
FURTHER READING: Saito & Womersley (1974); Womersley (2003).

Chondrophycus brandenii (Cape Peron, WA).

(above) *Chondrophycus brandenii* – branch detail (Cape Peron, WA). Scale = 1 mm.

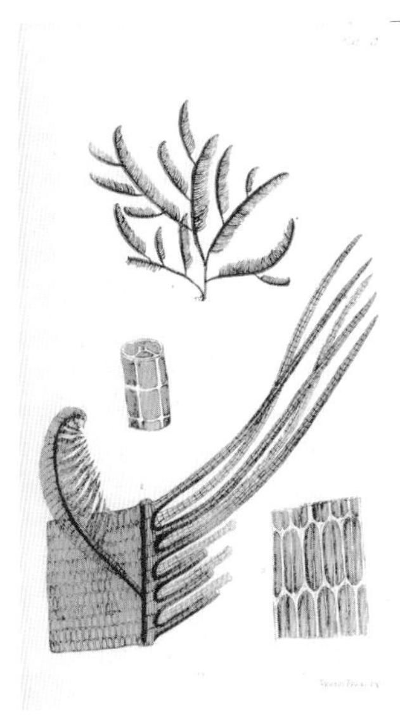

Cliftonaea

Cliftonaea includes the single species *Cliftonaea pectinata*, known only from southern and south-western Australia. The species is recognisable by its thallus form, in which the axes have inrolled apices and two closely set rows of short, undivided branches arising from one surface, giving the axes a comb-like appearance. The opposite surface has a membranous keel.

Cliftonaea pectinata

TYPE LOCALITY: Garden Island, Western Australia.
DISTRIBUTION: From the Montebello Islands, Western Australia, around south-western and southern Australia to Port Phillip Heads, Victoria.
FURTHER READING: Edgar (1997); Womersley (2003); Huisman (2018).

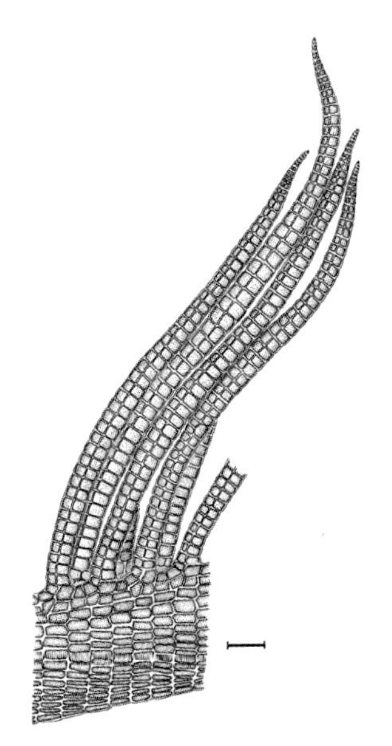

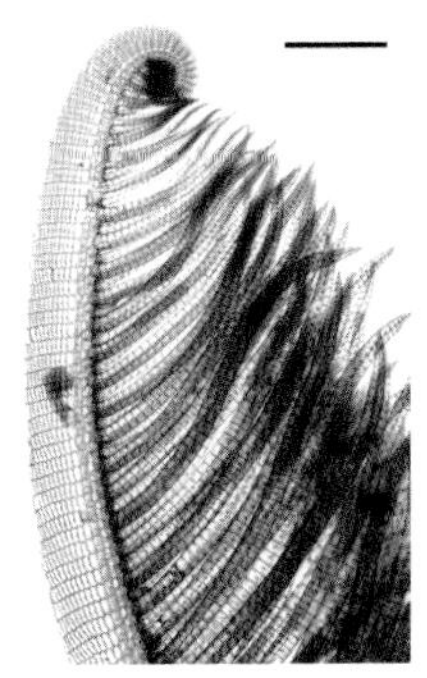

Cliftonaea pectinata (Cape Peron, WA).

(top) *Cliftonaea pectinata* – as illustrated by Harvey (1859a, pl. 100, as *Cliftonia pectinata*).

(above) *Cliftonaea pectinata* – detail of thallus (Map Reef, WA). Scale = 200 µm.

(right) *Cliftonaea pectinata* – inrolled branch apex (Cape Peron, WA). Scale = 1 mm.

Coeloclonium

Coeloclonium includes several species from southern and south-western Australia, of which *Coeloclonium umbellatum* and *Coeloclonium verticillatum* are included here. The thallus has a primary stem that bears whorls of short lateral branches. All branches are constricted at their bases and are soft but turgid. Structurally the thallus is mostly hollow, with a central axial filament ringed by five pericentral cells. Branched lateral filaments traverse the central cavity and join with the outer cortex. The pericentral cells have a distinctive shape, in which the distal region curves transversely to form the basal portion of the lateral filament. *Coeloclonium umbellatum* typically grows as an epiphyte on the seagrasses *Posidonia* and *Amphibolis*.

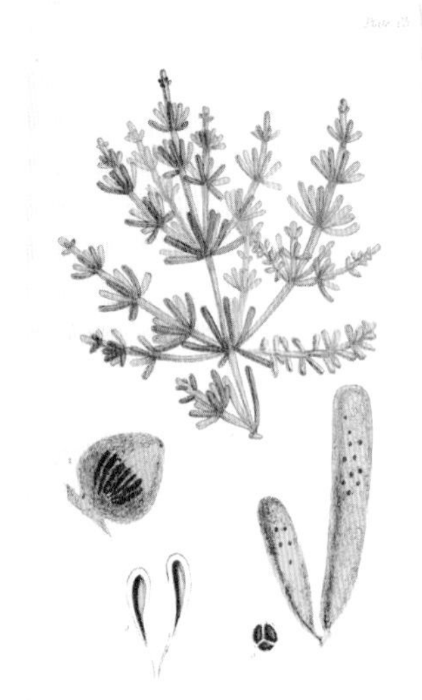

Coeloclonium umbellatum

TYPE LOCALITY: Garden Island, Western Australia.
DISTRIBUTION: From Rottnest Island, Western Australia, around southern Australia to Portland, Victoria.

Coeloclonium verticillatum

TYPE LOCALITY: Garden Island, Western Australia.
DISTRIBUTION: From the Perth region, Western Australia, around southern mainland Australia to Port Phillip Heads, Victoria, and around Tasmania.
FURTHER READING: Womersley (2003).

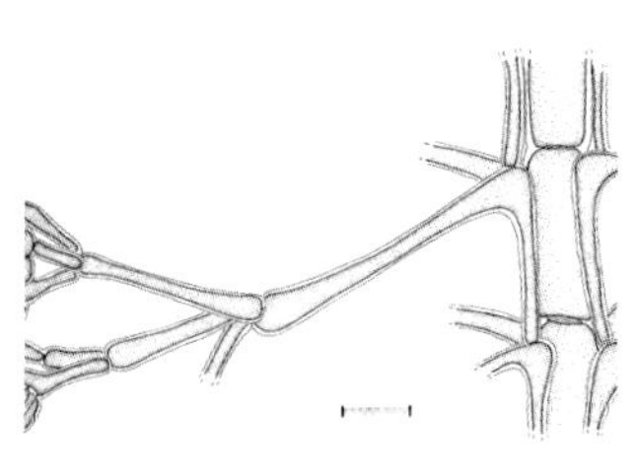

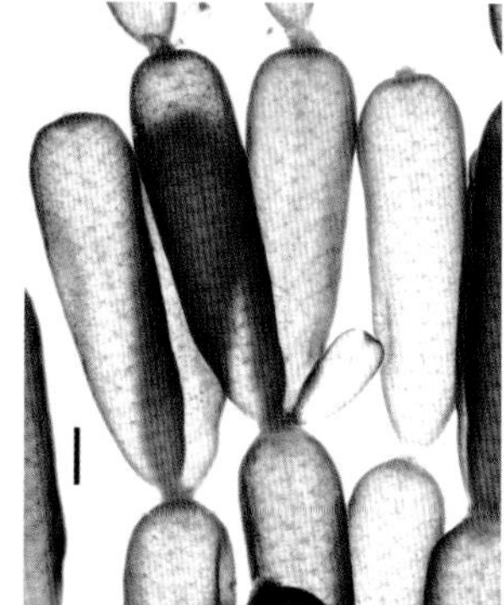

Coeloclonium umbellatum (Bremer Bay, WA).

Coeloclonium verticillatum (Map Reef, WA).

(top) *Coeloclonium verticilatum* – as illustrated by Harvey (1859a, pl. 102, as *Chondria verticillata*).

(far right) *Coeloclonium verticillatum* – branch detail (Cape Peron, WA). Scale = 1 mm.

(right) *Coeloclonium verticillatum* – partial longitudinal section showing central axis and one lateral filament (Rottnest Island, WA). Scale = 100 µm.

Coryneclαdia

Corynecladia is similar in appearance to *Laurencia* (p. 237), and the species included here, *Corynecladia elata*, was previously treated as a species of *Laurencia*. Recognition of the genus was based primarily on DNA sequence studies, and the only morphological feature that distinguishes *Corynecladia* is the presence of a secondary cortex. At present, the genus includes two species, the widespread *Corynecladia elata*, and *Corynecladia nova* (not included), which is restricted to New South Wales. *Corynecladia elata* has slightly compressed axes similar to those of *Laurencia brongniartii*, but has a different branching pattern.

Corynecladia elata

TYPE LOCALITY: King Island, Bass Strait, Tasmania.
DISTRIBUTION: From Port Denison, Western Australia, around southern Australia and Tasmania to Mossy Point, New South Wales. New Zealand.
FURTHER READING: Metti et al. (2015); Cassano et al. (2019).

Corynecladia elata (Rottnest Island, WA).

(top) *Corynecladia elata* – as illustrated by Harvey (1847–1849, pl. 33, as *Laurencia elata*).

Dasyclonium

A distinctive genus in which the polysiphonous main axes bear very regular, short lateral branches in an alternate-distichous pattern. The lateral branches arise from every second axial segment and themselves bear a row of undivided branches from their upper surface. *Dasyclonium incisum* is a common species in most Australian seas, often growing on seagrasses or other algae. Mature thalli are generally only a few centimetres in height, so the species is often overlooked. *Dasyconium flaccidum* has ultimate branches that are monosiphonous, whereas those of *Dasyconium incisum* are polysiphonous.

Dasyclonium flaccidum

TYPE LOCALITY: King George Sound, Western Australia.

DISTRIBUTION: Houtman Abrolhos, Western Australia, to Robe, South Australia, and northern Tasmania. Japan.

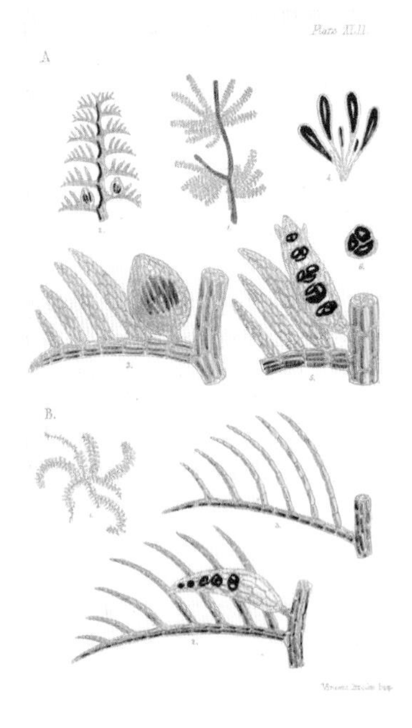

Dasyclonium incisum

TYPE LOCALITY: Australia.

DISTRIBUTION: From the Houtman Abrolhos, Western Australia, around southern Australia and Tasmania to Byron Bay, New South Wales. New Zealand. South Africa.

FURTHER READING: Millar (1990); Womersley (2003); Nelson (2013); Scott (2017).

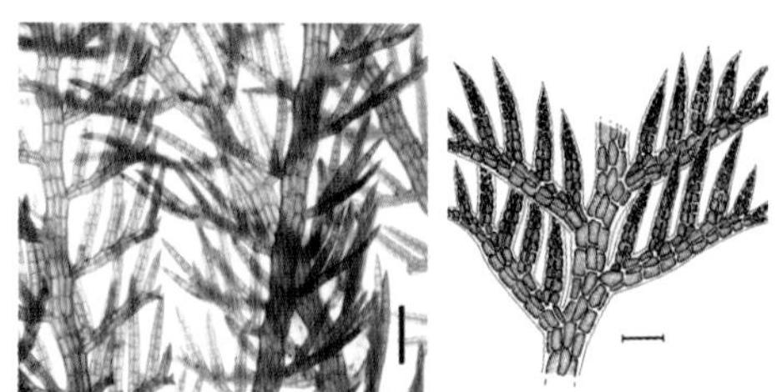

(above) *Dasyclonium incisum* (A) and *Dasyclonium flaccidum* (B) – as illustrated by Harvey (1858, pl. 42, as *Polyzonia*).

Dasyclonium incisum – growing on the stem of *Sargassum* (Cape Peron, WA).

(above left) *Dasyclonium incisum* – closer view of thallus (Cape Peron, WA). Scale = 1 mm.

(above right) *Dasyclonium incisum* – detail of thallus (Houtman Abrolhos, WA). Scale = 200 µm.

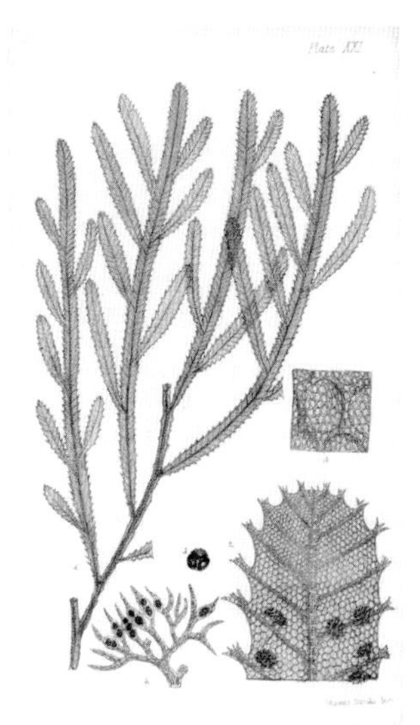

Dictyomenia

Thalli of *Dictyomenia* have flattened branches that are a relatively uniform width throughout and have a central midrib and membranous 'wings'. Lateral branches arise in an alternating sequence – some eventually form axes similar to the main axis, but the majority remain short and are visible as veins in the wings, terminated by a short tuft of filaments. Structurally the thallus is uniaxial, with the central axis ringed by six pericentral cells. *Dictyomenia* includes two relatively common species found in south-western waters. *Dictyomenia sonderi* is a large plant with a very conspicuous midrib and broad branches (to 15 mm) that taper near their point of attachment, while *Dictyomenia tridens* is a smaller plant with narrower branches.

Dictyomenia sonderi

TYPE LOCALITIES: Garden Island and Fremantle, Western Australia.
DISTRIBUTION: North of Dongara, Western Australia, to Backstairs Passage, South Australia.

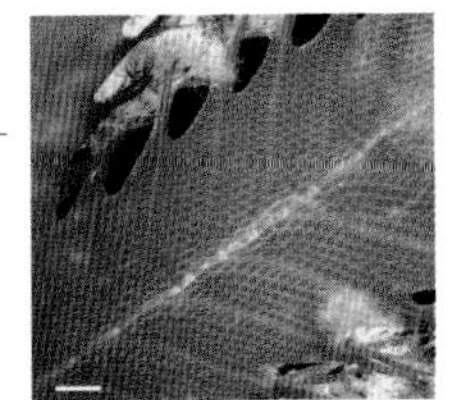

Dictyomenia sonderi (Cape Peron, WA).

(right) *Dictyomenia sonderi* – detail of thallus (Carnac Island, WA). Scale = 1 mm.

(top) *Dictyomenia sonderi* – as illustrated by Harvey (1858, pl. 21).

Dictyomenia tridens (Map Reef, WA).

(far right) *Dictyomenia tridens* – partial section of thallus showing central axis ringed by six pericentral cells (Jurien Bay, WA). Scale = 100 µm.

Dictyomenia tridens

TYPE LOCALITY: Western Australia.
DISTRIBUTION: From the Houtman Abrolhos, Western Australia, to San Remo, Victoria, Deal Island, Bass Strait, and Twofold Bay, New South Wales.
FURTHER READING: Womersley (2003).

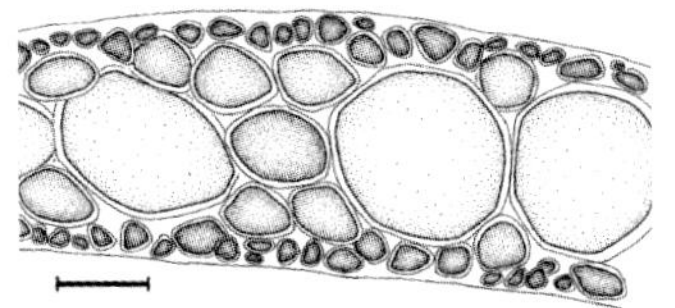

Digenea

Digenea is represented in Australia by the single species *Digenea simplex*, which grows on rock in the intertidal and shallow subtidal in warmer seas. Plants are often a dark purple, almost black colour, but become pale yellow in upper branches, and are often covered by epiphytic algae. The thallus is dichotomously or irregularly branched, with the thick cartilaginous stems hidden under short, slender lateral branches. Structurally the thallus is uniaxial, with each axial cell in the main axes ringed by nine or 10 pericentral cells, while those of lateral branches have six to nine pericentral cells.

Digenea simplex

TYPE LOCALITY: Trieste, Italy.
DISTRIBUTION: From Geraldton, Western Australia, around northern Australia to southern Queensland. Widespread in warmer seas.
FURTHER READING: Huisman (2018).

Digenea simplex (Barrow Island, WA).

Ditria

The genus *Ditria* is known in Australia by two species: the widespread *Ditria zonaricola* (not shown), and *Ditria expleta*, which is known from the western and southern coasts. Thalli of *Ditria expleta* appear to be exclusively epiphytic on the brown algae *Lobophora* and *Zonaria*, where they creep across the upper surface of the host. Structurally the thallus is uniaxial and each axial cell is surrounded by five pericentral cells. Lateral branches of limited growth arise from each axial cell in a spiral pattern, but because of the prostrate habit, they are displaced and appear to be in distichous lateral pairs. The dorsal lateral branch never develops beyond a branch primordium.

Ditria expleta

TYPE LOCALITY: Goss Passage, Houtman Abrolhos, Western Australia.
DISTRIBUTION: From the Houtman Abrolhos and Rottnest Island, Western Australia, and from Cootes Hill, Sturt Bay, Yorke Peninsula, to Sellicks reef, and Rocky Point, Kangaroo Island, South Australia, but probably more widespread.
FURTHER READING: Yoshida & Yoshida (1983); Huisman (1994), (2018); Womersley (2003).

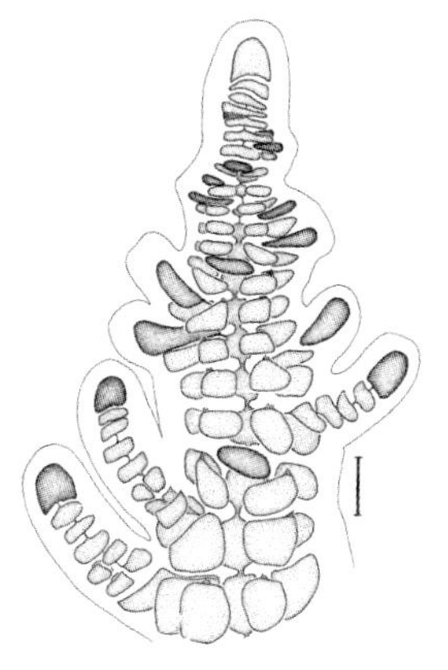

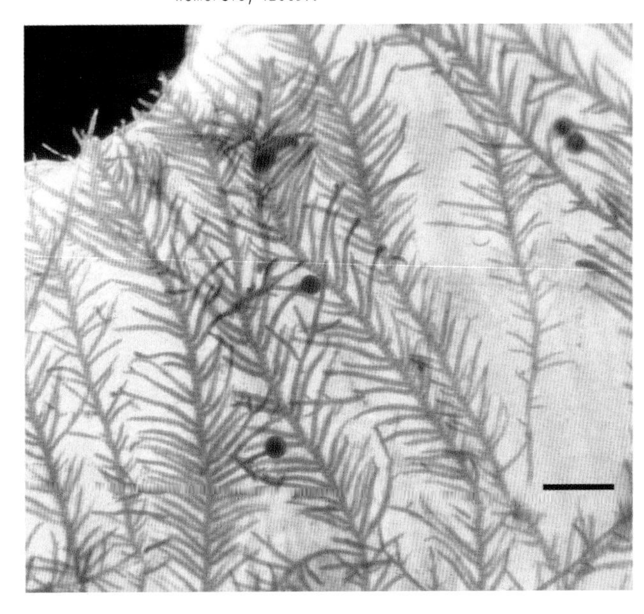

Ditria expleta – growing on *Lobophora* (Houtman Abrolhos, WA). Scale = 600 µm.

(above) *Ditria expleta* – detail of apical region (Houtman Abrolhos, WA). Scale = 10 µm.

Echinothamnion

Echinothamnion is a distinctive genus characterised by a thallus in which the main axes are densely corticated but bear uncorticated short lateral branches in a spiral pattern. In *Echinothamnion hystrix*, the lateral branches remain discrete and can be seen as short tufts, while in *Echinothamnion hookeri* they overlap and give the thallus a spongy texture; the latter also has several orders of indeterminate branches compared to the irregular branches of *Echinothamnion hystrix*. Structurally the thallus is uniaxial, with each axial cell ringed by four pericentral cells.

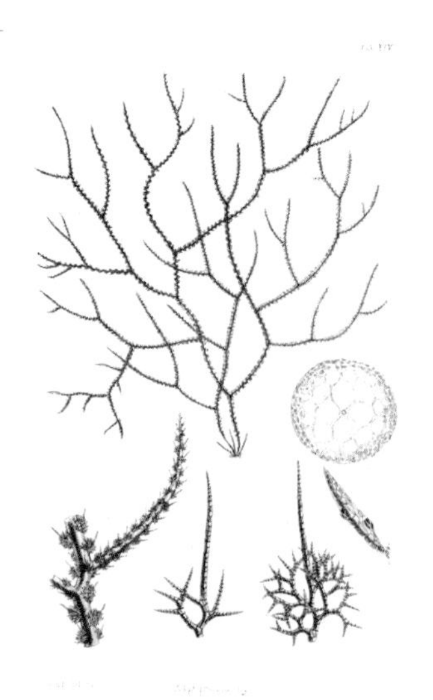

Echinothamnion hystrix

TYPE LOCALITY: Tasmania.
DISTRIBUTION: Nichol Bay, Western Australia, to Walkerville, Victoria, and around Tasmania. New Zealand.
FURTHER READING: Adams (1994); Womersley (2003).

Echinothamnion hookeri

TYPE LOCALITY: Georgetown, Tasmania.
DISTRIBUTION: Albany, Western Australia, to Walkerville, Victoria, and around Tasmania.

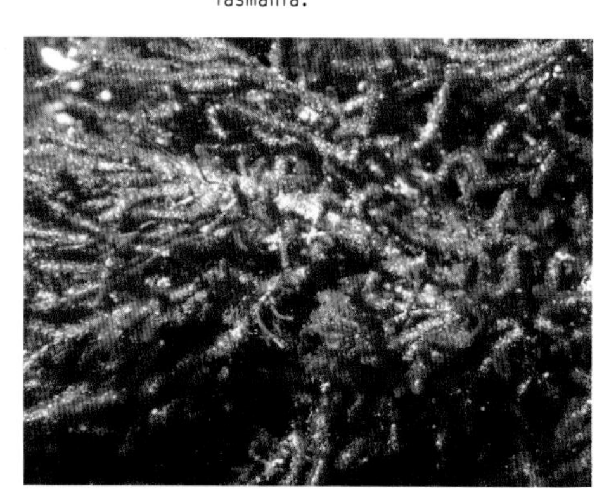

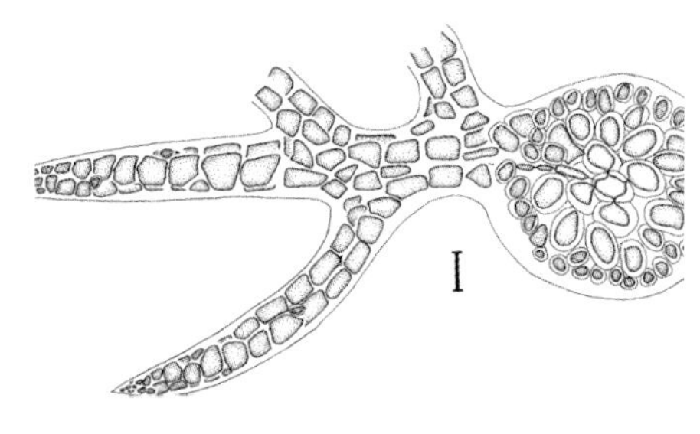

Echinothamnion hookeri (Eglinton Rocks, WA).

(right) *Echinothamnion hookeri* – section of thallus with lateral branch (Jurien Bay, WA). Scale = 300 µm.

Echinothamnion hystrix (Esperance, WA).

(top) *Echinothamnion hystrix* – as illustrated by Harvey (1847–1849, pl. 14, as *Polysiphonia hystrix*).

Endosiphonia

Endosiphonia includes four species, mostly distributed in warmer waters of the Indo-Pacific. The genus has terete branches that bear short spines, somewhat reminiscent of *Acanthophora*. *Endosiphonia* differs, however, in having four pericentral cells (as opposed to five), and with the medullary cells surrounding the pericentral cells being of a similar length to the pericentral cells. As a result, a regular banding pattern can often be seen from the surface. *Endosiphonia spinuligera* is a rare species, known in Australia only from the Houtman Abrolhos north to Broome, Western Australia, and Lord Howe Island, New South Wales.

Endosiphonia spinuligera

TYPE LOCALITY: Wokam Island, Aru Islands, Indonesia.
DISTRIBUTION: From the Houtman Abrolhos, Western Australia, probably around northern Australia to Lord Howe Island, New South Wales. Indonesia. Philippines. Marshall Islands.

Endosiphonia spinulosa

TYPE LOCALITY: Garden Island, Western Australia.
DISTRIBUTION: Rottnest Island, Western Australia, to Thistle Island, South Australia.
FURTHER READING: Dawson (1956); Huisman (2018).

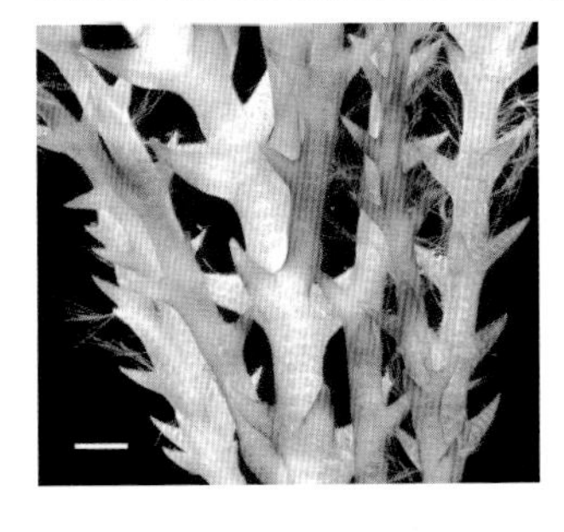

Endosiphonia spinuligera (Coral Bay, WA).

(above) *Endosiphonia spinuligera* – section of thallus (Houtman Abrolhos, WA). Scale = 100 µm.

Endosiphonia spinulosa (Rottnest Island, WA).

(right) *Endosiphonia spinulosa* – detail of thallus (Rottnest Island, WA). Scale = 1 mm.

(top) *Endosiphonia spinulosa* – as illustrated by Harvey (1860, pl. 130, as *Rhodomela spinulosa*).

Epiglossum

Epiglossum includes two species endemic to southern Australia. The genus was previously regarded as synonymous with *Lenormandia* (see p. 240), but was resurrected following DNA sequence analyses. Thalli have flattened, linear axes and are branched from the midrib, with rounded, recurved apices. Structurally the axial cells are surrounded by five pericentral cells and two to four pseudopericentral cells. The medulla has several layers of larger cells and the cortex is one to two cells thick. Reproductive structures are borne in short erect branches arising from the surface of the thallus. *Epiglossum smithiae* is a deep-water alga, distinguished by its dense surface coating of small proliferations and commonly a sponge coating.

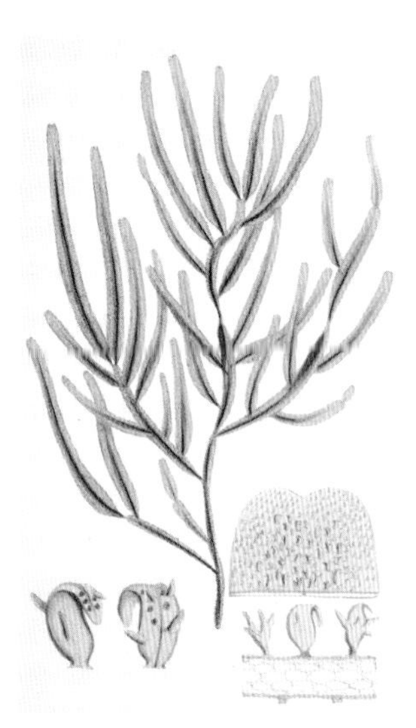

Epiglossum smithiae

TYPE LOCALITY: Circular Head, Tasmania.
DISTRIBUTION: Memory Cove, South Australia, to Green Cape, New South Wales, and around Tasmania.
FURTHER READING: Phillips (2002b); Womersley & Phillips (2003b).

Epiglossum smithiae (Kangaroo Island, SA; photo G.J. Edgar).

(top) *Epiglossum smithiae* – as illustrated by Harvey (1847, pl. 3, as *Polyphacum smithiae*).

Exophyllum

The genus *Exophyllum* includes the single species *Exophyllum wentii*, an extremely rare plant known only from a few sporadic collections worldwide. In Australia, the species has been collected on one occasion from the Montebello Islands, Western Australia. Thalli of *Exophyllum wentii* are tough and cartilaginous, with flattened, blade-like branches that are decumbent and attached by several short stipes. Structurally the thallus is uniaxial, although the central axis is quickly obscured by secondary cortication and cannot be recognised in mature branches. Reproductive structures occur only on the dorsal surface of the thallus: tetrasporangia are borne in club-like stichidia, cystocarps are globose and sessile, and spermatangia arise in flattened plates.

Exophyllum wentii

TYPE LOCALITIES: Savu Island and Borneo Bank, Indonesia. North Ubian Island, Sulu Province, Sulu Archipelago.
DISTRIBUTION: In Australia, known only from the Montebello Islands, Western Australia. Japan. Philippines. Indonesia.
FURTHER READING: Indy et al. (2006); Huisman (2018).

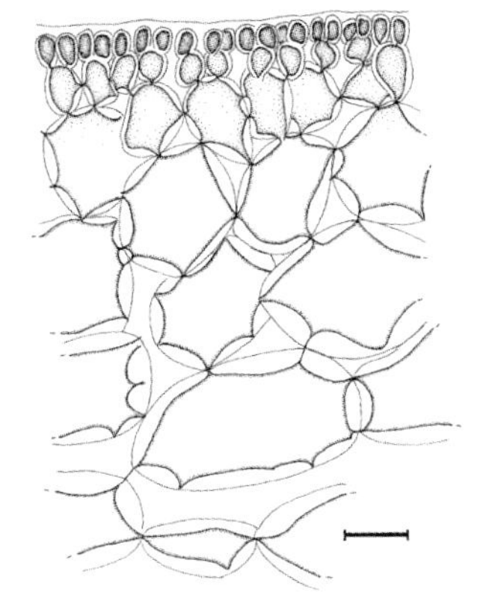

Exophyllum wentii (Montebello Islands, WA).

(above) *Exophyllum wentii* – partial section of thallus (Montebello Islands, WA). Scale = 30 µm.

Herposiphonia

Herposiphonia includes several species of small, creeping plants. The main axes bear two types of branches: determinate (which are short, unbranched and erect) and indeterminate (which are prostrate and eventually branch in a pattern identical to the main axis). The indeterminate branches generally arise on every fourth segment on alternating sides, while the determinate branches arise from the remaining segments. In this way, a precise branching pattern is produced, although in many specimens this is not fully realised. Attachment to the substratum is by unicellular, colourless rhizoids that arise from the ventral surface. Structurally the thallus is uniaxial, with each axial cell ringed by varying numbers of pericentral cells, depending on the species. In *Herposiphonia elongata* (one of the more common species) between six and nine pericentral cells have been reported. Colourless trichoblasts are generally present at the tips of upright branches, which are usually inrolled, and cortication is absent. *Herposiphonia tenella* has a similar structure but differs in the position of reproductive structures. *Herposiphonia pectinata* grows exclusively on the red alga *Amphiroa* and has between 11 and 13 pericentral cells.

Herposiphonia elongata

TYPE LOCALITY: Pulau Sampadi, Santubong, Kuching, Sarawak, Malaysia.
DISTRIBUTION: Widespread in north-western Australia. Malaysia. Japan.

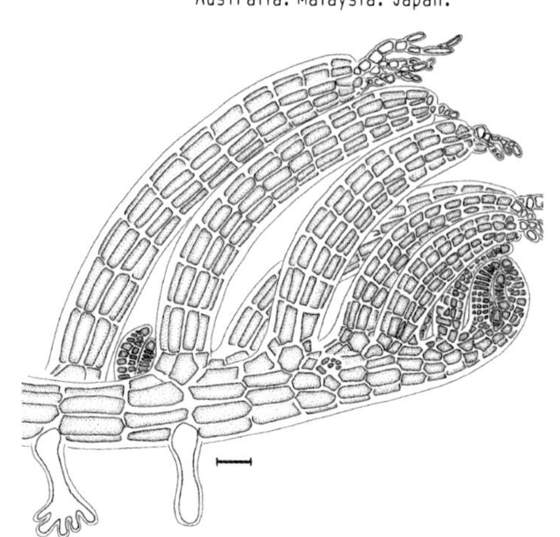

Herposiphonia pectinata

TYPE LOCALITY: Western Australia.
DISTRIBUTION: South-western Australia.

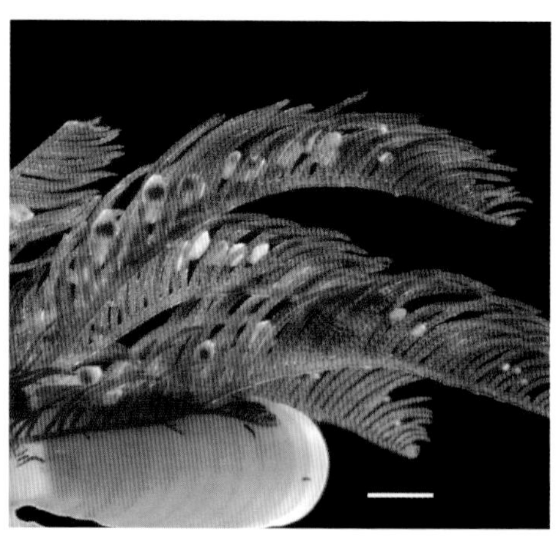

Herposiphonia elongata – detail of thallus (Montebello Islands, WA). Scale = 30 µm.

Herposiphonia pectinata (Cape Peron, WA). Scale = 1 mm.

Herposiphonia tenella

TYPE LOCALITY: Sicily.
DISTRIBUTION: Widespread in warmer seas.
FURTHER READING: Womersley (2003); Huisman et al. (2015); Huisman (2018).

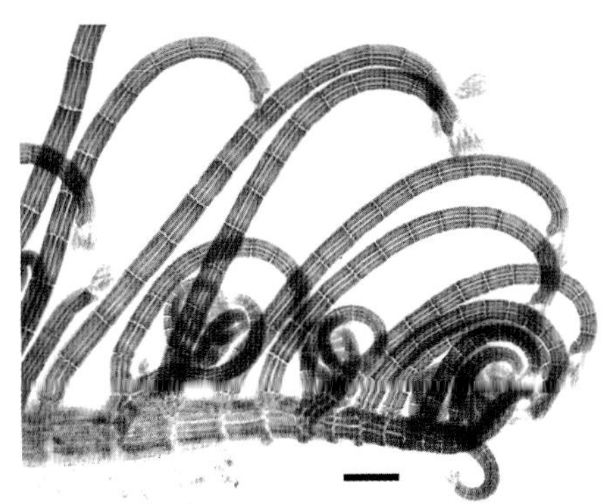

Herposiphonia tenella (Masthead Island, Qld).

(right) *Herposiphonia tenella* – closer view (Masthead Island, Qld). Scale = 100 µm.

Kuetzingia

A genus of several species found in south-western Australia and in Africa, *Kuetzingia* has flattened axes that bear opposite, lateral branches, in contrast to the alternate branches of several closely related genera (e.g. *Dictyomenia*). Structurally the thallus is uniaxial, with six pericentral cells, of which the lateral cells divide further to form wings. In section, the medulla is composed of a single layer of large cells, bordered by a smaller celled cortex. *Kuetzingia canaliculata* is a common alga in south-western Australia.

Kuetzingia canaliculata

TYPE LOCALITY: Perth, Western Australia.
DISTRIBUTION: From the Houtman Abrolhos to Point D'Entrecasteaux, Western Australia, and Wanna, South Australia.
FURTHER READING: Womersley et al. (2003b).

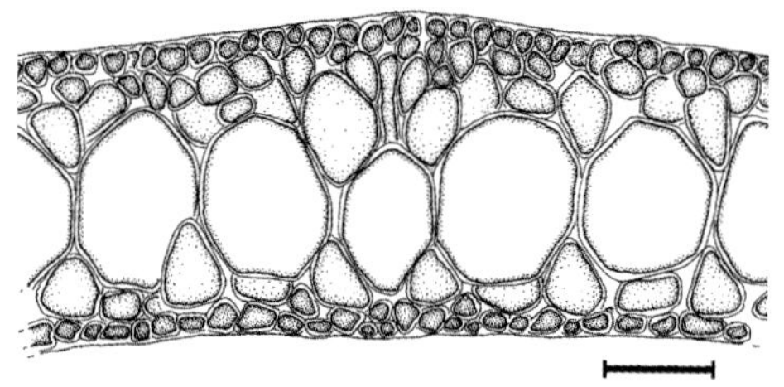

Kuetzingia canaliculata (Carnac Island, WA).

(top) *Kuetzingia canaliculata* – as illustrated by Harvey (1862, pl. 232).

(left) *Kuetzingia canaliculata* – section of thallus in the region of the central axis (Jurien Bay, WA). Scale = 100 µm.

Laurencia

Laurencia is a common genus in both tropical and temperate Australian seas. Thalli are cartilaginous and much branched, with axes that are either terete or flattened. A distinctive feature of the genus is the presence of a sunken apex, usually with emergent colourless filaments. Structurally the thallus is uniaxial, although the central axis is visible only near the apex and is generally not discernible in transverse sections. This feature is helpful in distinguishing *Laurencia* from the occasionally superficially similar *Chondria*, in which the central axis is always visible and surrounded by a ring of five pericentral cells. Separation of the species of *Laurencia* can be difficult and sometimes requires fertile material. The species included here can generally be recognised by their appearance; confirmation of the common *Laurencia dendroidea* (formerly known in Australia as *Laurencia majuscula*), however, requires an examination of the cortical cells near the apex, which show a distinctive convex surface (see figure).

Laurencia brongniartii

A distinctive species due to its markedly flattened branches and a prominent central axis.

TYPE LOCALITY: Martinique, West Indies.
DISTRIBUTION. Throughout Australia, but rarely encountered in southern waters. Widespread in warmer seas.

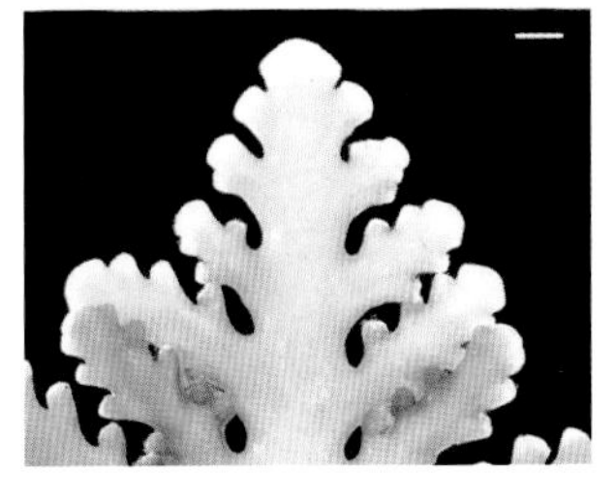

Laurencia brongniartii (Cape Peron, WA).

Laurencia dendroidea

A common species in most Australian seas.

TYPE LOCALITY: Brazil.
DISTRIBUTION: Widely distributed in warmer waters of the Indian and Pacific Oceans, including most Australian coasts.

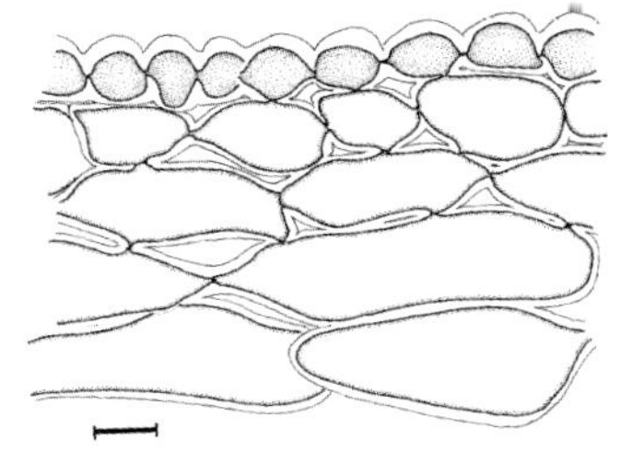

(left) *Laurencia brongniartii* – detail of thallus (Cape Peron, WA). Scale = 1 mm.

(top) *Laurencia brongniartii* – as illustrated by Harvey (1858, pl. 15, as *Laurencia grevilleana*).

Laurencia dendroidea (Little Turtle Island, WA).

(above) *Laurencia dendroidea* – partial longitudinal section of thallus (Houtman Abrolhos, WA). Scale = 30 µm.

Laurencia elegans

TYPE LOCALITY: Lord Howe Island, New South Wales.
DISTRIBUTION: Known from Cape Londonderry, northern Western Australia, to Adelaide, South Australia; also in Jervis Bay, Lord Howe Island and Norfolk Island.

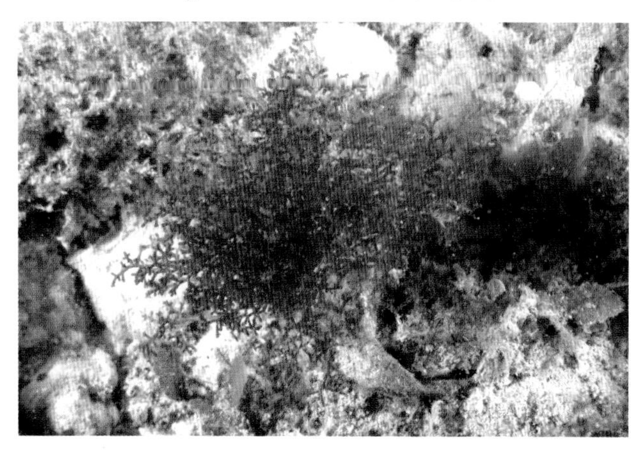

Laurencia similis

Laurencia similis has densely aggregated short branches, giving the thallus a warty appearance. It is similar in appearance to *Palisada perforata*.

TYPE LOCALITY: Low Isles, Queensland.
DISTRIBUTION: From Exmouth to Adele Island, northern Western Australia, and the Low Isles, near Port Douglas, north-eastern Queensland.

Laurencia forsteri

A common species, most often found as an epiphyte on the seagrass *Posidonia*.

TYPE LOCALITY: Australia (probably King George Sound, Western Australia).
DISTRIBUTION: From the Houtman Abrolhos, Western Australia, to Wilsons Promontory, Victoria, and northern Tasmania.

Laurencia elegans (Cassini Island, WA).

Laurencia forsteri (Cape Peron, WA).

Laurencia similis (Lizard Island, Qld).

Laurencia snackeyi

A coarse tropical species.

TYPE LOCALITY: Haingsisi (Hansisi), Samau Island (near Timor), Indonesia.
DISTRIBUTION: Widespread in warmer waters of the Indian and Pacific Oceans.

Laurencia filiformis

A widespread species that can vary in appearance.

TYPE LOCALITY: Western Australia.
DISTRIBUTION: In Australia from the Houtman Abrolhos, Western Australia, to Tilba, New South Wales and possibly Queensland, and around Tasmania.
FURTHER READING: Saito & Womersley (1974); Womersley (2003); Metti et al. (2018).

Laurencia snackeyi (Coral Bay, WA).

Laurencia filiformis – epiphytic on *Amphibolis griffithii* (Cape Peron, WA).

Lenormandia

Lenormandia is represented in Australia by five species, all largely confined to southern Australia. Plants of *Lenormandia latifolia* and *Lenormandia spectabilis* commonly occur as epiphytes on the seagrass *Amphibolis*. Thalli are broad and flattened, and can be branched (*Lenormandia spectabilis*) or mostly unbranched (*Lenormandia latifolia*). Structurally the thallus is uniaxial, but in *Lenormandia latifolia* the central axis is quickly obscured and is not visible as a midrib (as occurs in other species of the genus). The medulla is composed of large colourless cells, bordered by a cortex of smaller, pigmented cells. Reproductive structures are borne in specialised branches on the surface of the thallus. *Lenormandia latifolia* was previously included in the genus *Lenormandiopsis*, but was transferred following a DNA sequence analysis and morphological study.

Lenormandia latifolia

TYPE LOCALITY: Perth, Western Australia.
DISTRIBUTION: From the Houtman Abrolhos, Western Australia, to Cape Northumberland, South Australia.

Lenormandia marginata

TYPE LOCALITY: Mouth of the Tamar, Tasmania.
DISTRIBUTION: From the Houtman Abrolhos, Western Australia, to Cape Northumberland, South Australia.

Lenormandia spectabilis

TYPE LOCALITY: Western Australia.
DISTRIBUTION: From Geraldton, Western Australia, to Robe, South Australia.
FURTHER READING: Norris (1987); Phillips (2002a, 2006); Womersley & Phillips (2003a); Scott (2017).

Lenormandia latifolia (Seabird, WA).

Lenormandia marginata (Low Head, Tas; photo G.J. Edgar).

Lenormandia spectabilis (Cape Peron, WA).

(opposite, top) *Lenormandia spectabilis* – as illustrated by Harvey (1862, pl. 181).

Leveillea

Leveillea is a distinctive genus that is readily recognised. Its most widespread species, *Leveillea jungermannioides*, is an extremely common epiphyte of a wide variety of macroalgae. Plants have prostrate axes that bear small, leaf-like lateral branches in an alternating sequence. Both the tips of the axes and the lateral branches terminate in colourless trichoblasts, but these can be secondarily lost and are not always present. The species is generally too small to be seen with the unaided eye, but once recognised, its unusual appearance ensures that it is never forgotten.

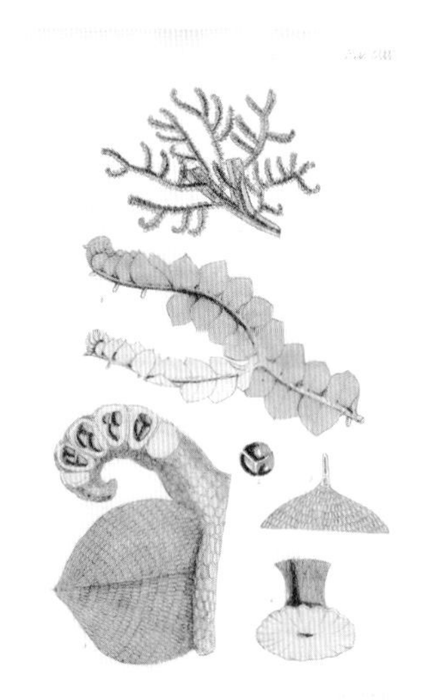

Leveillea jungermannioides

TYPE LOCALITY: El Tur, Gulf of Suez.

DISTRIBUTION: From Cape Peron, Western Australia, around northern Australia to northern New South Wales and Lord Howe Island. Widespread in warmer seas of the Indo-West Pacific.

FURTHER READING: Price & Scott (1992); Wynne (2003); Huisman (2018).

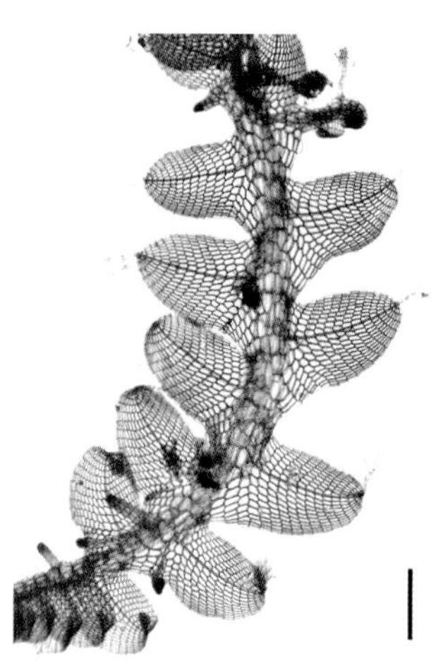

Leveillea jungermannioides – epiphytic on other algae (Cape Peron, WA).

(top) *Leveillea jungermannioides* – as illustrated by Harvey (1860, pl. 171, as *Leveillea schimperi*).

(above) *Leveillea jungermannioides* – thallus detail (Cape Peron, WA). Scale = 500 µm.

Lophocladia

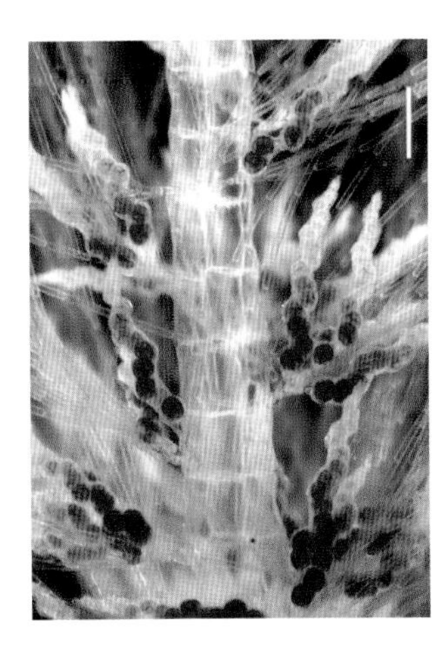

Lophocladia includes several species of upright, tufted plants. The primary axes are polysiphonous, with each axial cell ringed by four pericentral cells and with a persistent pigmented trichoblast arising in a spiral pattern and giving the thallus a furry appearance (in many other genera these are shed). The axes are corticated by descending rhizoids originating from the pericentral cells. Spermatangia and tetrasporangia are borne on specialised branches of the trichoblasts. *Lophocladia* is very similar to *Spirocladia* (see p. 254) and fertile specimens are often needed for positive identification. In spermatangial and tetrasporangial specimens, monosiphonous laterals arise from the fertile region of the trichoblast in *Spirocladia*, whereas these are absent in *Lophocladia*.

Lophocladia kuetzingii

TYPE LOCALITY: Fremantle, Western Australia.
DISTRIBUTION: Known from Port Denison, Western Australia, around southern Australia to Kangaroo Island, South Australia.
FURTHER READING: Womersley & Parsons (2003); Huisman (2018).

Lophocladia kuetzingii (Cape Peron, WA).

(top) *Lophocladia kuetzingii* – detail of branches with tetrasporangia (Cape Peron, WA). Scale = 250 µm.

Melanothamnus

Melanothamnus is one of several genera previously regarded as synonymous with *Polysiphonia* but now treated as independent genera following DNA sequence analyses (see also *Vertebrata*, p. 258). Morphologically, most species of *Melanothamnus* can be recognised by the presence of plastids restricted to radial walls of cells, but despite this being regarded as a defining feature of the genus it is not found in all species. Other features of *Melanothamnus* include spermatangial branches arising as one fork of a trichoblast and tetrasporangia forming in spiral series in upper branches. *Melanothamnus forfex* is commonly found as an epiphyte on seagrasses and some robust algae.

Melanothamnus blandii

TYPE LOCALITY: Brighton, Port Phillip, Victoria.
DISTRIBUTION: From Geraldton south to Geographe Bay, Western Australia, and Elliston, South Australia, to North Walkerville, Victoria. Tasmania.

Melanothamnus infestans

TYPE LOCALITY: Princess Royal Harbour, King George Sound, Western Australia.
DISTRIBUTION: Widespread in Australian seas. Indonesia.

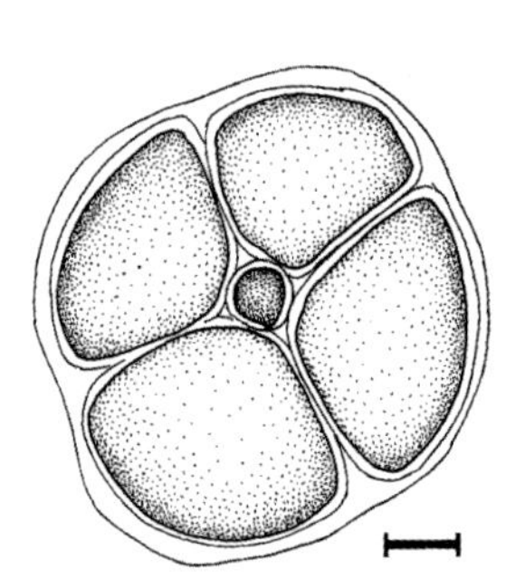

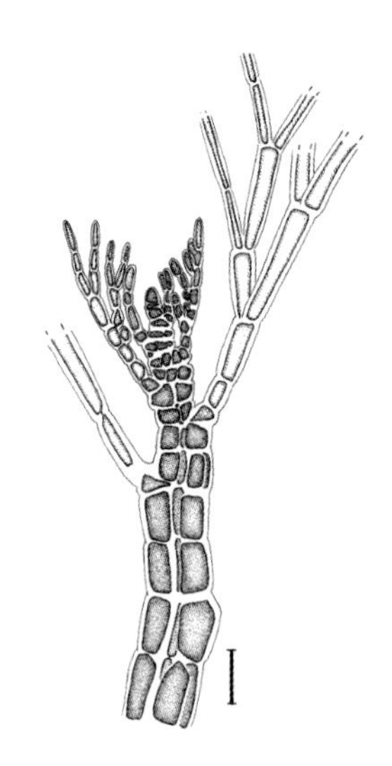

Melanothamnus blandii (Eglinton Rocks, WA).

(left) *Melanothamnus blandii* – section of thallus (Eglinton Rocks, WA). Scale = 50 µm.

(top) *Melanothamnus blandii* – as illustrated by Harvey (1862, pl. 184, as *Polysiphonia blandii*).

(right) *Melanothamnus infestans* – branch apex (Carbla, WA). Scale = 30 µm.

Melanothamnus forfex

TYPE LOCALITY: Rottnest Island, Western Australia.
DISTRIBUTION: Perth region to King George Sound, Western Australia.
FURTHER READING: Womersley (1979, as *Polysiphonia*, 2003, as *Polysiphonia*); Díaz-Tapia et al. (2017); Huisman (2018).

Melanothamnus forfex (Cape Peron, WA).

Neurymenia

A genus of two species, with the widespread *Neurymenia fraxinifolia* found in Australian seas. Thalli are large and coarse, with long stipes supporting flattened blades. The blades are somewhat leaflike and have a central midrib with lateral veins that arise alternately and extend to the margins, where they form branched teeth. Structurally the thallus is uniaxial, and the central axis is surrounded by a ring of five pericentral cells. The wings of mature blades have a medulla of two cell layers and a single layered cortex. *Neurymenia fraxinifolia* is readily recognised by its robust thallus with crisp 'leaves' and marginal teeth. It is regularly epiphytised by crustose coralline red algae that form conspicuous pale pink patches on the blade surface.

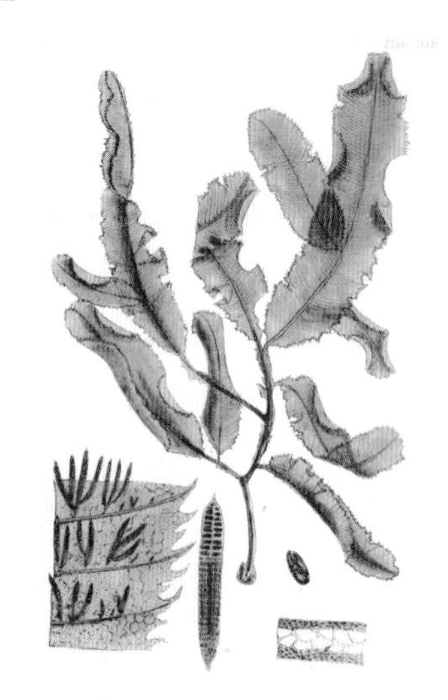

Neurymenia fraxinifolia

TYPE LOCALITY: East Indies.
DISTRIBUTION: From Cape Peron north to the Montebello Islands, Western Australia (at least). Warmer waters of the Indo-Pacific.
FURTHER READING: Tanaka & Itono (1969); Huisman (2018).

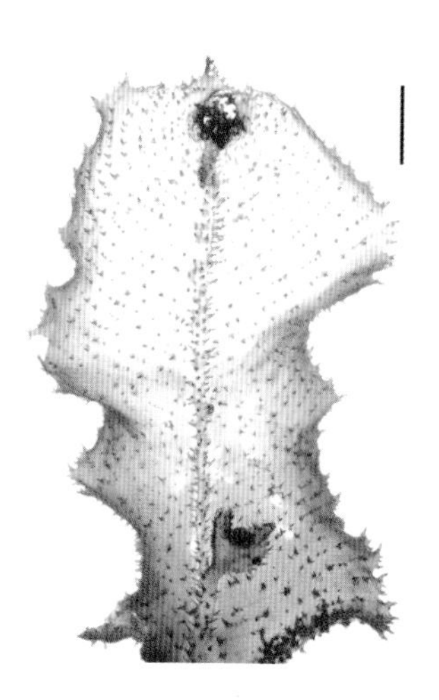

Neurymenia fraxinifolia (Cape Peron, WA).

(top) *Neurymenia fraxinifolia* – as illustrated by Harvey (1860, pl. 124, as *Dictymenia fraxinifolia*).

(above) *Neurymenia fraxinifolia* – detail of branch (Cape Peron, WA). Scale = 5 mm.

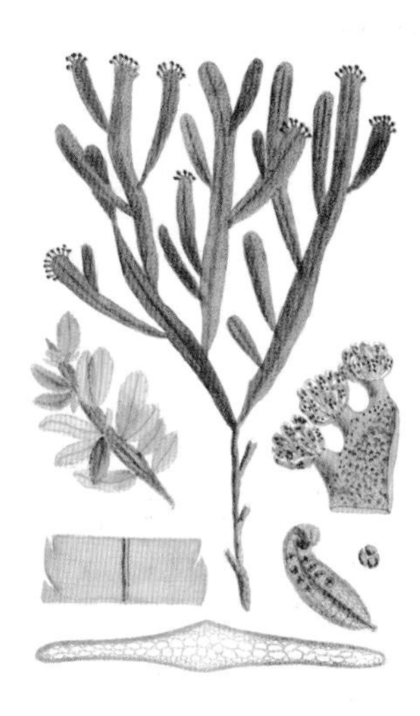

Osmundaria

Osmundaria includes a single species in Australia, the type of the genus, *Osmundaria prolifera*, a coarse plant whose branches are densely covered with short proliferations that give the thallus a rough texture. These proliferations are thought to be a response to a thin sponge layer that grows on the surface of all branches. The thallus has flattened branches, generally of a uniform width, with a central midrib and lateral wings with serrated margins. A vein system (prominent in the related genus *Vidalia*) is absent or inconspicuous. Structurally the genus is uniaxial, with each axial cell surrounded initially by five pericentral cells, those on the dorsal surface cutting off two pseudopericentral cells in the mature axis. The wings are formed by outgrowths of the lateral pericentral cells and the entire thallus is secondarily corticated.

Osmundaria prolifera

TYPE LOCALITY: Western Australia.
DISTRIBUTION: Kalbarri, Western Australia, to Victor Harbor and Kangaroo Island, South Australia.
FURTHER READING: Norris (1991); Womersley (2003).

Osmundaria prolifera (Cape Leeuwin, WA).

(top) *Osmundaria prolifera* – as illustrated by Harvey (1862, pl. 188, as *Polyphacum proliferum*).

Palisada

Palisada is another genus that was segregated from *Laurencia* and shares many morphological similarities with that genus. It differs from *Laurencia* in producing two (as opposed to four) pericentral cells per axial cell, and in lacking secondary pit connections between epidermal cells. *Palisada perforata* (formerly known as *Chondrophycus papillosa*) is found in most tropical and subtropical seas and can be recognised by its dense covering of short, stumpy branches. It is similar in appearance to *Laurencia similis*, differing in the absence of secondary pit connections between epidermal cells. In general, *Laurencia similis* is taller, more slender, and often more branched than *Palisada perforata*. *Palisada tumida* is a robust species that occurs near low tide level on moderately rough coasts in south-eastern Australia.

Palisada perforata

TYPE LOCALITY: Santa Cruz de Tenerife, Canary Islands.

DISTRIBUTION: From the Houtman Abrolhos, Western Australia, probably around northern Australia to Jervis Bay, New South Wales. Widespread in tropical seas.

Palisada tumida

TYPE LOCALITY: Robe, South Australia.

DISTRIBUTION: Elliston, South Australia, to San Remo, Victoria, and around Tasmania.

FURTHER READING: Garbary & Harper (1998); Nam & Saito (1991); Womersley (2003, as *Chondrophycus*); Metti et al. (2018).

Palisada perforata (Scott Reef, WA).
Palisada tumida (Point Lonsdale, Vic).

Periphykon

Periphykon forms small blade-like thalli that are typically prostrate on the surfaces of other algae, in Australia most commonly on species of the calcified green alga *Halimeda*. The thalli attach by clusters of rhizoids forming multicellular holdfasts on the ventral surface. Structurally the blades are composed of laterally coherent polysiphonous axes, these initially with each axial cell ringed by four pericentral cells, of which the dorsal two divide once longitudinally, then twice transversely to form eight small pericentral cells from each original cell. *Periphykon beckerae* is the only species of the genus known in Australia.

Periphykon beckerae

TYPE LOCALITY: Nusa Kembangan [Kambangan], Java, Indonesia.
DISTRIBUTION: Barrow Island and the Montebello Islands, north-western Australia. Indonesia.
FURTHER READING: Huisman (2018).

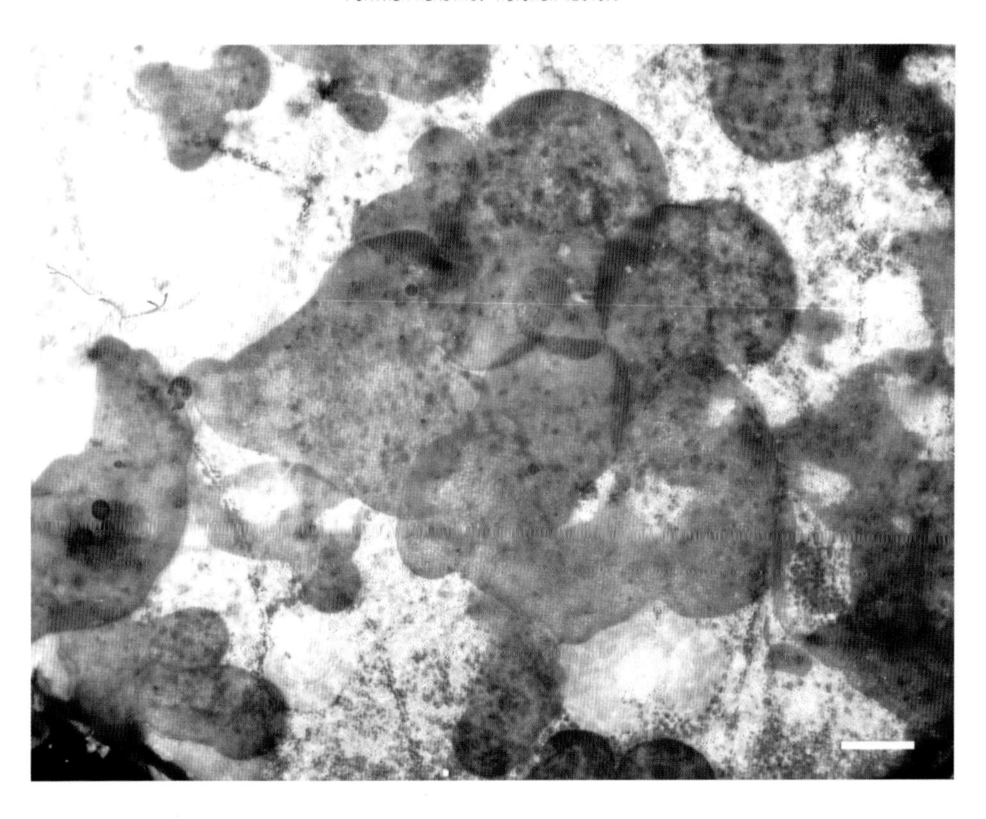

Periphykon beckerae – growing on *Halimeda discoidea* (Barrow Island, WA). Scale = 1 mm.

Pollexfenia

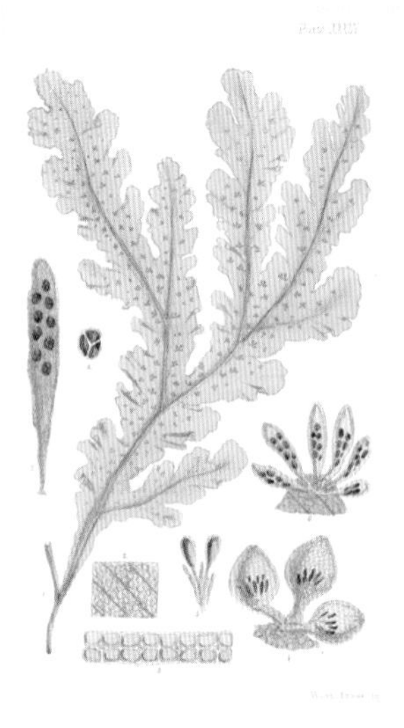

The genus *Pollexfenia* includes several species wholly restricted to southern Australia. Plants are erect, membranous, and complanately branched, with lobed, rounded lateral branches. Fine veins can be seen in lateral branches. Structurally the thallus is multiaxial with the axes congenitally fusing laterally to form the flattened branches. The primary axes are polysiphonous, with each axial cell initially cutting off four pericentral cells that eventually divide transversely several times. The thallus is mostly two cells thick, but in some species has a well-defined thickened midrib. Trichoblasts are formed on the surface and margins of the thallus and are associated with reproductive structures in fertile specimens. *Pollexfenia lobata* can be common and is often found as an epiphyte on the seagrass *Amphibolis*.

Pollexfenia lobata

TYPE LOCALITY: Port Arthur, Tasmania.
DISTRIBUTION: From the Houtman Abrolhos, Western Australia, to Walkerville, Victoria, and around Tasmania.
FURTHER READING: Womersley (2003); Scott (2017).

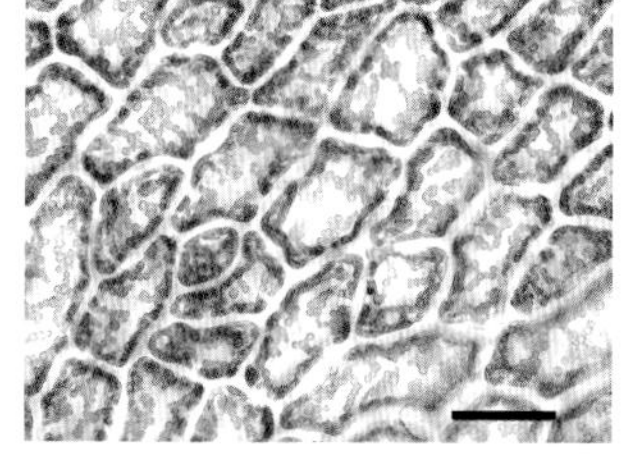

Pollexfenia lobata – epiphytic on *Amphibolis antarctica* (Cape Peron, WA).

(top) *Pollexfenia lobata* – as illustrated by Harvey (1858, pl. 33, as *Jeannerettia lobata*).

(above) *Pollexfenia lobata* – surface view of thallus (Cape Peron, WA). Scale = 50 µm.

Polysiphonia

Species of *Polysiphonia* are common in most seas and grow in a variety of situations, from epiphytic on seagrasses to epilithic on intertidal rocky shores. Thalli are usually profusely branched, with axes composed of a central filament whose cells are surrounded by a ring of cells of similar length. These are known as pericentral cells and are found in some form in all members of the families Rhodomelaceae, Delesseriaceae and Dasyaceae. In *Polysiphonia*, the number of pericentral cells can vary from four to many, depending on the species. Branch apices in most species bear colourless trichoblasts. These are deciduous and are generally not found in lower branches. Of the species included here, *Polysiphonia decipiens* has a relatively large thallus that is often a dark brown colour and epiphytic on seagrasses and other algae. *Polysiphonia sertularioides* is a common species. Many other species are likely to be encountered, however, and more detailed studies should be consulted if an accurate identification is required.

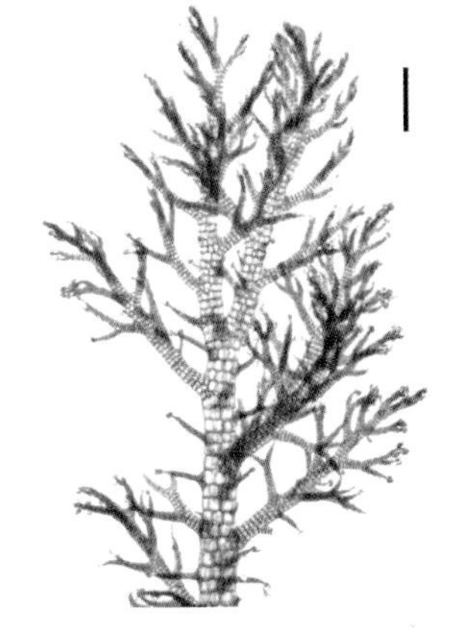

Polysiphonia decipiens

TYPE LOCALITY: Uncertain, possibly Otago or Akaroa Harbour, South Island, New Zealand.

DISTRIBUTION: Shark Bay and Geraldton, Western Australia, around southern Australia and Tasmania to Wilsons Promontory, Victoria, and to Newcastle, New South Wales. New Zealand. Tierra del Fuego.

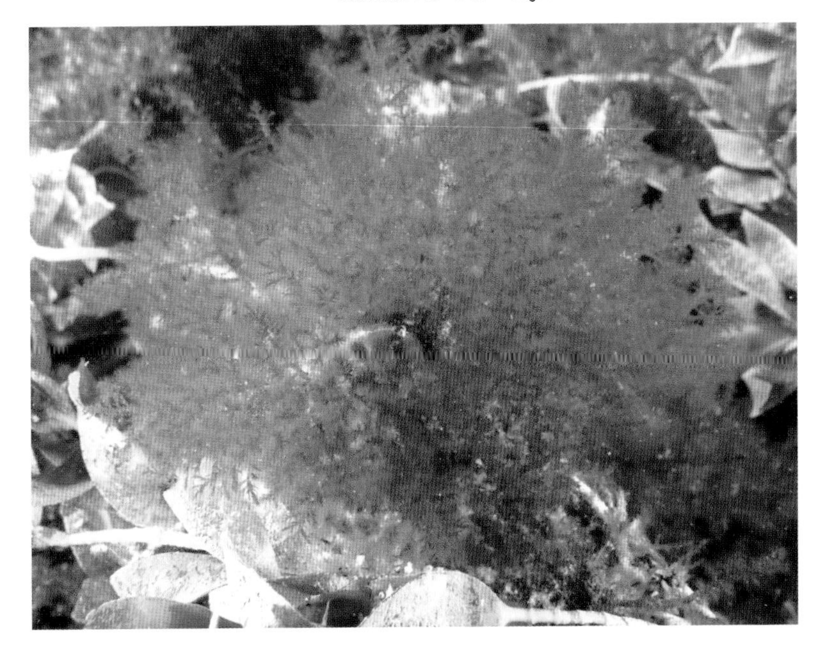

Polysiphonia decipiens (Cape Peron, WA).

(top) *Polysiphonia decipiens* – detail of thallus (Cape Peron, WA). Scale = 1 mm.

Polysiphonia beltoniorum

TYPE LOCALITY: Montgomery Reef, Kimberley, Western Australia.
DISTRIBUTION: Known only from Montgomery Reef and King and Conway Islands.

Polysiphonia sertularioides

TYPE LOCALITY: Cette, Golfe Du Lion, France.
DISTRIBUTION: Widespread in most seas.
FURTHER READING: Womersley (1979), (2003); Nelson (2013); Huisman (2018).

Polysiphonia beltoniorum (King and Conway Islands, WA).

Polysiphonia sertularioides (Cape Peron, WA).

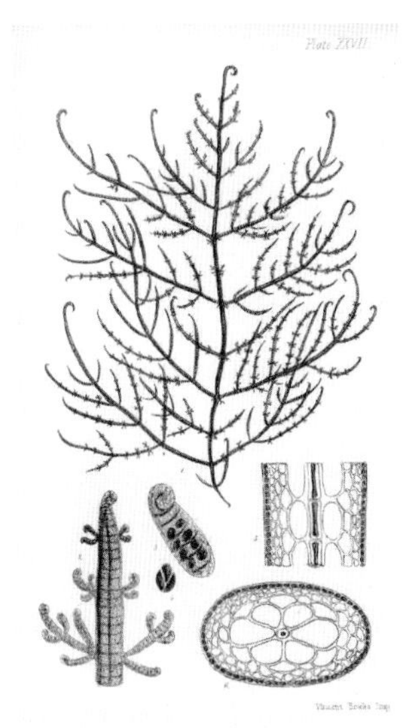

Protokuetzingia

Protokuetzingia includes the single species *Protokuetzingia australasica*, which can be common in shallow water on calm to rough-water coasts in southern Australia. Plants grow to about 20 cm tall, with terete to slightly compressed axes that can be regularly oppositely or irregularly alternately branched, but with regularly spaced opposite short, determinate branchlets. Branch apices are curved. Structurally the thallus is uniaxial and polysiphonous, with each axial cell ringed by six pericentral cells. Reproductive structures are borne on the determinate branchlets.

Protokuetzingia australasica

TYPE LOCALITY: Storm Bay, Tasmania.
DISTRIBUTION: From Port Denison, Western Australia, to San Remo, Victoria, and around Tasmania.
FURTHER READING: Womersley et al. (2003a); Phillips & De Clerck (2005).

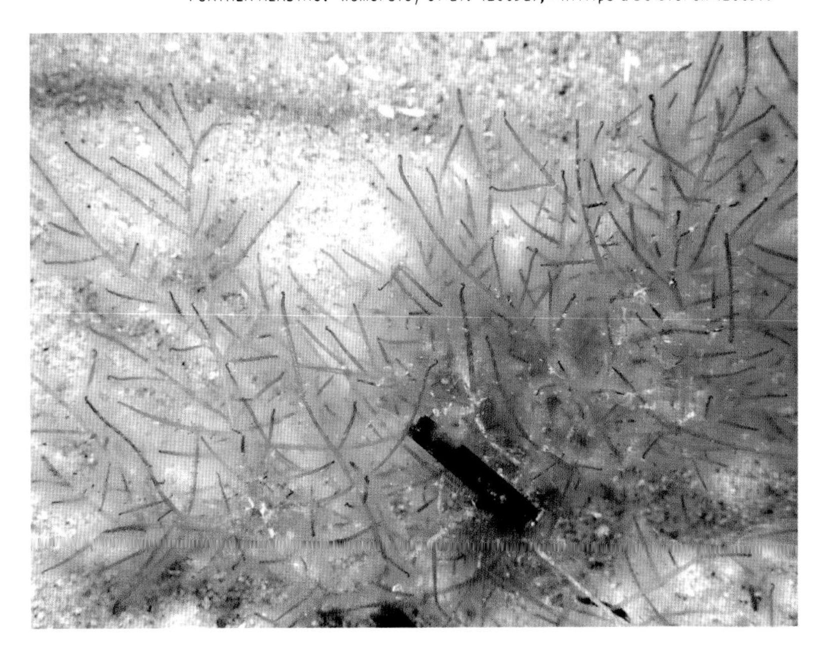

Protokuetzingia australasica (Busselton Jetty, WA).

(top) *Protokuetzingia australasica* – as illustrated by Harvey (1858, pl. 27, as *Rytiphloea australasica*).

Spirocladia

Three species of *Spirocladia* are known in Australia, primarily from warmer waters, of which *Spirocladia barodensis* is included here. Thalli are profusely branched and the terete axes are clothed with pigmented, filamentous lateral branches. Structurally the axes are uniaxial, with each axial cell bearing four pericentral cells, the latter then lightly corticated by descending rhizoidal filaments. Tetrasporangia and spermatangia are borne in specialised branches (known as stichidia) that in *Spirocladia* arise on monosiphonous stalks. A distinguishing feature of *Spirocladia* is the presence of sterile filaments arising laterally from the stichidia. This feature serves to separate the genus from the superficially similar *Lophocladia kuetzingii* (see p. 243).

Spirocladia barodensis

TYPE LOCALITY: Baroda, India.
DISTRIBUTION: From the Houtman Abrolhos, Western Australia, around northern Australia to Jervis Bay, New South Wales. India.
FURTHER READING: Millar (1990); Huisman (2018).

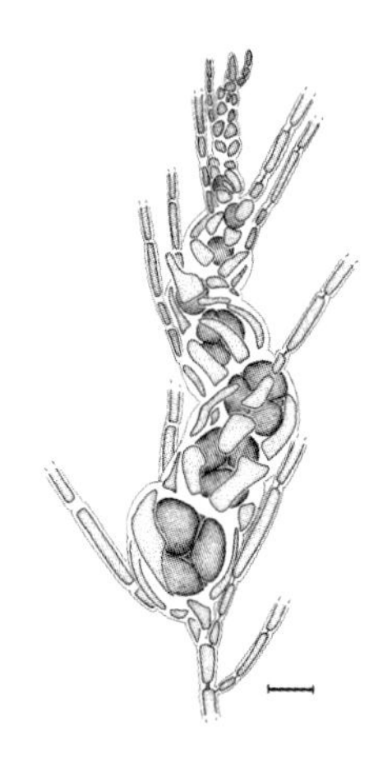

Spirocladia barodensis (Houtman Abrolhos, WA).

(above) *Spirocladia barodensis* – detail of tetrasporangial stichidium with lateral filaments (Houtman Abrolhos, WA). Scale = 30 µm.

Symphyocladia

Symphyocladia includes small prostrate or sprawling plants that have ribbon-like branches generally with irregular marginal lobes. The unusual appearance is due to the growth pattern in which the lateral branches remain congenitally attached. Structurally the thallus is mostly two cells thick and can have branched trichoblasts arising from the margins. Tetrasporangia are formed in straight rows, one per segment, in stichidia that are also laterally fused. The plant in the photograph was growing on shell debris in a shaded sandy habitat, but epiphytic plants are also known.

Symphyocladia marchantioides

TYPE LOCALITY: Cape Kidnappers, Hawkes Bay, New Zealand.
DISTRIBUTION: Houtman Abrolhos, Western Australia, around southern and eastern Australia to Gladstone, Queensland. Widespread in most seas.
FURTHER READING: Womersley (2003).

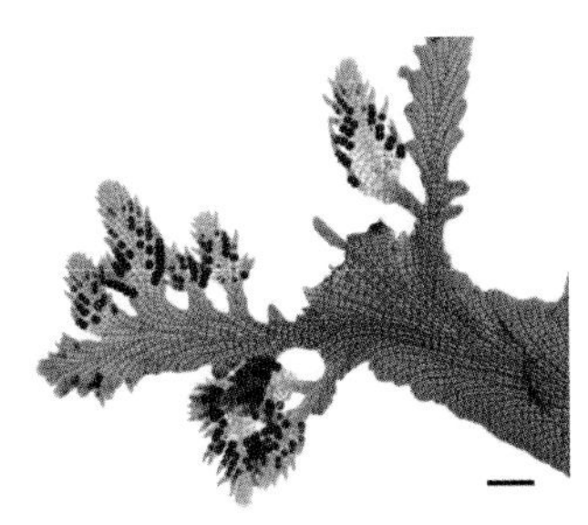

Symphyocladia marchantioides (Cape Peron, WA).

(above) *Symphyocladia marchantioides* – detail of branches with tetrasporangia (Cape Peron, WA). Scale = 500 µm.

Tolypiocladia

Tolypiocladia is a widespread genus with two species recorded from Australia, of which *Tolypiocladia glomerulata* is commonly encountered in warmer waters. Thalli are profusely branched, with a dense covering of short lateral branches that almost obscures the main axes. Structurally the thallus is uniaxial with each axial cell surrounded by a ring of four pericentral cells. The short lateral branches arise from each segment and are themselves profusely branched in their upper regions, with spinous apices. No secondary cortication occurs. In *Tolypiocladia calodictyon*, the lateral spinous branches coalesce to form a loose outer layer, but these remain free in *Tolypiocladia glomerulata*.

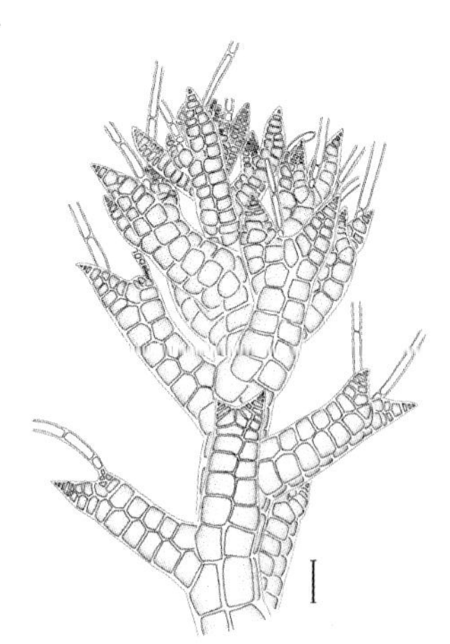

Tolypiocladia calodictyon

TYPE LOCALITY: Tonga.
DISTRIBUTION: Widespread in the Indian and western Pacific Oceans.

Tolypiocladia glomerulata

TYPE LOCALITY: Shark Bay, Western Australia.
DISTRIBUTION: From Rottnest Island, Western Australia, around northern Australia to Lord Howe Island, New South Wales. Warmer waters of the Indo-Pacific.
FURTHER READING: Cribb (1983); Price & Scott (1992); Wynne (1999); Huisman (2018).

Tolypiocladia calodictyon (Barrow Island, WA).

Tolypiocladia glomerulata (Houtman Abrolhos, WA).

(top) *Tolypiocladia glomerulata* – detail of apex (Houtman Abrolhos, WA). Scale = 100 µm.

Trichidium

One of several red algal parasites that occur in Australia, *Trichidium* includes the single species *Trichidium pedicellatum*, which was first discovered in 1971 at Port Denison in Western Australia but then not seen again until 2019, when a population was found at Cape Peron. This population has subsequently also seemingly disappeared, despite the host *Lophocladia kuetzingii* being commonly found. The species occurs as small white tufts in the branch axils of the host, with erect radiating branches from a basal hemispherical cushion. As with many red algal parasites, *Trichidium* is closely related to its host *Lophocladia*.

Trichidium pedicellatum

TYPE LOCALITY: Port Denison, Western Australia
DISTRIBUTION: Known only from Cape Peron and Port Denison, Western Australia.
FURTHER READING: Noble & Kraft (1983); Huisman (2019).

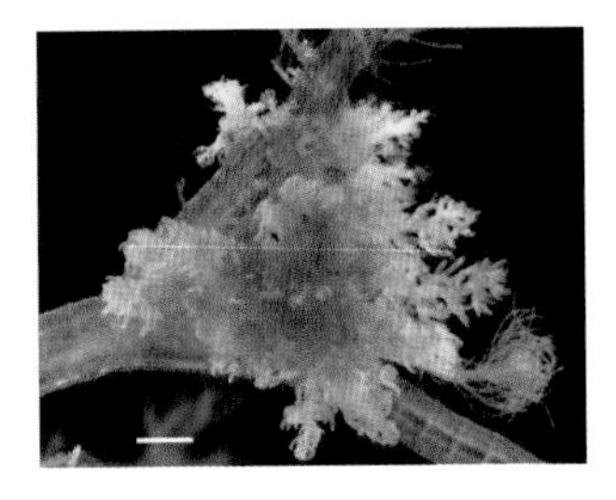

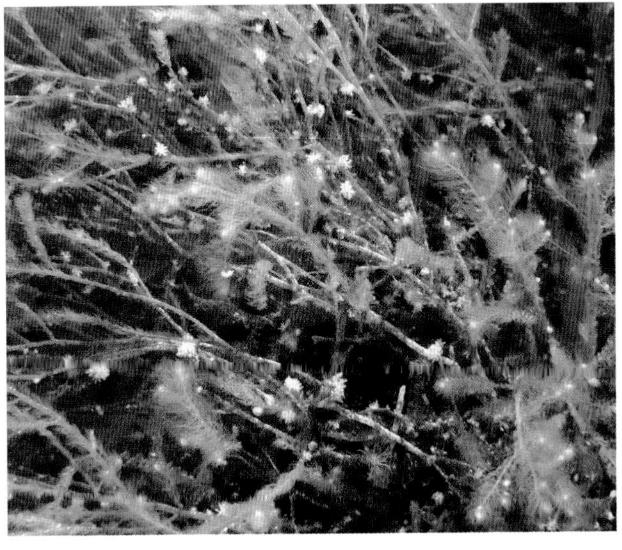

Trichidium pedicellatum - visible as small white tufts growing on *Lophocladia kuetzingii* (Cape Peron, WA).

(above) *Trichidium pedicellatum* - detail of thallus (Cape Peron, WA). Scale = 500 µm.

Vertebrata

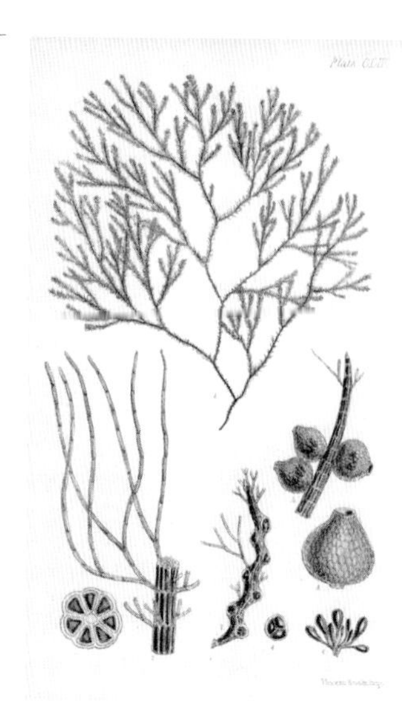

Vertebrata was recently resurrected following DNA sequences analyses to accommodate several species previously included in *Polysiphonia* and *Brongniartella* (and other genera). The only morphological feature unique to species of *Vertebrata* is the presence of colourless trichoblasts found at the apices of branches with multinucleate cells, those of other genera having uninucleate cells. *Vertebrata australis* is a common species on rough-water to sheltered coasts in southern Australia, ranging from low tide level to deep water. Plants are dark red to purple and can be up to 50 cm tall. The thallus is irregularly branched and is structurally polysiphonous, with the central axis ringed by seven pericentral cells and a dense covering of pigmented trichoblasts, although these can be shed from older axes.

Vertebrata australis

TYPE LOCALITY: Geographe Bay, Cape Leeuwin, or King George Sound, Western Australia.

DISTRIBUTION: From Whitfords Beach, Western Australia, around southern Australia to Bemm Reef, Victoria, and around Tasmania. New Zealand, from Wellington south to Stewart Island.

FURTHER READING: Parsons (1980, as *Brongniartella*); Womersley & Parsons (2003, as *Brongniartella*); Díaz-Tapia et al. (2017).

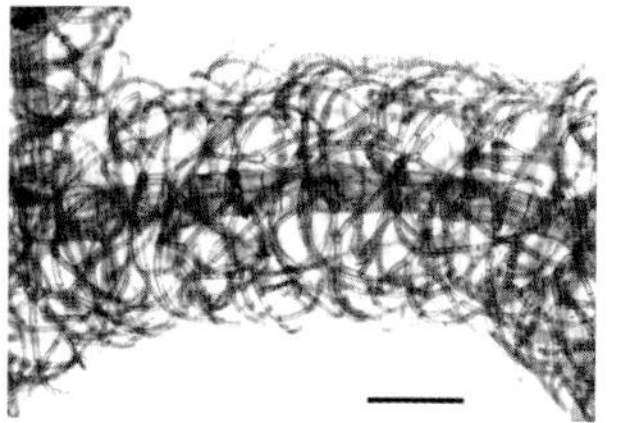

Vertebrata australis – epiphytic on *Sargassum* (Cape Peron, WA).

(top) *Vertebrata australis* – as illustrated by Harvey (1860, pl. 154, as *Polysiphonia cladostephus*).

(above, middle) *Vertebrata australis* (Point Roadknight, Vic).

(above) *Vertebrata australis* – detail of thallus (Cape Peron, WA). Scale = 2 mm.

Vidalia

Vidalia includes three species in Australia and is represented here by *Vidalia spiralis*, which, as the name suggests, has regularly spiralled branches; it is readily recognised. It can be common in shallow but shaded habitats on the western and southern coasts. The thallus has flattened branches with a central midrib and lateral wings with serrated margins. Structurally the genus is uniaxial, with each axial cell surrounded initially by five pericentral cells, those on the dorsal surface cutting off two pseudopericentral cells in the mature axis. The wings are formed by outgrowths of the lateral pericentral cells and the entire thallus is secondarily corticated. The serrations at the thallus margins are formed alternately, as are faint veins (which correspond to lateral axes of limited growth) that can be seen running from the midrib to the margins. Species of *Vidalia* have been included in *Osmundaria*, but recent studies have recognised the two as independent genera.

Vidalia spiralis

TYPE LOCALITY: Western Australia.
DISTRIBUTION: From the Houtman Abrolhos, Western Australia, to Port Phillip Heads, Victoria.
FURTHER READING: Womersley (2003).

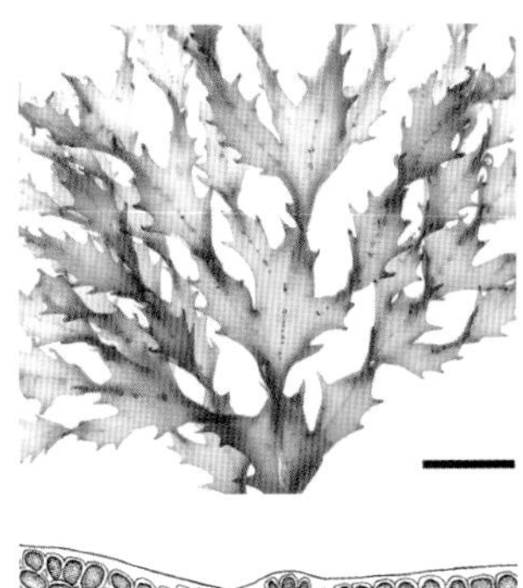

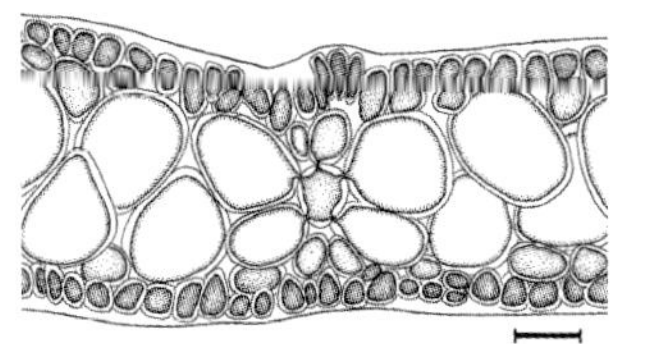

Vidalia spiralis (Cape Peron, WA).

(top) *Vidalia spiralis* – detail of thallus (Cape Peron, WA). Scale = 1 cm.

(above) *Vidalia spiralis* – section of thallus (Eglinton Rocks, WA). Scale = 50 µm.

Dasya

A relatively common genus, recognisable by its corticated main axes bearing numerous filaments that give the thallus a hairy appearance. Structurally the thallus is uniaxial, with the central axis ringed by five pericentral cells, although these are not always obvious in sections. A distinctive feature of *Dasya* and the Dasyaceae is the occurrence of sympodial growth, a process in which the apical cell regularly relinquishes its position to the lateral initial on the subapical cell, which then becomes the new apical cell. This pattern can be difficult to observe, but can be the only way to separate *Dasya* from other superficially similar genera. Fortunately, the process results in a distinctive pattern at the junction of cells in the filaments – the basal region of the distal cells remain laterally attached, such that their walls (and that of the bearing cell) form a 'Y' pattern. This feature allows ready recognition of the Dasyaceae. Cystocarps in *Dasya* are urn-shaped and often borne on a short stalk. Spermatangia and tetrasporangia occur in modified branches known as stichidia. Numerous species of *Dasya* occur in Australia, distinguished by features of their vegetative and reproductive structures.

Dasya frutescens

TYPE LOCALITY: Rottnest Island, Western Australia.
DISTRIBUTION: Safety Bay (near Perth), north to the Maret Islands, Western Australia.

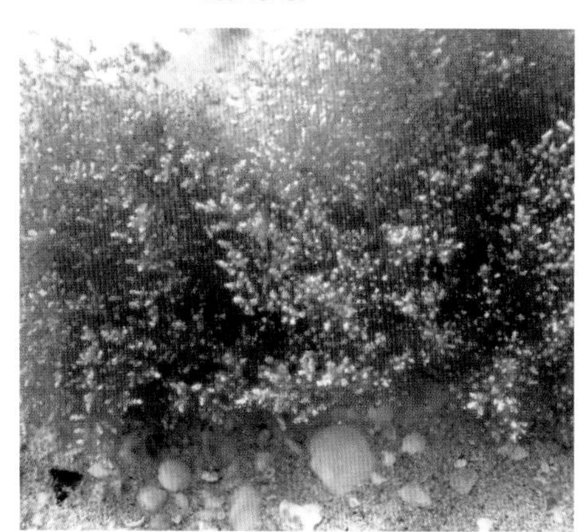

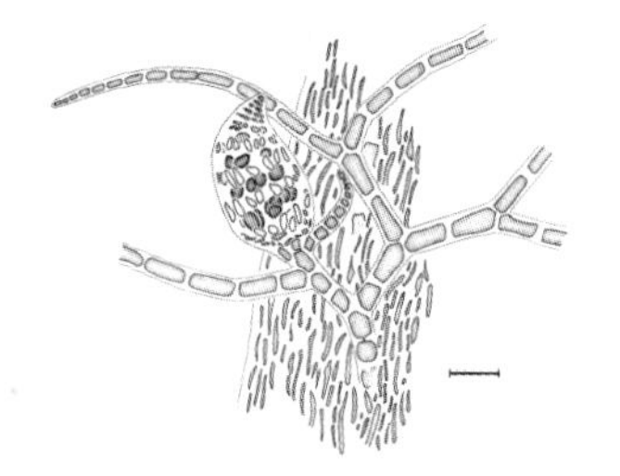

Dasya villosa

TYPE LOCALITY: Georgetown, Tasmania.
DISTRIBUTION: From the Perth region, Western Australia, around southern Australia to Western Port, Victoria, and the north coast of Tasmania.
FURTHER READING: Parsons (1975); Parsons & Womersley (1998); Huisman (2018).

Dasya frutescens (Woodman Point, WA).

Dasya villosa (Cape Peron, WA).

(top) *Dasya villosa* – as illustrated by Harvey (1847–1849, pl. 20).

(left) *Dasya villosa* – portion of thallus with a tetrasporangial stichidium. Note the 'Y' junctions at dichotomies in the filament (Rottnest Island, WA). Scale = 100 µm.

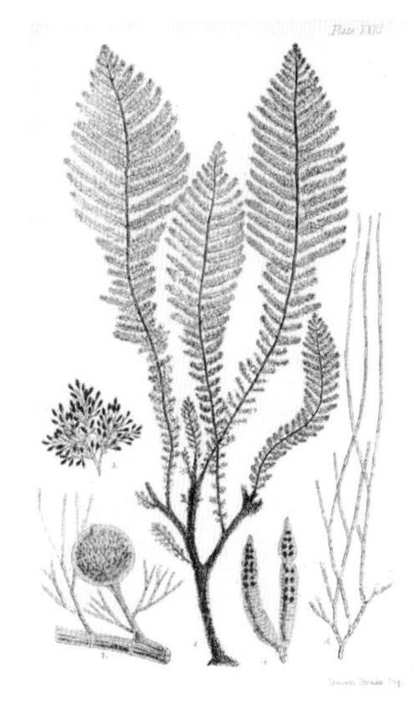

Heterosiphonia

Heterosiphonia has upright thalli with the main axis ringed by varying numbers of pericentral cells, depending on the species, and often secondarily corticated. It is somewhat similar to *Dasya* (p. 260), as the lateral branches are also produced in a sympodial manner and tetrasporangia are produced in stichidia. *Heterosiphonia* differs, however, in its lateral branches arising in an alternate-distichous pattern and being separated by internodes of between two and nine cells. This pattern is well displayed in *Heterosiphonia muelleri*, one of several species found in Australia. In *Heterosiphonia muelleri*, seven or eight pericentral cells are formed and the lateral branches are separated by two-celled internodes. The species is readily recognised by its feather-like upper axes and densely corticated lower axes. Another heavily corticated species is *Heterosiphonia lawrenciana*, in which the laterals are separated by 4-7- celled internodes. *Heterosiphonia callithamnium* is a smaller species that commonly grows as an epiphyte on the seagrass *Amphibolis*.

Heterosiphonia callithamnium

TYPE LOCALITY: South-west Western Australia.
DISTRIBUTION: Champion Bay, Western Australia, to Kangaroo Island, South Australia.

Heterosiphonia callithamnium (Cape Peron, WA).

(top) *Heterosiphonia muelleri* – as illustrated by Harvey (1858, pl. 31, as *Dasya muelleri*).

Heterosiphonia crassipes

TYPE LOCALITY: Jetty Reef, Rottnest Island, Western Australia.
DISTRIBUTION: Known from the Indian Ocean and several localities in eastern Australia, including Coffs Harbour, New South Wales, and Lord Howe Island.

Heterosiphonia muelleri

TYPE LOCALITY: Port Phillip, Victoria.
DISTRIBUTION: From Lancelin, Western Australia, around southern mainland Australia to Walkerville, Victoria, and the north coast of Tasmania.

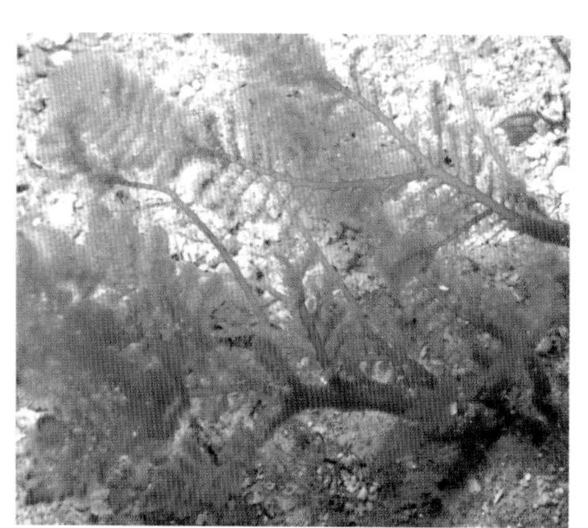

Heterosiphonia lawrenciana

TYPE LOCALITY: Georgetown, Tasmania.
DISTRIBUTION: Cottesloe, Western Australia, to Port Phillip Heads, Victoria, and northern Tasmania.
FURTHER READING: Parsons (1975); Parsons & Womersley (1998).

Heterosiphonia crassipes (Cape Peron, WA).

Heterosiphonia muelleri (Rottnest Island, WA).

Heterosiphonia lawrenciana (Cape Peron, WA).

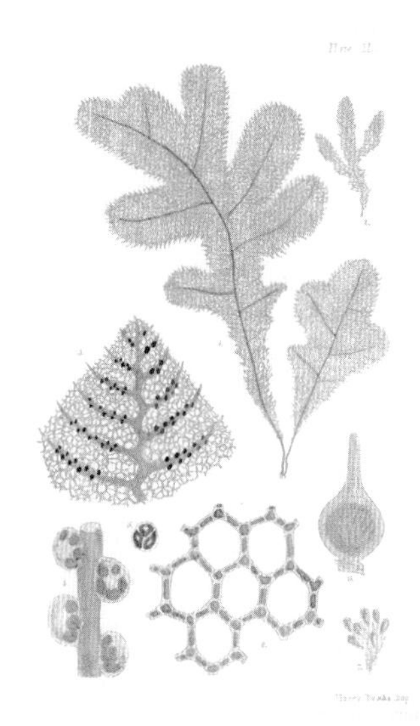

Thuretia

Thuretia includes two species in Australia, with *Thuretia quercifolia* the most commonly encountered. Plants are erect, with terete axes producing lateral branches that link together to form a flat network. Cystocarps are urn-shaped, usually with a prominent neck, and tetrasporangia occur in stichidia that develop from the lateral branches. *Thuretia quercifolia* is a distinctive species on southern rough-water coasts, usually occurring in deeper water. *Thuretia australasica*, the second Australian species, has a terete or only slightly compressed network (not illustrated).

Thuretia quercifolia

TYPE LOCALITY: Australia.
DISTRIBUTION: Dongara, Western Australia, to Walkerville, Victoria, and northern Tasmania.
FURTHER READING: Parsons (1975); Parsons & Womersley (1998).

Thuretia quercifolia (Augusta, WA).

(top) *Thuretia quercifolia* – as illustrated by Harvey (1858, pl. 40).

Division Ochrophyta

THE BROWN ALGAE

Canistrocarpus cervicornis – detail of thallus. Scale = 1 mm.

The Ochrophyta (from the Greek *ochros* meaning 'yellow' and *phyton* meaning 'plant') includes the most conspicuous algae in marine habitats. The shallow reef systems of southern Australia are regularly dominated by the kelp *Ecklonia radiata* or any number of species of the order Fucales (including the genera *Sargassum* and *Cystophora*). The brown algae contain the green pigments chlorophylls *a* and *c*, but these are masked by the brown accessory pigment fucoxanthin. As in the green algae, a motile unicellular stage is present at some stage of the life history – those stages of the brown algae, however, are 'heterokont', which means that the two flagella present on each motile cell differ in structure from one another. The brown algae are almost wholly restricted to the marine environment and include around 1,500 to 2,000 species worldwide, of which approximately 350 occur in Australian seas.

The large size and colouration of the brown algae are often sufficient to allow easy recognition. Perhaps the only colour variations (other than shades of brown) are found in the iridescent members of the order Dictyotales, which often appear blue underwater.

Feldmannia

Feldmannia is one of several genera of small, tufted brown algae that are commonly found as epiphytes on other algae and seagrasses and also growing on rocks. The thallus is composed of profusely branched filaments that are only a single cell wide and often taper into a hair-like structure at the apices. Growth is via an intercalary meristem that is usually situated at the base of the branches and can be recognised as a cluster of smaller cells. Reproductive structures are borne directly on the branches. *Feldmannia mitchelliae* has previously been included in a variety of genera, including *Giffordia* and *Hincksia*. *Feldmannia indica* is similar in appearance, but is less profusely branched.

Feldmannia mitchelliae

TYPE LOCALITY: Nantucket, Massachusetts, USA.
DISTRIBUTION: Throughout Australia. Widespread in temperate and subtropical seas.

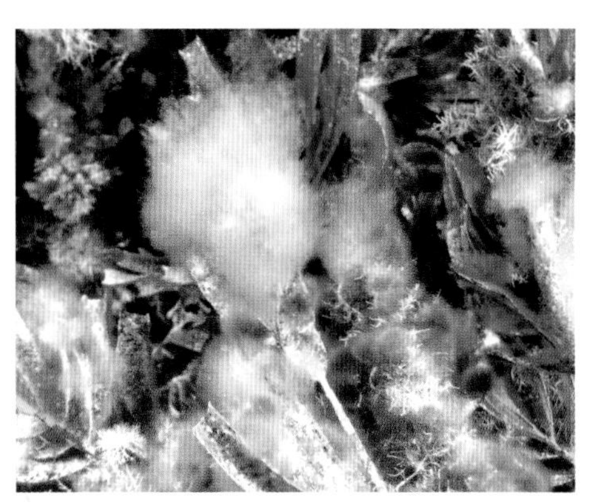

Feldmannia mitchelliae – growing on *Amphibolis antarctica*. The darker brown tufts are *Sphacelaria rigidula* (Cape Peron, WA).

Feldmannia indica

TYPE LOCALITY: Bima Bay, Sumbawa, Indonesia.
DISTRIBUTION: Widespread in tropical seas.
FURTHER READING: Womersley (1987, as *Giffordia*); Huisman (2015).

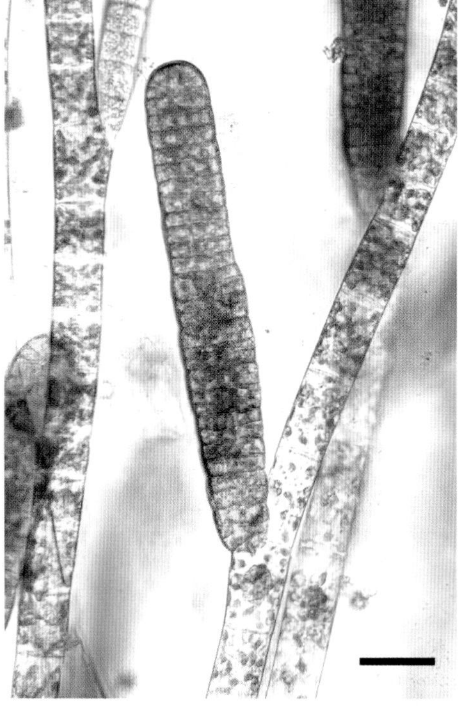

Feldmannia indica – detail of reproductive structures (plurangia) (Heron Island, Qld). Scale = 40 µm.

Asperococcus

Asperococcus has undivided, saccate or tubular thalli that are generally covered in numerous colourless hairs. The cellular layer is three to four cells thick, with a small-celled cortex and a larger-celled medulla. Reproductive structures are borne in clusters on the surface of the cortex. *Asperococcus bullosus* is a common epiphyte on the seagrasses *Posidonia* (as in the photograph) and *Amphibolis*. It is usually found in sheltered areas.

Asperococcus bullosus

TYPE LOCALITY: Mediterranean coast of France.
DISTRIBUTION: From the Houtman Abrolhos, Western Australia, around southern mainland Australia and the north and east coasts of Tasmania, to Port Stephens, New South Wales. Widely distributed in temperate seas.
FURTHER READING: Womersley (1987); Scott (2017).

Asperococcus bullosus (Houtman Abrolhos, WA).

Cladosiphon

Thalli of *Cladosiphon* have cylindrical branches that are flaccid and mucilaginous. Structurally they have a filamentous medulla that can become hollow as it matures. Towards the cortex, the medullary filaments become more closely aligned. The cortex is composed of unbranched, pigmented filaments that arise perpendicular to the medullary filaments and have curved apices. Colourless hairs are present and project above the surface of the cortex. Reproductive structures are borne either on the outer cells of cortical filaments (plurangia) or on subcortical cells (unangia). *Cladosiphon filum* is a common epiphyte on the seagrasses *Amphibolis* and *Posidonia* in shallow waters.

Cladosiphon filum

TYPE LOCALITY: King George Sound, Western Australia.
DISTRIBUTION: From the Houtman Abrolhos, Western Australia, around southern Australia to Nowra, New South Wales.
FURTHER READING: Womersley (1987).

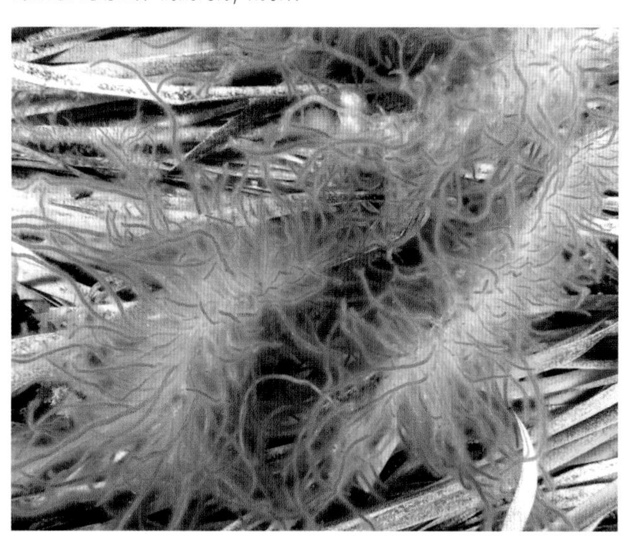

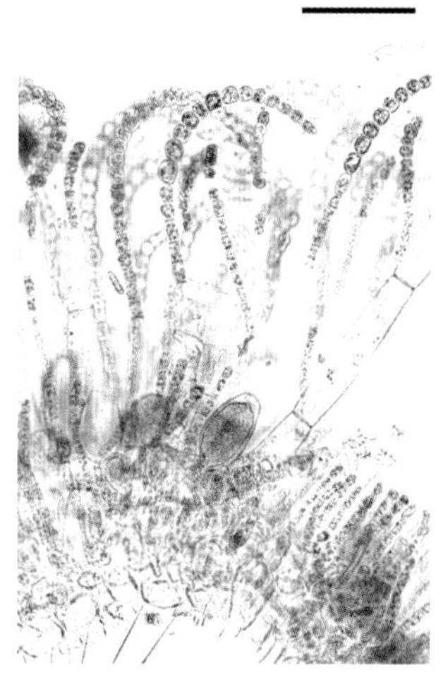

Cladosiphon filum (Cape Peron, WA).

(above) *Cladosiphon filum* – detail of cortex (Cape Peron, WA). Scale = 100 µm.

Elachista

Plants of *Elachista nigra* occur as small tufts to 2 cm high, in Australia growing epiphytically on the kelp *Ecklonia radiata*. Erect branches grow from a basal layer that is firmly attached to the host and consists of an unpigmented lower portion of inflated cells and an upper portion of pigmented barrel-shaped cells. Both plurangia and obovate unangia arise within the cortex from the outer medullary cells. Surface hairs are absent. *Elachista nigra* is possibly an introduced species, as there are no records of it in Australia prior to 1976. In its native range, it grows on *Undaria pinnatifida*.

Elachista nigra

TYPE LOCALITY: North-east Honshu, Japan.
DISTRIBUTION: In southern Australia, known from Rottnest Island, Western Australia, from Port Noarlunga to Port Elliot, South Australia, and from Garie Beach, New South Wales.
FURTHER READING: Womersley (1987); Uwai et al. (2002).

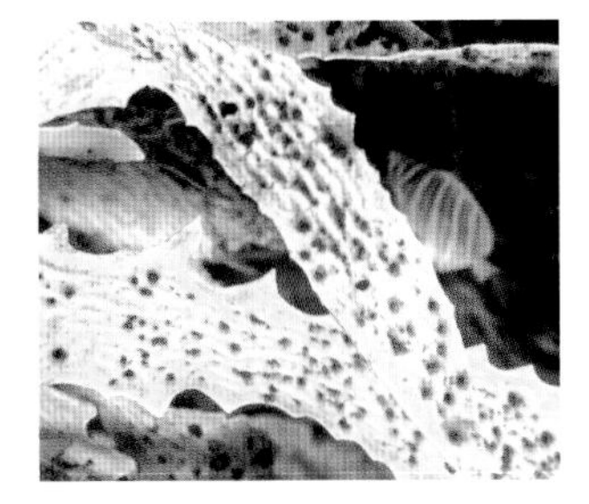

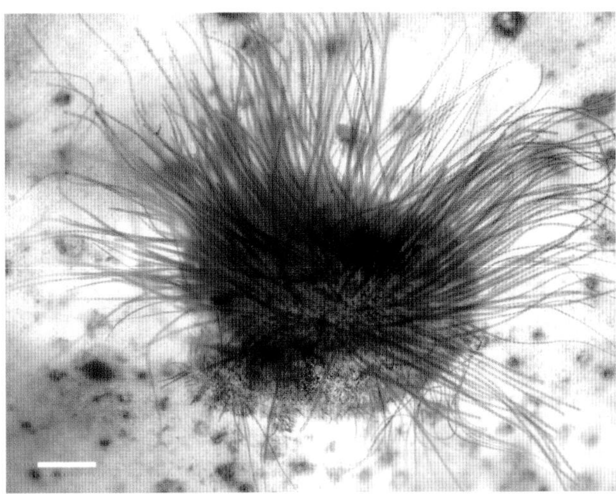

Elachista nigra – detail of thallus (Cape Peron, WA). Scale = 1 mm.

(above) *Elachista nigra* – growing on *Ecklonia radiata* (Cape Peron, WA).

Myriogloea

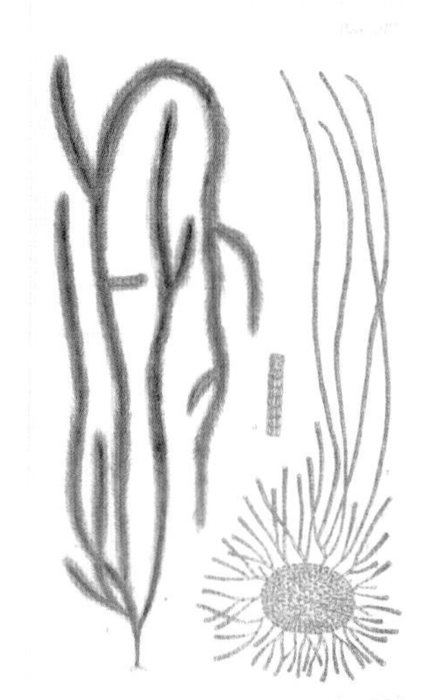

Thalli of *Myriogloea* are worm-like and very mucilaginous, with branched or unbranched (depending on the level of exposure) terete axes that can grow to over a metre in length. The genus is represented in Australia by *Myriogloea sciurus*, which can be common in the low intertidal of rough-water coasts. Structurally the thallus has a medulla of primarily longitudinally oriented filaments that give rise to a subcortex of radially oriented, branched, non-pigmented filaments. These then produce a cortex with a mix of short and long, unbranched, straight pigmented filaments. Hairs are absent, a feature that distinguishes *Myriogloea* from most similar looking genera with the exception of *Papenfussiella* (see p. 273), which differs in having only a slight subcortex, if any. The life history of *Myriogloea* includes a filamentous gametophyte stage and, in Australia, the sporophytes of *Myriogloea sciurus* mostly occur as a summer annual.

Myriogloea sciurus

TYPE LOCALITY: Port Fairy, Victoria.
DISTRIBUTION: From Point Drummond, South Australia, to Newcastle, New South Wales, and around Tasmania.
FURTHER READING: Womersley & Bailey (1987).

Myriogloea sciurus (Cressy Beach, Tas; photo M.D. Guiry).

(top) *Myriogloea sciurus* – as illustrated by Harvey (1858, pl. 58, as *Myriocladia sciurus*).

Papenfussiella

Papenfussiella includes some 10 species worldwide, mostly found in colder seas. Plants are medium to dark brown and cord-shaped, up to 50 cm tall, and branched with very long laterals. They have a layer of surface filaments and look hairy as a result, but are also very mucilaginous and slimy to touch. Two species are found in southern Australia, *Papenfussiella lutea* known from the east coast of Tasmania, and *Papenfussiella extensa* from Cape Peron in Western Australia. The two species are separated mainly on robustness and dimensions of the long surface filaments.

Papenfussiella extensa

TYPE LOCALITY: Point Peron, Western Australia.
DISTRIBUTION: Currently known only from Cape Peron [Point Peron], Western Australia.
FURTHER READING: Womersley & Bailey (1987).

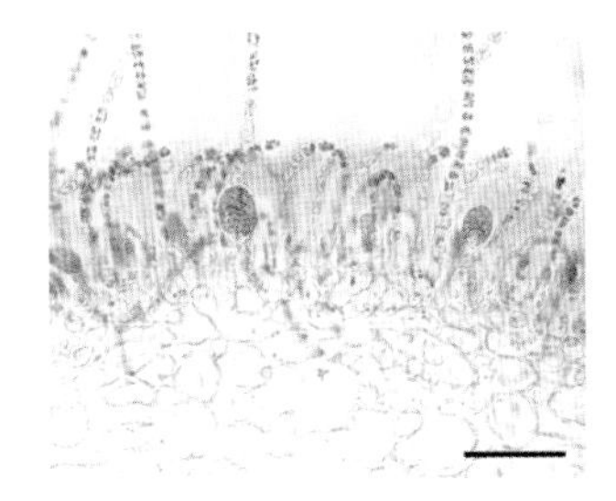

Papenfussiella extensa (Cape Peron, WA).

(above) *Papenfussiella extensa* – section of thallus showing cortex with reproductive structures (Cape Peron, WA). Scale = 50 µm.

Polycerea

Polycerea includes two species of worm-like, mucilaginous brown algae that are endemic to southern and south-western Australia, with *Polycerea zostericola* known from only scattered locations. Plants are typically epiphytic on the seagrasses *Posidonia* and *Amphibolis* and are summer annuals. Branches are cylindrical; structurally they have a multiaxial medulla and a cortex of unbranched, straight assimilatory filaments with an enlarged terminal cell. Hairs are also present in the cortex.

Polycerea zostericola

TYPE LOCALITY: King George Sound, Western Australia.
DISTRIBUTION: From Rottnest Island, Cape Peron and King George Sound, Western Australia.
FURTHER READING: Womersley & Bailey (1987).

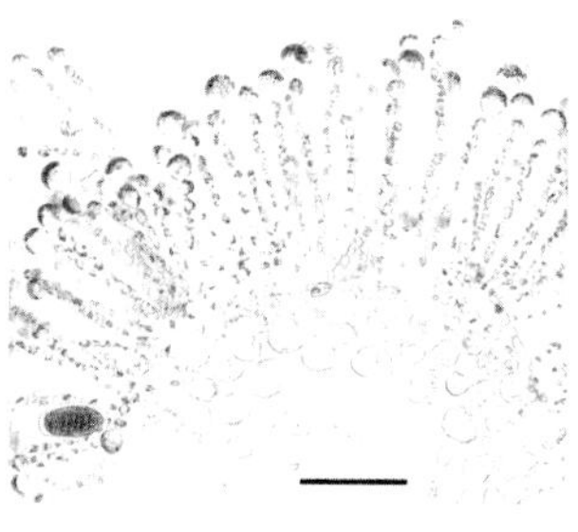

Polycerea zostericola – epiphytic on *Posidonia sinuosa* (Cape Peron, WA).

(above) *Polycerea zostericola* – section of thallus (Cape Peron, WA). Scale = 100 µm.

Tinocladia

Tinocladia is a genus of worm-like, mucilaginous brown algae, represented in Australia by the single species *Tinocladia australis*. The genus is distinguished from similar-looking algae by the presence of an extensive subcortex, which radiates from a medulla of longitudinal filaments and bears groups of cortical assimilatory filaments. *Tinocladia australis* generally grows on sand-covered lower intertidal rock during summer. The thallus can be up to 30 cm in height and has cylindrical branches, usually branched from near the base, although some plants can be profusely branched throughout. A related species, *Tinocladia crassa* (not shown), is eaten in Japan, where it and other similar looking algae are known as *somen-nori*.

Tinocladia australis

TYPE LOCALITY: Georgetown, Tasmania.
DISTRIBUTION: From Sceale Bay, South Australia, to Port Phillip, Victoria, and around Tasmania.
FURTHER READING: Womersley & Bailey (1987).

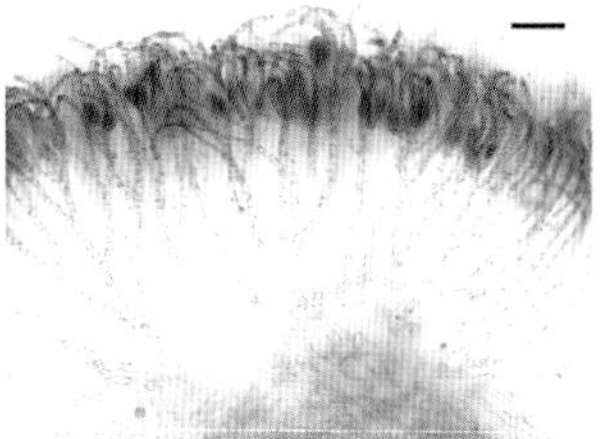

Tinocladia australis (Point Roadknight, Vic).

(above) *Tinocladia australis* – section of thallus (Point Roadknight, Vic). Scale = 100 µm.

Petrospongium

Petrospongium includes three species, of which *Petrospongium rugosum* is found in south-eastern Australia. Plants grow on intertidal rock, where they form red-brown to dark-brown crusts, with broad marginal lobes and a rugose surface. The species was first observed at Point Lonsdale, Victoria, in 1954 and may have been introduced, as it was apparently not present prior to this time. *Petrospongium* has a heteromorphic life history, with the macroscopic sporophyte alternating with a gametophytic microthallus.

Petrospongium rugosum

TYPE LOCALITY: Japan.
DISTRIBUTION: From Cape Marengo, Victoria, to Bilgola, New South Wales. Japan. Pacific coast of North America. New Zealand.
FURTHER READING: Womersley (1987); Racault et al. (2009); Nelson (2013).

Petrospongium rugosum (Point Roadknight, Vic).

Colpomenia

Colpomenia is a common genus in most regions of Australia. Thalli are hemispherical or irregularly lobed and consist of a crisp outer membrane encasing a hollow interior. The membrane is from three to 10 cells thick, with an inner medulla of large colourless cells and an outer cortex of smaller pigmented cells. *Colpomenia* resembles the closely related *Hydroclathrus* (see next page), differing in the lack of perforations in the membranous layer. Several species are known in Australia and require reproductive material to be accurately identified. Reproductive structures are borne in variously shaped patches (depending on the species) on the surface of the thallus, usually associated with clusters of hairs.

Colpomenia peregrina

TYPE LOCALITY: Brittany, France.
DISTRIBUTION: Throughout eastern, western and southern Australia. Widespread in temperate seas.

Colpomenia sinuosa

TYPE LOCALITY: Near Cádiz, Spain
DISTRIBUTION: Widespread in tropical to temperate seas.
FURTHER READING: Womersley (1987); Kraft (2009); Huisman (2015); Scott (2017).

Colpomenia peregrina (Cape Peron, WA).

Colpomenia sinuosa (Heron Island, Qld).

Hydroclathrus

Only six species of *Hydroclathrus* are recognised, of which the common and widely distributed *Hydroclathrus clathratus* is known in Australia. Thalli are globular or irregularly lobed, spreading over the substratum and attaching at several points. Structurally the thallus consists of a hollow interior, bordered by an outer cellular layer, the latter being composed of a medulla of large, clear cells and an outer cortex of smaller, pigmented cells. A distinctive feature of the genus is the presence of numerous holes of various sizes in the cellular layer. Reproductive structures are borne in patches on the surface of the thallus.

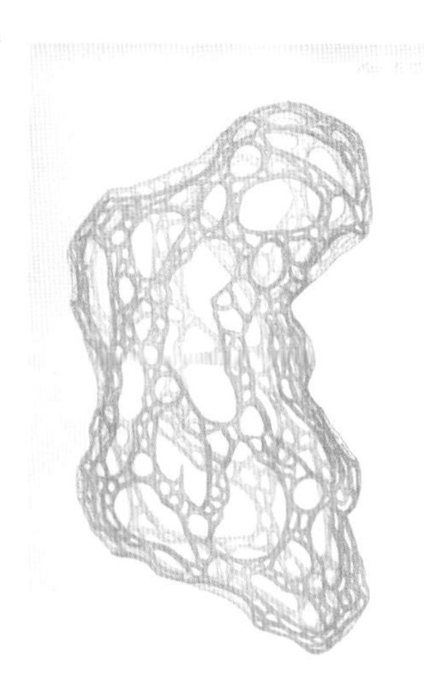

Hydroclathrus clathratus

TYPE LOCALITY: Belle Isle, France.
DISTRIBUTION: Throughout Australia except Tasmania. Widespread in tropical and temperate seas.
FURTHER READING: Womersley (1987); Kraft (2009); Huisman (2015).

Hydroclathrus clathratus (Coral Bay, WA).

(top) *Hydroclathrus clathratus* – as illustrated by Harvey (1859a, pl. 98, as *Hydroclathrus cancellatus*).

Petalonia

Plants of *Petalonia* are golden brown and are almost always found in clusters on intertidal rock. Each plant can grow to about 10 cm in height and is composed of a flattened blade attached by a small disc-shaped holdfast. The blades are shaped a little like an elongate leaf, or slightly curved like a scythe. *Petalonia binghamiae* is locally common during winter and spring on intertidal rocks. The species is also known by the name Habanori, which comes from the Japanese and means 'wide nori', a reference to the more commonly eaten red seaweed, nori, and familiar to most as the dark wrapping around sushi rolls. Habanori is eaten in Asia, where it is collected from the wild and is much prized, sometimes commanding high prices at the markets. Plants are dried and used to flavour soups, or roasted with a little soy sauce and added to boiled rice. *Petalonia fascia* is very similar in appearance to *Petalonia binghamiae*, but differs in having a central medulla that is composed of tightly packed cells, rather than the loosely arranged filaments of the latter.

Petalonia binghamiae

TYPE LOCALITY: Vicinity of Santa Barbara, California, USA.
DISTRIBUTION: Worldwide in temperate and tropical seas.

Petalonia fascia

TYPE LOCALITY: Near Kristiansand, Norway.
DISTRIBUTION: Widespread in temperate and colder seas.
FURTHER READING: Boo (2010).

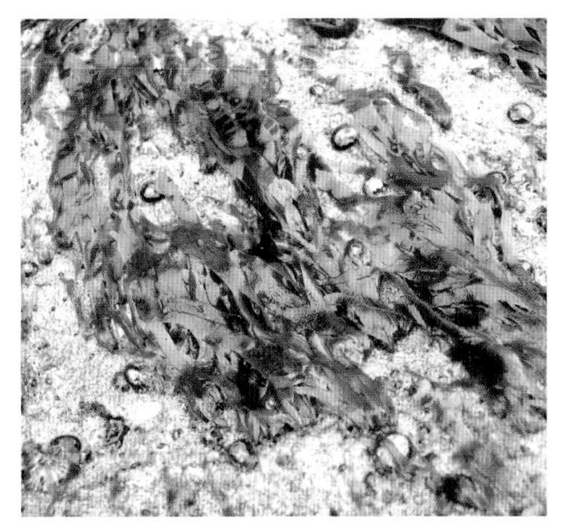

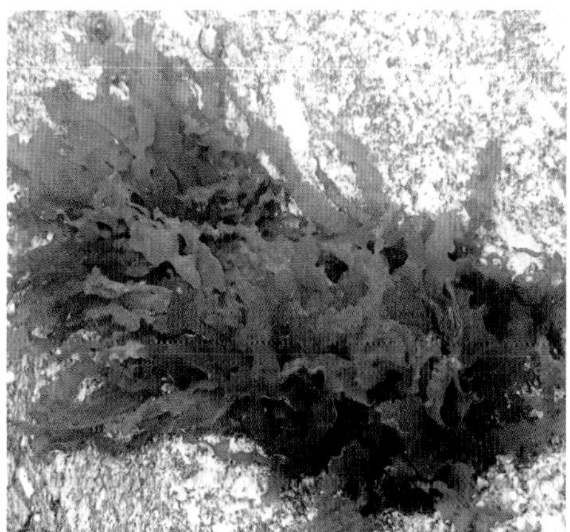

Petalonia binghamiae (Cottesloe, WA).

Petalonia fascia (Albany, WA).

Pseudochnoospora

Pseudochnoospora includes the single species *Pseudochnoospora implexa*, which is often found growing on tropical reef flats in shallow water. Plants have solid branches that are profusely and irregularly divided, often resulting in a hemispherical thallus that can be dislodged from the reef and appear as tumbleweed. Superficially *Pseudochnoospora* is similar in appearance to some species of *Rosenvingea* (p. 281), but that genus has hollow branches whereas those of *Pseudochnoospora* are solid.

Pseudochnoospora implexa

TYPE LOCALITY: Near Tor, Sinai Peninsula, Egypt.
DISTRIBUTION: Widespread in tropical seas.
FURTHER READING: Kraft (2009, as *Chnoospora*); Huisman (2015, as *Chnoospora*); Santiañez et al. (2018).

Pseudochnoospora implexa (Heron Island, Qld).

Rosenvingea

The genus *Rosenvingea* is widespread in tropical to subtropical seas, with four species known from Australia. Thalli of *Rosenvingea* have branched, hollow axes with an inner cavity and an outer cellular layer. Plants of *Rosenvingea australis* form dense, spreading tufts, with numerous anastomoses between branches. In contrast, thalli of *Rosenvingea orientalis* are essentially upright. *Rosenvingea nhatrangensis* is a rare species with broad branches to 2 cm in diameter; in Australia, it has been recorded only from the north-west.

Rosenvingea australis

TYPE LOCALITY: Cape Peron, Western Australia.
DISTRIBUTION: Presently known only from Cape Peron, Western Australia.

Rosenvingea nhatrangensis

TYPE LOCALITY: Cua Bé near Truong Dông, Vietnam.
DISTRIBUTION: North-western Australia. Vietnam. India.

Rosenvingea orientalis

TYPE LOCALITY: Manilla, Philippines.
DISTRIBUTION: Widespread in tropical and warm seas.
FURTHER READING: Kraft (2009); Huisman (2015); Huisman et al. (2018b).

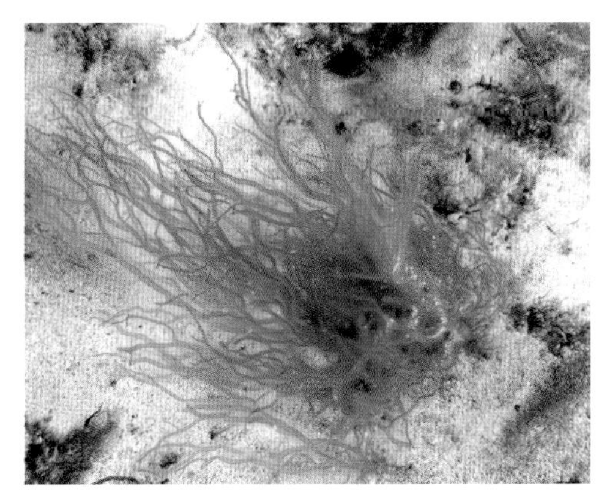

Rosenvingea australis (Cape Peron, WA).

Rosenvingea nhatrangensis (Maret Islands, WA).

Rosenvingea orientalis (Cape Peron, WA).

Scytosiphon

Scytosiphon is similar to *Rosenvingea* (see previous page) in having hollow branches, but differs in being unbranched, with clusters of fronds arising from a small discoid holdfast. The fronds are generally cylindrical or slightly compressed, and are often constricted at intervals. Only the single species, *Scytosiphon lomentaria*, occurs in Australia and typically grows on rock in the lower intertidal to shallow subtidal. It is mainly found during winter and occurs as a crustose stage in summer.

Scytosiphon lomentaria

TYPE LOCALITY: Quivig, Faeroe Islands, Denmark.
DISTRIBUTION: Widespread in temperate and colder waters. In Australia from Port Gregory, Western Australia, around southern Australia to Moreton Bay, Queensland.
FURTHER READING: Womersley (1987).

Scytosiphon lomentaria (Woodman Point, WA).

Cladostephus

Thalli of *Cladostephus* have one to several subdichotomous main axes, each with numerous short lateral branches that are arranged in whorls. In some plants, particularly those from shallow water, the laterals can become crowded along the axis and give the thallus a spongy texture. Reproductive structures are borne on the lateral branches. Two species, *Cladostephus hirsutus* and *Cladostephus kuetzingii*, are known from Australia, the latter only from Victoria.

Cladostephus hirsutus

TYPE LOCALITY: Probably Trondheim, Norway.
DISTRIBUTION: From Yanchep, Western Australia, around southern Australia and Tasmania to Keppel Bay, Queensland. Widespread in temperate seas.
FURTHER READING: Heesch et al. (2020).

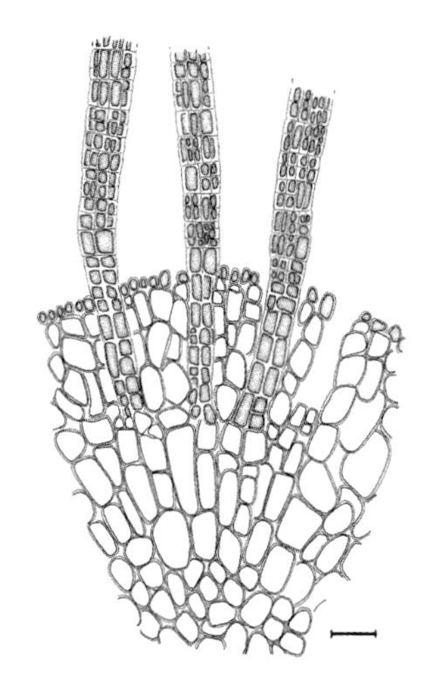

Cladostephus hirsutus (Carnac Island, WA).

(above) *Cladostephus hirsutus* – partial section of thallus, showing the bases of several lateral branches and the cavity from which one has been shed (Carnac Island, WA). Scale = 30 µm.

Sphacelaria

Sphacelaria forms dense brown tufts, growing on rock or a variety of algal and seagrass hosts. It is commonly present on the axes of larger brown algae, particularly *Sargassum*. Many of the species are a dark brown colour and are quite rigid. These features are helpful in recognising *Sphacelaria* in the field, but a microscopic examination is required for accurate identification of both genus and species. Growth is from a single large apical cell that initially divides transversely; this division is followed by a variable number (depending on the species) of longitudinal divisions. The axis therefore appears as tiers of two to several elongate cells. Several species of *Sphacelaria* produce distinctive dwarf branches that break away from the parent thallus and grow into new individuals. These structures are known as propagules, as they serve to vegetatively propagate the plant. Sexual reproduction also occurs.

Sphacelaria rigidula

TYPE LOCALITY: Red Sea.
DISTRIBUTION: Widespread in temperate and warmer seas.

Sphacelaria tribuloides

TYPE LOCALITY: Gulf of Spezia, northern Italy.
DISTRIBUTION: Widespread in temperate and tropical seas.
FURTHER READING: Womersley (1987); Kraft (2009); Huisman (2015).

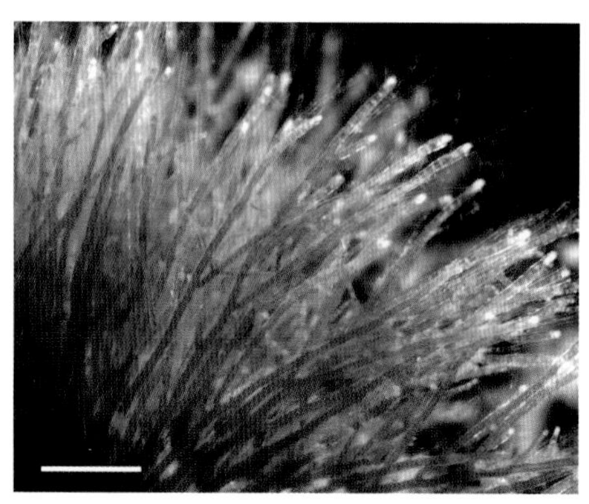

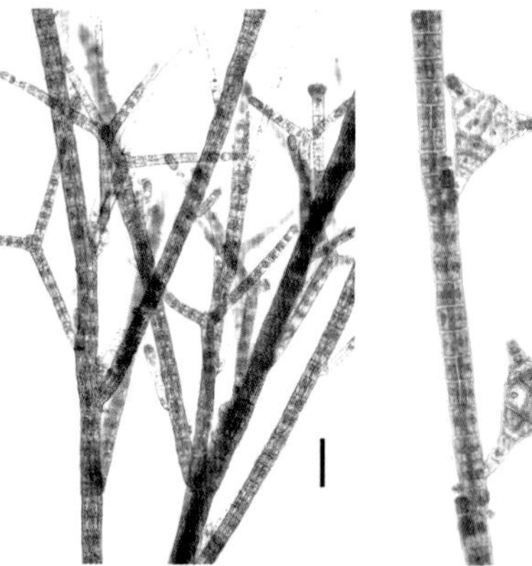

Sphacelaria rigidula – thallus detail (Cape Peron, WA). Scale = 500 µm.

(right) *Sphacelaria rigidula* – propagules (Cape Peron, WA). Scale = 100 µm.

Sphacelaria tribuloides (Cape Peron, WA).

(far right) *Sphacelaria tribuloides* – propagules (Cape Peron, WA). Scale = 100 µm.

Canistrocarpus

Canistrocarpus is represented in Australia by two species, the widespread and common *Canistrocarpus cervicornis* and the less regularly seen *Canistrocarpus crispatus*. Both species were previously included in *Dictyota*, but were segregated based on DNA sequence analyses and the presence of a ring of sterile cells surrounding the reproductive structures. Thalli are ribbon-like and growth is via a single apical cell. Structurally the thallus has a one-cell-thick medulla surrounded by a one-cell-thick cortex, with reproductive structures scattered on the surface. *Canistrocarpus cervicornis* often has twisted and recurved branches.

Canistrocarpus cervicornis

TYPE LOCALITY: Key West, Florida, USA.
DISTRIBUTION: From Cape Peron, Western Australia, around northern Australia to southern Queensland. A pantropical alga.

Canistrocarpus crispatus

TYPE LOCALITY: Antilles, Caribbean Sea.
DISTRIBUTION: Widespread in tropical and subtropical seas.
FURTHER READING: De Clerck et al. (2006); Kraft (2009); Huisman & Phillips (2015).

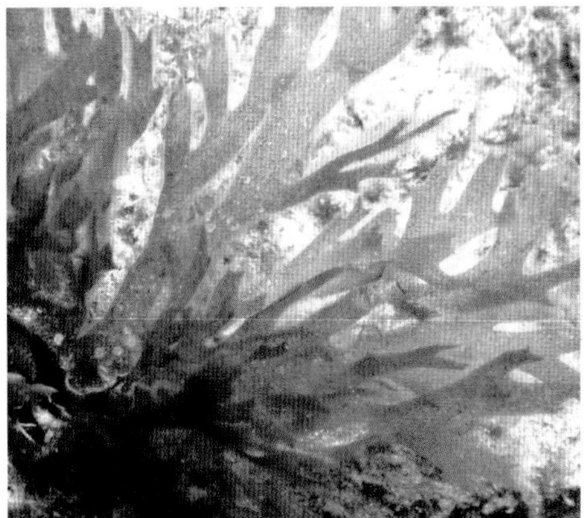

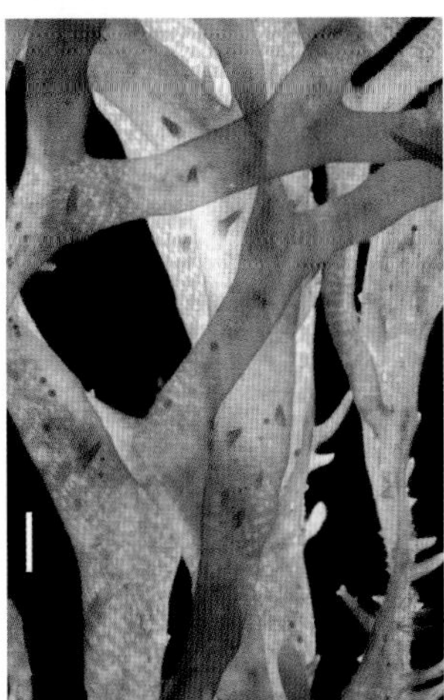

Canistrocarpus cervicornis (Cape Peron, WA).

(left) *Canistrocarpus cervicornis* – lower branches with surface proliferations (Cape Peron, WA). Scale = 1 mm.

Canistrocarpus crispatus (Broome, WA).

Dictyopteris

Species of *Dictyopteris* are common in most Australian seas. They have a distinctive flattened thallus that can be subdichotomously or laterally branched, depending on the species. All have a conspicuous midrib and lateral wings, with the latter in some species traversed by fine veins. Branches are generally of a uniform width throughout, although in many species the lateral wings can be denuded in older parts, leaving the midrib bare. Branch margins can be smooth (e.g. *Dictyopteris muelleri* and *Dictyopteris australis*) or beset with small spines (e.g. *Dictyopteris serrata*). Fine veins are present in some species (e.g. *Dictyopteris plagiogramma*), and *Dictyopteris secundispiralis* has tightly twisted branches. Structurally, the thallus grows from a cluster of cells in an apical depression, and lateral wings of mature branches are several cells thick, becoming thicker near the midrib. Hair tufts occur on the surface of the lateral wings – these can be arranged randomly, or, as is the case in *Dictyopteris australis*, in reflexed lines from the midrib to the branch margin. Reproductive structures are borne on the surface of the lateral wings.

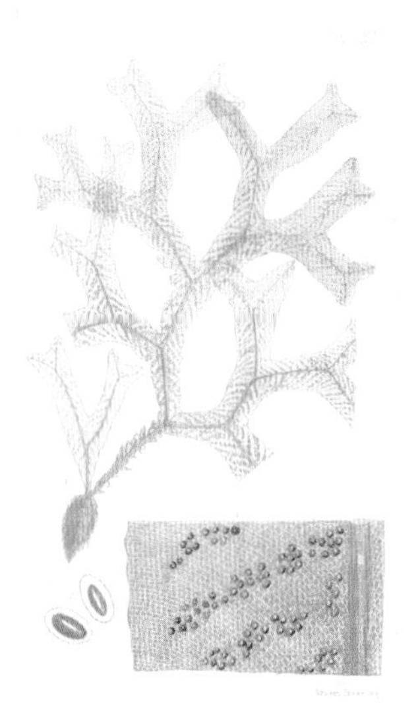

Dictyopteris australis

TYPE LOCALITY: Lefevre Peninsula, South Australia.
DISTRIBUTION: From the Dampier Archipelago, Western Australia, to Port Noarlunga, South Australia, Queensland, Lord Howe Island, New South Wales. Pakistan. India. Hawaiian Islands.

Dictyopteris muelleri

TYPE LOCALITY: Lefevre Peninsula, South Australia.
DISTRIBUTION: From the Houtman Abrolhos, Western Australia, around southern mainland Australia and Tasmania to Port Jackson, New South Wales.

Dictyopteris australis (Cape Peron, WA).

(top) *Dictyopteris australis* – as illustrated by Harvey (1858, pl. 29, as *Haliseris pardalis*).

Dictyopteris muelleri (Rottnest Island, WA).

Dictyopteris plagiogramma

TYPE LOCALITY: Havana, Cuba.
DISTRIBUTION: From Rottnest Island, Western Australia, around northern Australia to Lord Howe Island, New South Wales.

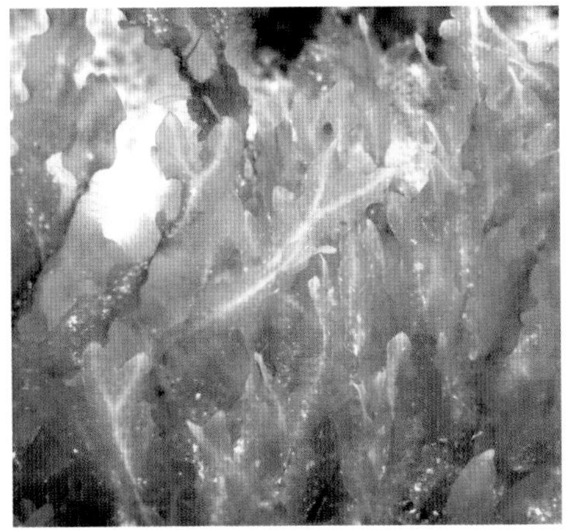

Dictyopteris secundispiralis

TYPE LOCALITY: Horrocks Beach, near Northampton, Western Australia.
DISTRIBUTION: From Hamersley Shoals, Dampier Archipelago, south to Five Fathom Bank (between Fremantle and Mandurah), Western Australia.

Dictyopteris serrata

TYPE LOCALITY: Port Natal, South Africa.
DISTRIBUTION: From the North West Cape region to Dongara, Western Australia. South Africa. Mauritius. Mozambique. Réunion.

Dictyopteris woodwardia

TYPE LOCALITY: North coast of Australia.
DISTRIBUTION: Ningaloo Reef, near Exmouth, Western Australia, across northern Australia to Magnetic Island, Queensland. Seychelles. India. Sri Lanka. Indonesia.
FURTHER READING: Allender & Kraft (1983); Womersley (1987); Phillips & Huisman (1998); Phillips (2000); Kraft (2009); Huisman & Phillips (2015); Scott (2017).

Dictyopteris plagiogramma (Rottnest Island, WA).
Dictyopteris secundispiralis (Stragglers Reef, near Perth, WA).
Dictyopteris serrata (Coral Bay, WA).
Dictyopteris woodwardia (Port Douglas, Qld; photo G.J. Edgar).

Dictyota

Thalli of *Dictyota* have flattened axes that are generally dichotomously branched and of a similar width throughout. Growth is via a single apical cell per axis and mature branches have a medulla composed of a single layer of large, colourless cells bordered on either side by a one-cell-thick cortex. Occasionally extra cell layers can occur in the medulla or cortex. Reproductive structures are borne on the surface of the thallus. Separation of species is often difficult as many are poorly defined and variation in the thallus can be significant.

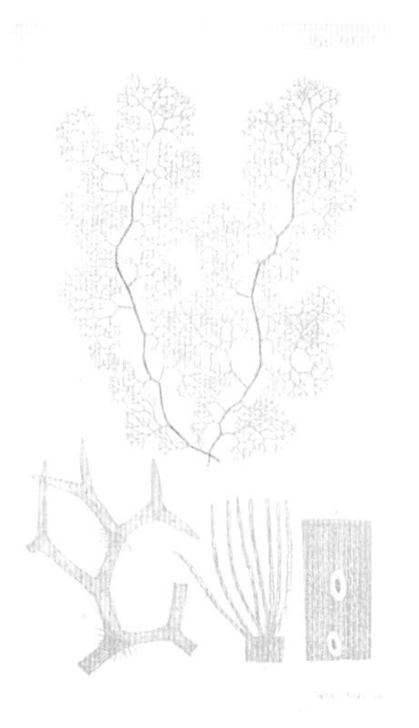

Dictyota canaliculata

Dictyota canaliculata is an eastern Australian species that can be identified by its semi-prostrate habit and slightly channelled branches (i.e., with upwardly curved margins).

TYPE LOCALITY: Patch Reef east of Loloata Island, near Port Moresby, Papua New Guinea.

DISTRIBUTION: Lizard Island, Queensland. Indonesia. Papua New Guinea. Palau.

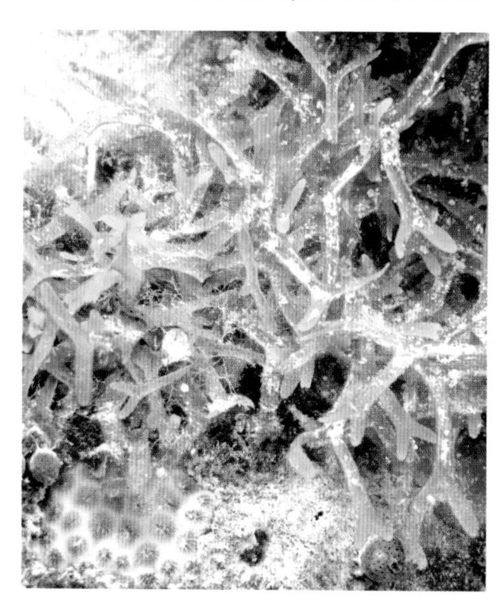

Dictyota ciliolata

Dictyota ciliolata is characterised by the presence of small spines along the margins of the thallus.

TYPE LOCALITY: La Guayra, Venezuela.

DISTRIBUTION: From the Houtman Abrolhos, Western Australia, around northern Australia to Queensland. Widely distributed in tropical and subtropical waters.

(top) *Dictyota polyclada* – as illustrated by Harvey (1858: pl. 38, as *Dictyota furcellata*).

Dictyota canaliculata (Lizard Island, Qld).

Dictyota ciliolata (Cape Peron, WA).

Opposite page:

Dictyota flagellifera (Heron Island, Qld).

Dictyota friabilis (Coral Bay, WA).

Dictyota nigricans (Port Lincoln, SA; photo G.J. Edgar).

Dictyota furcellata (Cape Peron, WA)

Dictyota flagellifera

Dictyota flagellifera often has elongate outgrowths from the surfaces of the blades, often near the base, and these serve to anchor the thallus.

TYPE LOCALITY: Wistari Reef, southern Great Barrier Reef, Queensland.
DISTRIBUTION: Known only from Heron Island, Wistari Reef, and One Tree Island in the southern Great Barrier Reef, Queensland.

Dictyota friabilis

Thalli of *Dictyota friabilis* are typically small and creep along the surface of rocks or are entangled in other algae or invertebrates, but they often catch the eye because of their blue iridescence.

TYPE LOCALITY: Tafaa Point, Tahiti, Society Islands, French Polynesia.
DISTRIBUTION: Widespread in the tropical and subtropical Indo-Pacific.

Dictyota nigricans

On mature specimens, *Dictyota nigricans* typically has a dense covering of small branches that give the thallus a velvety texture. Based on this feature it was formerly placed in *Glossophora*, a genus that is no longer recognised.

TYPE LOCALITY: Orford, Tasmania.
DISTRIBUTION: From Dongara, Western Australia, around southern mainland Australia to Walkerville, Victoria, and around Tasmania.

Dictyota furcellata

A species with very narrow branches, similar to *Dictyota polyclada* but with branching less divaricate and without the multilayered cortex.

TYPE LOCALITY: Shark Bay, Western Australia.
DISTRIBUTION: Shark Bay, Western Australia around southern Australia to Westernport Bay, Victoria, and northern Tasmania.

Dictyota polyclada

Dictyota polyclada has very narrow axes that are typically widely branched. Lower branches have a multilayered cortex.

TYPE LOCALITY: Western Australia.
DISTRIBUTION: Champion Bay, Western Australia to Port Phillip, Victoria.

Dictyota naevosa

One of the largest species of *Dictyota* in Australia.

TYPE LOCALITY: Algoa Bay, South Africa.
DISTRIBUTION: In Australia, from the Houtman Abrolhos, Western Australia, to Port Willunga, South Australia. Southern coast of South Africa.

Dictyota robusta

This species was formerly placed in *Dilophus*, a genus separated from *Dictyota* based on the presence of a multilayered medulla but no longer recognised.

TYPE LOCALITY: Port Phillip Heads, Victoria.
DISTRIBUTION: Houtman Abrolhos, Western Australia, around southern Australia to Port Phillip Heads, Victoria.

Dictyota vittata

Dictyota vittata typically has longitudinal dark patches that look like miniature footprints on the surface of the thallus. The alga in the photo has small branchlets arising along the branch margins, but these are not always present.

TYPE LOCALITY: Coral Gardens, Heron Island, southern Great Barrier Reef, Queensland.
DISTRIBUTION: Known only from Heron Island, Wistari Reef, and One Tree Island in the southern Great Barrier Reef, Queensland.
FURTHER READING: De Clerck & Coppejans (1997); De Clerck (2003); De Clerck et al. (2006); Kraft (2009); Huisman & Phillips (2015); Scott (2017).

Dictyota polyclada – growing on the seagrass *Amphibolis antarctica* (Two Peoples Bay, WA).
Dictyota naevosa (Cape Peron, WA)..
Dictyota robusta (Houtman Abrolhos, WA).
Dictyota vittata (Heron Island, Qld).

Distromium

Distromium is characterised by a meristem composed of a row of apical cells and a thallus that remains two cells thick throughout. The genus has three Australian species, of which *Distromium flabellatum* is included here. *Distromium flabellatum* has a flabellate thallus and is somewhat similar in appearance to species of the genus *Lobophora*, but is a more delicate plant without the multi-layered thallus of that species.

Distromium flabellatum

TYPE LOCALITY: Port Willunga, South Australia.
DISTRIBUTION: From the Houtman Abrolhos, Western Australia, around southern Australia to Port Phillip, Victoria, and Bass Strait.
FURTHER READING: Womersley (1987); Scott (2017).

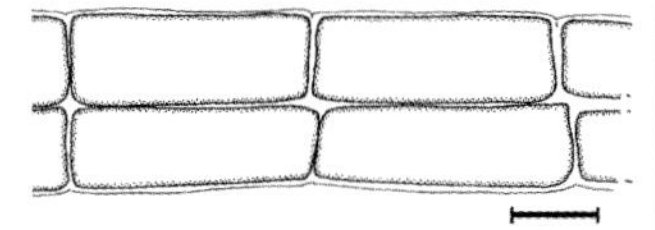

Distromium flabellatum (Bicheno, Tas; photo G.J. Edgar).

(above) *Distromium flabellatum* – section of thallus (Houtman Abrolhos, WA). Scale = 30 µm.

Lobophora

Thalli of *Lobophora* are flattened and mostly entire and fan-shaped, although some become secondarily divided. In these features *Lobophora* is similar to other genera such as *Stypopodium*, and the two are easily misidentified. *Lobophora* displays a variety of forms depending on the species and the conditions under which it is growing. In most situations, thalli grow close to the substratum, and upper regions are secondarily attached. Plants from deeper water, however, are often entirely upright. Growth is via a row of apical cells that lines the entire margin and produces a flattened blade that becomes seven to nine cells thick. A characteristic feature of the genus is a central row of distinctly larger medullary cells. Reproductive structures are borne in scattered patches on the surface of the thallus. Species of *Lobophora* are often abundant in tropical areas, particularly in association with coral reefs.

Lobophora sonderi

TYPE LOCALITY: South-west Western Australia.
DISTRIBUTION: Western Australia.
FURTHER READING: Womersley (1987); Kraft (2009); Sun et al. (2012); Huisman & Phillips (2015); Vieira et al. (2016).

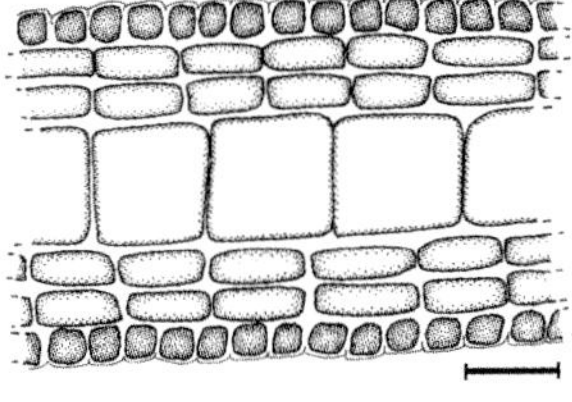

Lobophora sonderi (Cape Peron, WA).

(above) *Lobophora sonderi* – section of thallus (Houtman Abrolhos, WA). Scale = 30 µm.

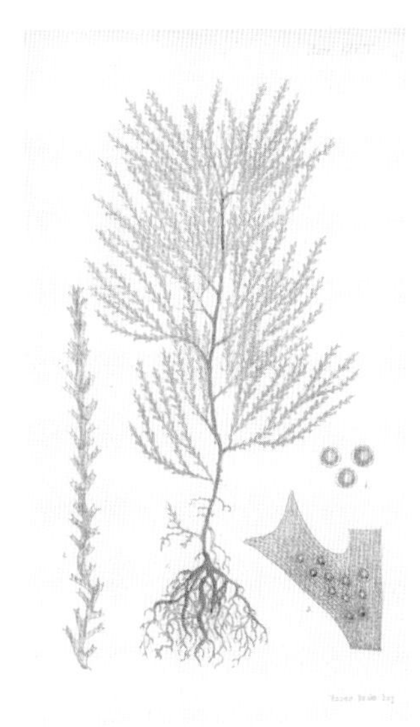

Lobospira

Lobospira includes the single species *Lobospira bicuspidata*. Thalli are much-branched, with narrow, spirally twisted axes that bear short, pointed or bicuspid branches. Structurally the thallus grows from a short row of apical cells and becomes four to many cells thick. Reproductive structures are embedded in the thallus. *Lobospira bicuspidata* is a distinctive species that is readily recognised. It can be very common on shallow reefs in south-western Australia, often just below the reef crest.

Lobospira bicuspidata

TYPE LOCALITY: Port Adelaide, South Australia.
DISTRIBUTION: From Nickol Bay, Western Australia, around southern Australia to Eden, New South Wales.
FURTHER READING: Womersley (1987).

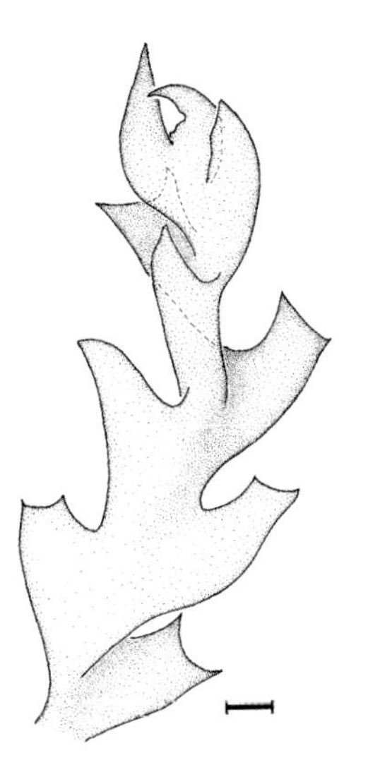

Lobospira bicuspidata (Cape Peron, WA).

(top) *Lobospira bicuspidata* – as illustrated by Harvey (1858, pl. 34).

(above) *Lobospira bicuspidata* – apex of thallus (Jurien Bay, WA). Scale = 200 µm.

Padina

Padina is unusual in being one of only two genera of brown algae that are calcified, albeit lightly and generally only on one surface. Thalli are flattened and usually fan-shaped, although some can be secondarily divided. A distinctive feature of *Padina* is its circinately inrolled apical margin. Growth is via a row of apical cells that lines the entire margin and produces the flattened blade that becomes several cells thick (from two to nine, depending on the species). Reproductive structures are borne in concentric rows on one or both sides of the thallus. Species of *Padina* are separated on habit, colour, the number of cell layers of the thallus, and features of the reproductive structures. *Padina gymnospora* is from four to nine cells thick and *Padina fraseri* is consistently three cells thick. All other species included here are two cell layers thick and their distinguishing features are given under the specific headings.

Padina elegans

Sporangia in *Padina elegans* occur on the ventral surface, usually with the remnants of a thin membranous cover.

TYPE LOCALITY: Mudurup Reef, Cottesloe, Western Australia.
DISTRIBUTION: From the Dampier Archipelago, Western Australia, around south-western Australia to Pearson Island, South Australia.

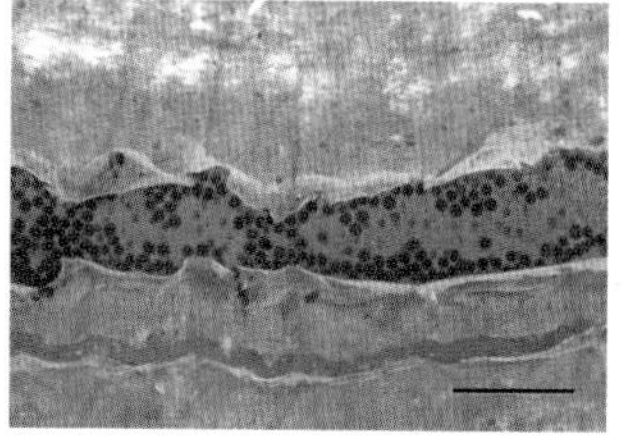

Padina elegans (Yanchep, WA).
Padina elegans – detail of sporangia with flaking membranous cover (Yanchep, WA). Scale = 1 mm.
Opposite page:
Padina boryana (Cape Peron, WA).
Padina australis (White Island, WA).
Padina fraseri (Point Roadknight, Vic).
Padina melemele (Heron Island, Qld).

Padina australis

Sporangia in *Padina australis* occur on the dorsal surface, in fertile zones that alternate with sterile zones. The dorsal surface is defined as the one without the inrolled margin.

TYPE LOCALITY: Cape York, Queensland.
DISTRIBUTION: Widespread in the Indo-Pacific.

Padina boryana

Thalli of *Padina boryana* grow in two forms: the typical fan-shaped calcified thallus, and also a second form that is much smaller, with pointed branch tips.

TYPE LOCALITY: Tongatapu, Tonga.
DISTRIBUTION: Widespread in the Indo-Pacific.

Padina fraseri

This species is often common in intertidal pools on the Victorian coast.

TYPE LOCALITY: Australia.
DISTRIBUTION: From Warrnambool, Victoria, around south-eastern Australia to the mid-north coast of New South Wales, and the north coast of Tasmania.

Padina melemele

The uncalcified surface of *Padina melemele* is bright yellow, strongly contrasting with the white calcified surface.

TYPE LOCALITY: Ilio Pt, Moloka'i, Hawaiian Islands.
DISTRIBUTION: In Australia known from the southern Great Barrier Reef. Papua New Guinea. The Solomon Islands. Hawaiian Islands. French Polynesia.

Padina gymnospora

TYPE LOCALITY: St Thomas, West Indies.
DISTRIBUTION: From Augusta, Western Australia, around northern Australia. Apparently not found on the east coast. Widespread in warmer waters.

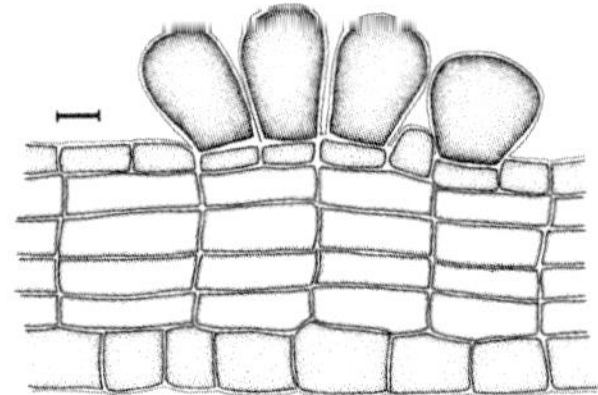

Padina sanctae-crucis

Sporangia in *Padina sanctae-crucis* occur on the dorsal surface; as in *Padina elegans* they usually have the remnants of a thin membranous cover.

TYPE LOCALITY: St Croix, West Indies.
DISTRIBUTION: Widespread in warmer waters.
FURTHER READING: Womersley (1987); Kraft (2009); Ni-Ni-Win et al. (2014); Huisman & Phillips (2015).

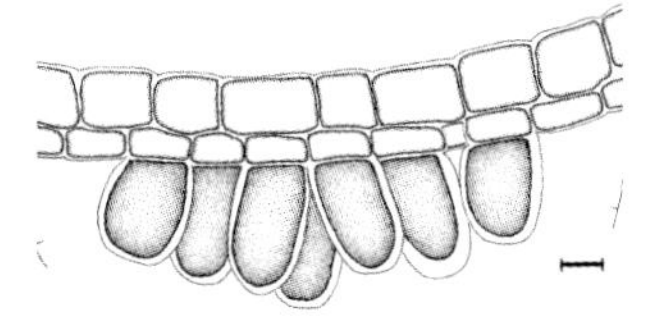

Padina gymnospora (Rottnest Island, WA).

(top) *Padina gymnospora* – section of thallus with sporangia (Rottnest Island, WA). Scale = 30 µm.

Padina sanctae-crucis (Houtman Abrolhos, WA).

(above) *Padina sanctae-crucis* – section of thallus with sporangia (Shark Bay, WA). Scale = 50 µm.

Spatoglossum

Spatoglossum is a widespread genus found in most tropical and cool-temperate southern hemisphere waters. The genus is characterised by its flattened thallus without a midrib and with an apical meristem composed of a localised cluster of cells. This contrasts with other genera of the Dictyotales that either have a continuous marginal meristem (e.g. *Padina*, *Lobophora*) or grow from a single apical cell (e.g. *Dictyota*). Structurally the thallus is from four to six cells thick, with a large-celled medulla and a slightly smaller-celled cortex. Reproductive structures are embedded in the thallus. Several species are known from Australia. *Spatoglossum macrodontum* is known from tropical to subtropical regions on both the east and west coasts. It often has proliferous growth from the margins. *Spatoglossum asperum* can be recognised by its stalked thallus anchored by a fibrous clump of rhizoids.

Spatoglossum asperum

TYPE LOCALITY: Sri Lanka.
DISTRIBUTION: In Australia, known from Queensland and Lord Howe Island. Widespread in the Indo-Pacific.

Spatoglossum macrodontum

TYPE LOCALITY: Port Denison, Queensland.
DISTRIBUTION: From the Houtman Abrolhos, Western Australia, around northern Australia to Lord Howe Island, New South Wales. Warmer waters of eastern and western Australia.
FURTHER READING: Allender & Kraft (1983); Kraft (2009); Huisman & Phillips (2015).

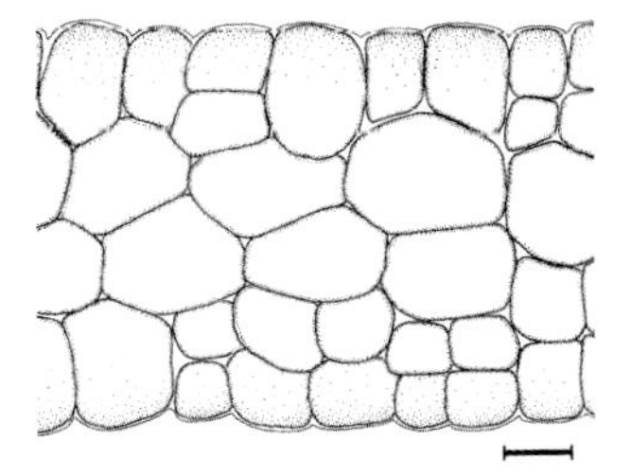

Spatoglossum asperum (Heron Island, Qld).

Spatoglossum macrodontum (James Price Point, WA).

(right) *Spatoglossum macrodontum* – section of thallus (Ningaloo Reef, WA). Scale = 30 µm.

Stoechospermum

Stoechospermum includes the single species *Stoechospermum polypodioides*, which in Australia is known only from the west coast. Plants are erect, tan to dark-brown in colour, and regularly dichotomously branched. They appear similar to some species of *Dictyota*, but differ in their mode of growth, with the apices of *Stoechospermum* having an inrolled margin and growing via a short row of three to five cells. In contrast, growth in *Dictyota* is via a single apical cell and the margins are flat. Structurally the thallus has four to eight medullary cell layers and a single-layered cortex. Reproductive sporangia arise in patches that lie close to, and parallel with, the branch margins. *Stoechospermum* is also similar to *Dictyopteris* (p. 286), but that genus is distinguished by having a central midrib, which is absent in *Stoechospermum*.

Stoechospermum polypodioides

TYPE LOCALITY: Unknown.
DISTRIBUTION: In Australia, known from Jurien Bay north to Broome, Western Australia. Warmer waters of the Indian Ocean.
FURTHER READING: Phillips et al. (1993); Huisman & Phillips (2015).

Stoechospermum polypodioides (Barrow Island, WA).

Stypopodium

Stypopodium is a common inhabitant of tropical regions. Thalli are fan-shaped or divided to some degree as a result of the splitting of the margins. Growth is via an apical row of cells and gives rise to a thallus around four to nine cells thick. The central medulla is composed of large colourless cells that are irregularly arranged and bordered by a cortex of a single layer of pigmented cells. Concentric lines of hairs arise on the surface of the thallus. *Stypopodium* is closely related to *Zonaria* but differs in its non-tiered medullary cells and features of the sporangia. *Stypopodium flabelliforme* is a distinctive species that frequently occurs in clusters of several fronds. Thalli are recumbent and often overlap with adjacent fronds. Young thalli often have a blue iridescence, but this can be lost as the thallus matures. *Stypopodium australasicum* differs in its more upright habit and by being attached by a single holdfast.

Stypopodium australasicum

TYPE LOCALITY: Lord Howe Island, New South Wales.
DISTRIBUTION: Lord Howe Island, New South Wales, and Rottnest Island to the Houtman Abrolhos, Western Australia.

Stypopodium flabelliforme

TYPE LOCALITIES: Rotti Island, Indonesia; Pearl Bank, Tawitawi Province, Sulu Archipelago, Philippines.
DISTRIBUTION: From Rottnest Island, Western Australia, around northern Australia to Lord Howe Island, New South Wales. Widespread in the warmer waters of the Indo-Pacific.
FURTHER READING: Allender & Kraft (1983); Kraft (2009); Huisman & Phillips (2015).

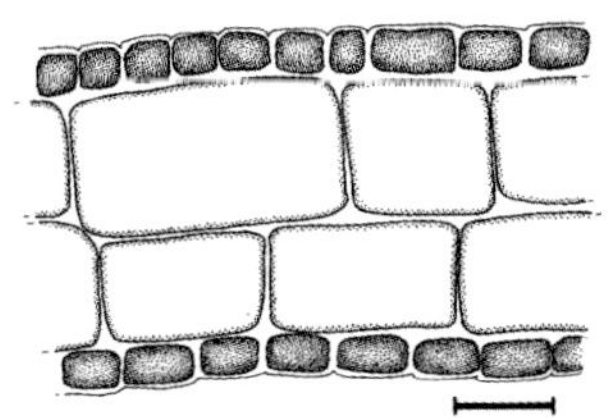

Stypopodium australasicum (Houtman Abrolhos, WA).

Stypopodium flabelliforme (Heron Island, Qld).

(right) *Stypopodium flabelliforme* – section of thallus (Montebello Islands, WA). Scale = 30 µm.

Zonaria

Thalli of *Zonaria* are fan shaped, as seen in *Zonaria diesingiana*, or secondarily branched as a result of splitting of the apices, as in *Zonaria turneriana*. All species grow from an apical row of cells that gives rise to a thallus with a many-celled medulla (with cells in tiers) and a single-celled cortical layer. Cortical cells are always in pairs above single outer medullary cells (when viewed in a transverse section). Reproductive structures are borne in patches on the surface of the thallus and are intermingled with sterile filaments.

Zonaria diesingiana

TYPE LOCALITY: Australia (probably near Sydney).
DISTRIBUTION: In Australia, from New South Wales, Lord Howe Island, Norfolk Island and the Perth region. Japan. China. Taiwan. St Paul Island. Philippines.

Zonaria turneriana

TYPE LOCALITIES: Warrington, Otago, New Zealand.
DISTRIBUTION: From the Houtman Abrolhos, Western Australia, around southern Australia to Port Phillip Heads, Victoria, and Tasmania.
FURTHER READING: Allender & Kraft (1983); Womersley (1987); Kraft (2009).

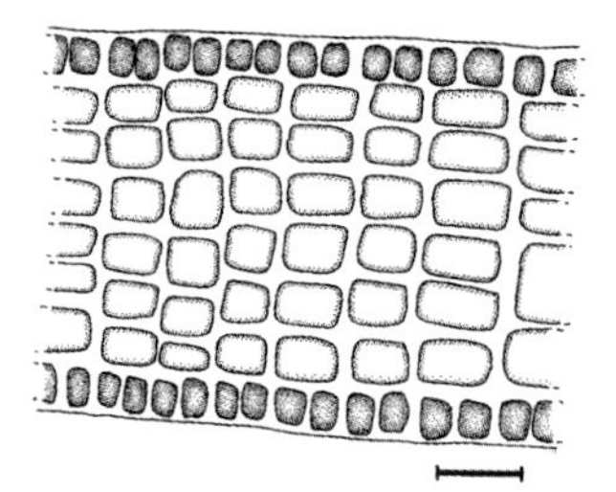

Zonaria diesingiana (Cape Peron, WA).

Zonaria turneriana (Gnarabup, WA).

(right) *Zonaria turneriana* – section of thallus (Houtman Abrolhos, WA). Scale = 30 µm.

(top) *Zonaria turneriana* – as illustrated by Harvey (1862, pl. 190, as *Zonaria interrupta*).

Cutleria

Cutleria displays one of the more unusual growth patterns found in the brown algae. By a process known as 'trichothallic growth', the dividing cells lay down a cellular thallus in one direction and filamentous tufts in the other. This pattern results in the unusual fringe formed at the edges of the thallus. Species of *Cutleria* can be much-branched (as in *Cutleria multifida* from southern Australia) or, as in *Cutleria kraftii* shown here, have a fan-shaped thallus superficially similar in appearance to several genera of the Dictyotales. *Cutleria* has a heteromorphic life history – the gametophyte is shown here while the sporophyte is an encrusting phase without the marginal fringe.

Cutleria kraftii

TYPE LOCALITY: Bynoe Island, Houtman Abrolhos, Western Australia.
DISTRIBUTION: Known from Rottnest Island north to the Dampier Archipelago, Western Australia.
FURTHER READING: Huisman (2000); Huisman (2015).

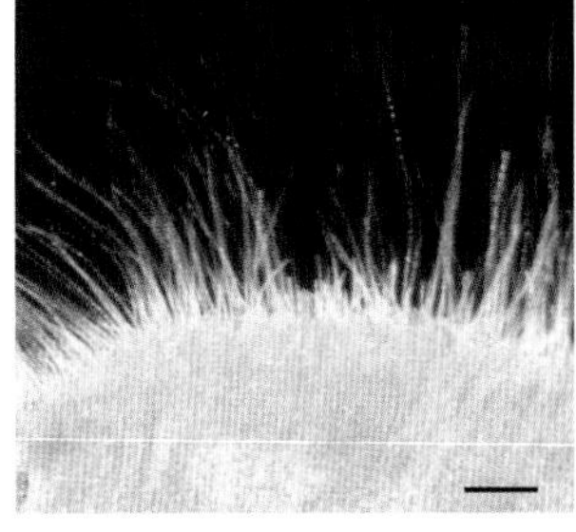

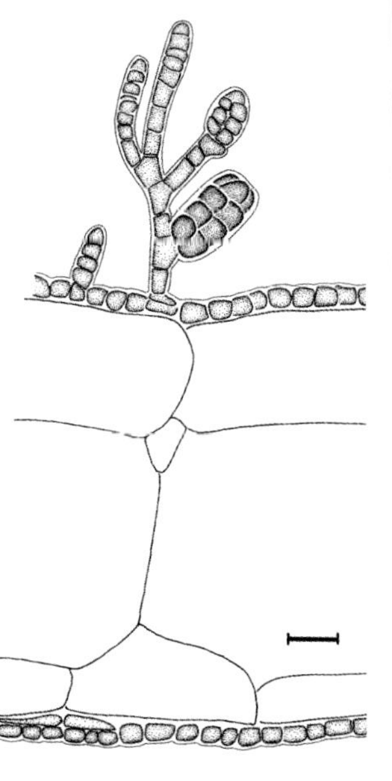

Cutleria kraftii (Rottnest Island, WA).

(top left) *Cutleria kraftii* – detail of marginal fringe (Rottnest Island, WA). Scale = 1 mm.

(left) *Cutleria kraftii* – section of thallus with female gametangia on surface hairs (Houtman Abrolhos, WA). Scale = 30 μm.

Splachnidium

Splachnidium includes only the single species *Splachnidium rugosum*, a relatively common plant found in the intertidal region of rough-water coasts. Thalli are very slimy to touch, with cylindrical branches that become very wrinkled. Structurally the branches have a hollow centre, a filamentous outer medulla, and a pseudoparenchymatous cortex. Reproductive structures are borne in cavities in the surface of the plant. The plant pictured is the sporophyte stage of the life history, the gametophyte being small and filamentous.

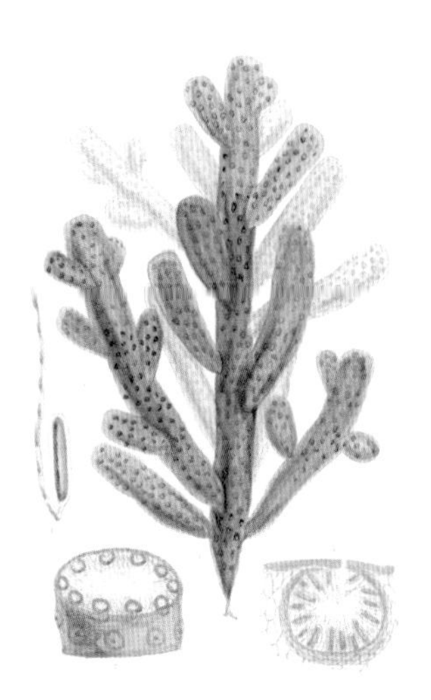

Splachnidium rugosum

TYPE LOCALITY: South Africa.
DISTRIBUTION: From Point Sinclair, South Australia, to Narrabeen, New South Wales, and around Tasmania. South Africa. Juan Fernandez Islands. New Zealand. Also several subantarctic islands.
FURTHER READING: Womersley (1987); Scott (2017).

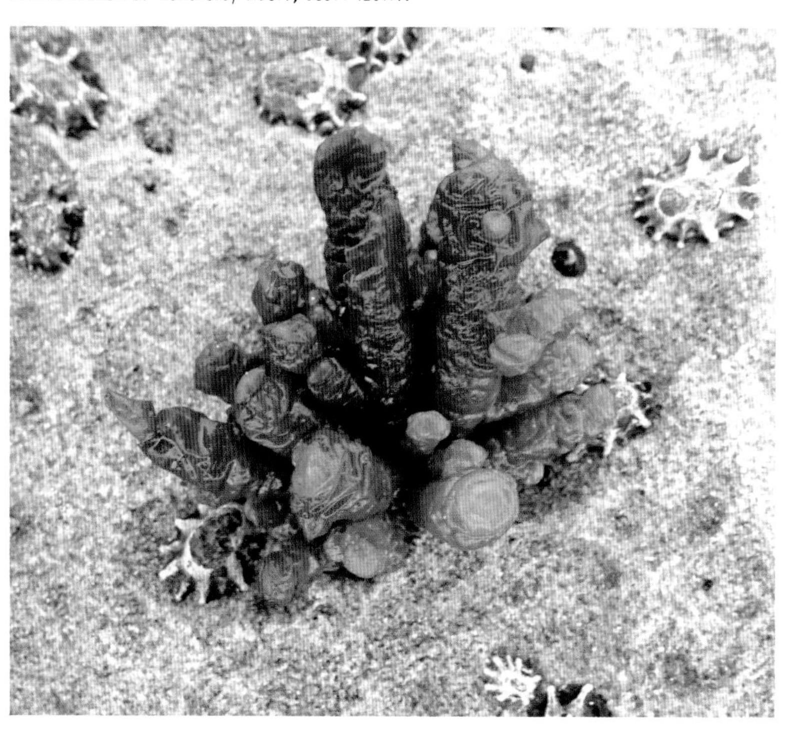

Splachnidium rugosum (Point Roadknight, Vic).

(top) *Splachnidium rugosum* – as illustrated by Harvey (1858, pl. 14).

Neoralfsia

Thalli of *Neoralfsia* form dark-brown crusts on intertidal and shallow subtidal rock. Only a single species, the apparently widespread *Neoralfsia expansa*, is recorded from Australia, but recent studies suggest this entity is different from the species as it is known in Mexico (the type locality) and it might represent a new species. Structurally the thallus is pseudoparenchymatous and composed of closely adherent filaments, with distinct cortical and medullary layers. The surface is marked by pits containing clusters of hairs.

Neoralfsia expansa

TYPE LOCALITY: Veracruz, Mexico.
DISTRIBUTION: Widespread in tropical seas.
FURTHER READING: Kraft (2009); Huisman (2015).

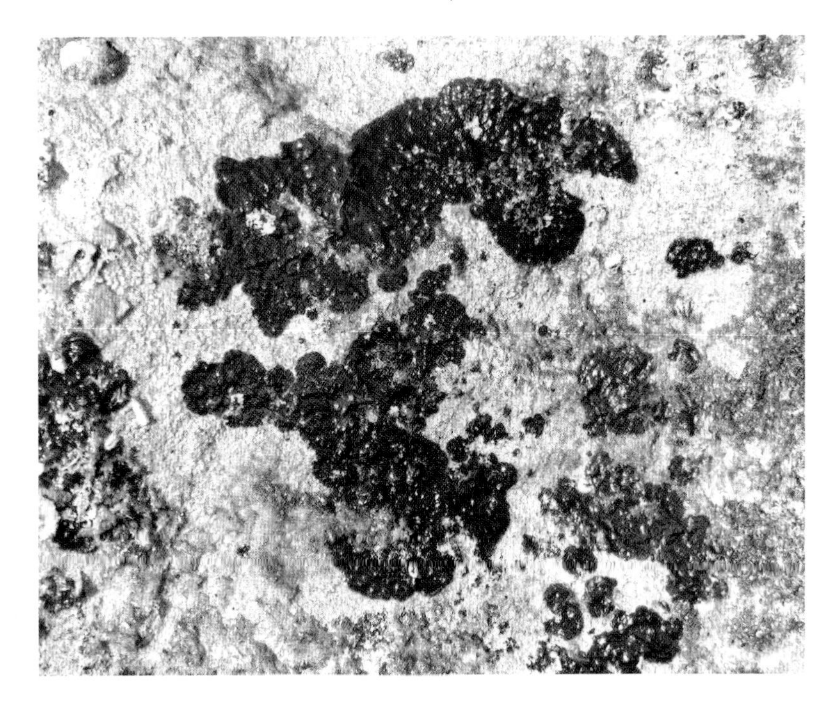

Neoralfsia expansa (James Price Point, WA).

Bellotia

Bellotia includes two species in Australia, *Bellotia eriophorum* (shown here) from the south-east coast and *Bellotia simplex* from the Great Barrier Reef, both generally found in deep water. Plants have one to several primary axes that can be unbranched or have successive series of umbellate secondary lateral branches arising from the apices of lower axes, with each branch terminating in a tuft of pigmented filaments. Reproductive structures are borne in swollen regions of the axes, below the apical tufts; the region between the fertile tissue and the apex is sterile.

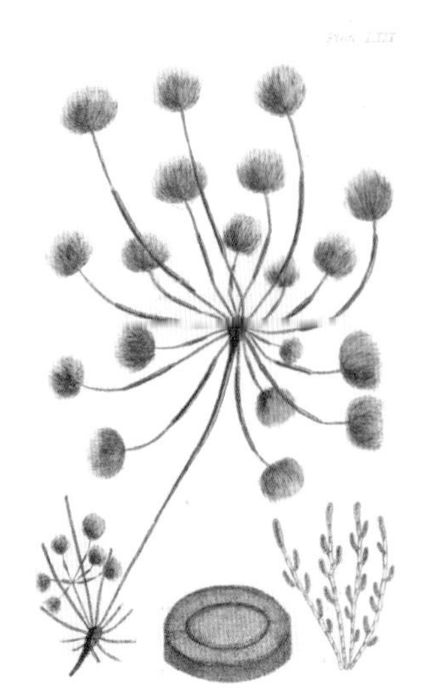

Bellotia eriophorum

TYPE LOCALITY: Port Phillip Heads, Victoria.
DISTRIBUTION: From Isles of St Francis, South Australia, to Walkerville, Victoria, and around Tasmania; Queensland.
FURTHER READING: Womersley (1987); Kraft (2007); Scott (2017).

Bellotia eriophorum (Flinders Island, Tas; photo G.J. Edgar).

(top) *Bellotia eriophorum* – as illustrated by Harvey (1859a, pl. 69).

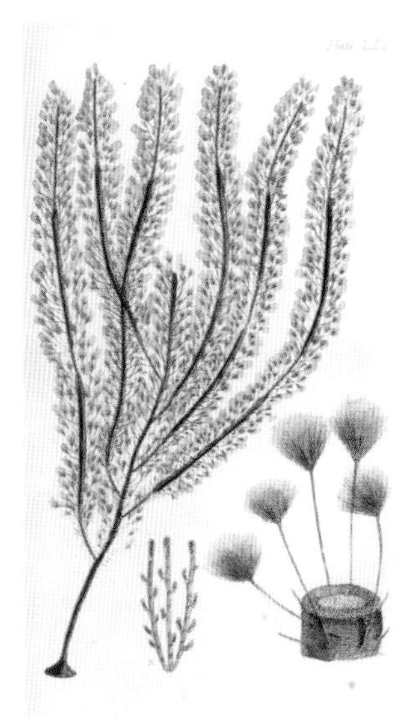

Encyothalia

Thalli of *Encyothalia* are radially branched, with the terete main axes bearing numerous short branches of equal length, each with a conspicuous apical tuft of filaments. Growth is via a trichothallic meristem that produces the apical tuft in one direction and the cellular branch in the other. A distinctive feature of the genus is the presence of reproductive structures in raised patches that are spread over the surface of the main axes and the bases of the short branches. *Encyothalia* includes only a single species, *Encyothalia cliftonii*.

Encyothalia cliftonii

TYPE LOCALITY: Fremantle, Western Australia.
DISTRIBUTION: From Kalbarri, Western Australia, around southern Australia to Walkerville, Victoria.
FURTHER READING: Womersley (1987).

Encyothalia cliftonii (Gnarabup, WA).

(top) *Encyothalia cliftonii* – as illustrated by Harvey (1859a, pl. 62).

(above) *Encyothalia cliftonii* – closer view of thallus (Gnarabup, WA).

Lucasia

Yet another genus with a single species, *Lucasia lasiocarpa* is known in Australia only from deep-water collections, primarily growing associated with sand at the base of reef slopes. The genus is morphologically very similar to *Sporochnus* (see p. 308) in having raised fertile regions immediately below the apical hair tufts, but DNA sequence studies showed *Lucasia* to be only distantly related to *Sporochnus* and prompted its recognition as an independent genus.

Lucasia lasiocarpa

TYPE LOCALITY: South-west Lagoon, Cape Woodin, Noumea, New Caledonia.
DISTRIBUTION: Capricorn group of the southern Great Barrier Reef, Queensland. New Caledonia.
FURTHER READING: Yee et al. (2009).

Lucasia lasiocarpa (Heron Island, Qld).

Perithalia

Perithalia includes two species, *Perithalia caudata* from south-eastern Australia and *Perithalia capillaris* from New Zealand (not shown). Plants of *Perithalia caudata* can be up to 100 cm long and are much branched, with each branch terminating in a rounded apex or a cup-shaped cap covering the apex. The cap is composed of the bases of trichothallic filaments that have adhered laterally. Tufts of free filaments occur only occasionally. *Perithalia* has a heteromorphic life history that involves a conspicuous sporophyte phase and a microscopic filamentous gametophyte. Reproductive structures in the sporophyte occur in sori that are formed some distance below the branch apices. *Perithalia caudata* generally occurs on rough-water coasts, in shaded areas just below low tide level to 15 m deep.

Perithalia caudata

TYPE LOCALITY: Cape van Diemen, south-east Tasmania.
DISTRIBUTION: From West Bay, Kangaroo Island, South Australia, to Wilsons Promontory, Victoria, and around Tasmania.
FURTHER READING: Womersley (1987); Nelson (2013).

Perithalia caudata (Flinders Island, Tas; photo G.J. Edgar).

(top) *Perithalia caudata* – as illustrated by Harvey (1862, pl. 238, as *Carpomitra inermis*).

Sporochnus

Sporochnus is a distinctive genus, readily recognised by its smooth, terete axes that bear numerous short, lateral branches topped with tufts of pigmented filaments. The genus has a heteromorphic life history, with the large sporophyte alternating with a microscopic, filamentous gametophyte. Only the sporophyte phase is known for all Australian species. Sporangia are borne in distinct raised patches that encircle the axes, in most cases immediately below the apical tufts.

Sporochnus comosus

TYPE LOCALITY: Australia.
DISTRIBUTION: From the Houtman Abrolhos, Western Australia, around southern Australia and Tasmania to the Calliope River, Queensland.
FURTHER READING: Womersley (1987); Yee et al. (2015).

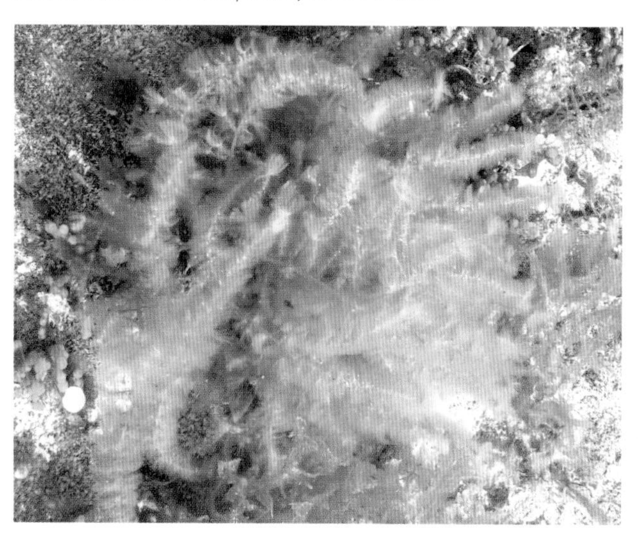

Sporochnus comosus (Barrow Island, WA).

(top) *Sporochnus comosus* – as illustrated by Harvey (1859a, pl. 104)

(above) *Sporochnus comosus* – detail of apical hair tufts and sporangial sorus (Barrow Island, WA). Scale = 1 mm.

Tomaculopsis

Tomaculopsis is known from a single species, *Tomaculopsis herbertiana*, found only in subtropical eastern Australia. The genus has the tufts of apical filaments typical of the Sporochnaceae, but differs from other genera in having an unbranched stem. *Tomaculopsis* is spectacular when viewed underwater, as it often occurs in clusters and the apical tufts are iridescent in shades of pale green to blue. Sporangia are borne in raised patches that encircle the axes, with a short sterile area projecting beyond the fertile region.

Tomaculopsis herbertiana

TYPE LOCALITY: Burleigh Heads, Queensland, Australia.
DISTRIBUTION: Known from Noosa Heads, Queensland, south to Byron Bay, New South Wales.
FURTHER READING: Cribb (1960).

Tomaculopsis herbertiana (Byron Bay, NSW; photo N. Coleman).

Undaria

Undaria pinnatifida, a medium to large kelp with flat brown blades, has been introduced to south-eastern Australian coasts, where it can be very common during spring. It is similar in appearance to the native common kelp *Ecklonia radiata*, the most distinguishing feature being the presence of a thick midrib in the centre of the blade in *Undaria*. No midrib occurs in *Ecklonia* (see p. 312). When fertile, *Undaria* produces sporangia in ruffled blades that surround the stipe. *Undaria pinnatifida* is edible and is harvested and sold under the name 'wakame'.

Undaria pinnatifida

TYPE LOCALITY: Shimoda, Shizuoka Prefecture, Japan.
DISTRIBUTION: Widely distributed in colder seas. Introduced to many regions worldwide.
FURTHER READING: Nelson (2013); Scott (2017).

Undaria pinnatifida (Safety Cove, Tas; photo M.D. Guiry).

(above) *Undaria pinnatifida* – ruffled reproductive tissue (photo CSIRO).

Macrocystis

Macrocystis – also known by the common name 'giant kelp' – is the tallest growing of the Australian marine plants and can occasionally form dense kelp forests up to 20 m high. Only a single species is known in the genus, *Macrocystis pyrifera*. Plants have terete stipes that bear unilateral, unbranched blades. Young blades arise at the apices of fronds by a process in which a juvenile frond progressively splits longitudinally. Mature blades have spinous margins and a slightly elongate, basal pneumatocyst ('float'). The genus was once harvested in many countries as a source of alginate, a cell-wall component that is used as a thickening and stabilising agent in a number of food and industrial applications.

Macrocystis pyrifera

TYPE LOCALITY: South Atlantic Ocean.
DISTRIBUTION: From Cape Jaffa, South Australia, to Walkerville, Victoria, and the north and north-west coasts of Tasmania. Widespread in colder seas.
FURTHER READING: Womersley (1987); Macaya & Zuccarello (2010); Scott (2017).

Macrocystis pyrifera – detail of thallus (Bicheno, Tas; photo G.J. Edgar).

(top) *Macrocystis pyrifera* – as illustrated by Harvey (1862, pl. 202).

(above) *Macrocystis pyrifera* (Point Lonsdale, Vic).

Ecklonia

Ecklonia includes some six species worldwide, of which one, possibly two, occurs in Australian seas. *Ecklonia radiata* forms large, conspicuous beds in subtidal waters on moderate- to rough-water coasts. It is the only true kelp (i.e. a member of the order Laminariales) found in the warmer seas of Western Australia and southern Queensland. Thalli have a terete basal stipe that bears a flattened blade with distichous lateral branches. The blade is often very spiny and has a corrugated surface. Structurally the thallus has a central filamentous medulla and an outer parenchymatous cortex, with only the peripheral cells of the latter conspicuously pigmented. The life history of *Ecklonia* – as is the case for all members of the Laminariales – includes an inconspicuous filamentous gametophyte phase that alternates with the more visible sporophyte. One population in south-western Australia has been observed reproducing vegetatively by the production of secondary holdfasts from the apices of lateral branches. These holdfasts attach to the substratum and, along with a portion of the lateral branch, become detached from the parent thallus. This type of vegetative propagation is also found in the New Zealand *Ecklonia brevipes,* and the Australian population was previously tentatively assigned to that species. However, DNA sequence analyses suggest that this entity is not *Ecklonia brevipes* and is possibly only a form of *Ecklonia radiata*. As its status is yet to be established it is included here as *Ecklonia* sp.

Ecklonia radiata

TYPE LOCALITY: Port Jackson, New South Wales.
DISTRIBUTION: From Kalbarri, Western Australia, around southern Australia and Tasmania to Caloundra, Queensland. New Zealand. South Africa.
FURTHER READING: Womersley (1987); Bolton & Anderson (1994); Nelson (2013); Rothman et al. (2015); Scott (2017).

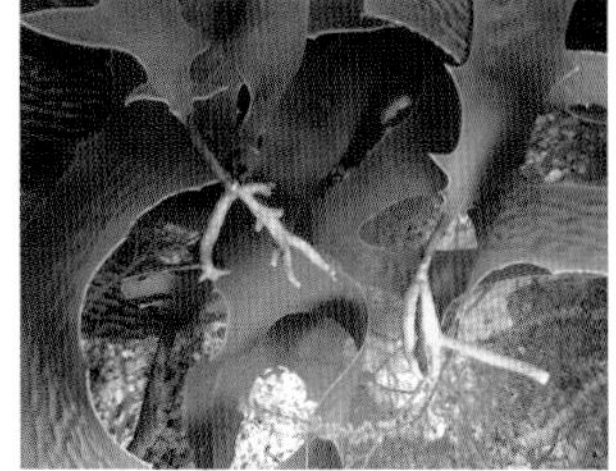

Ecklonia radiata (Cape Peron, WA).

(top) *Ecklonia radiata* – showing rugose surface of blades (Cape Peron, WA).

(above) *Ecklonia* sp. – secondary holdfasts formed at the apices of lateral branches (Gnarabup, WA).

Lessonia

The 11 species of the kelp genus *Lessonia* are found only in the southern hemisphere. Only *Lessonia corrugata* occurs in Australia, and this is virtually endemic to Tasmania, where it commonly occurs on rough-water coasts in the lower intertidal, but can also grow in deeper water. Plants are medium to dark brown and can be up to 2 m tall. The holdfast supports numerous branched stipes that bear terminal elongate, unbranched blades, which are generally around 3 cm broad, with marginal spines and distinctive longitudinal corrugations.

Lessonia corrugata

TYPE LOCALITY: Port Arthur, Tasmania.
DISTRIBUTION: Phillip Island, Victoria, and around Tasmania.
FURTHER READING: Womersley (1987); Scott (2017).

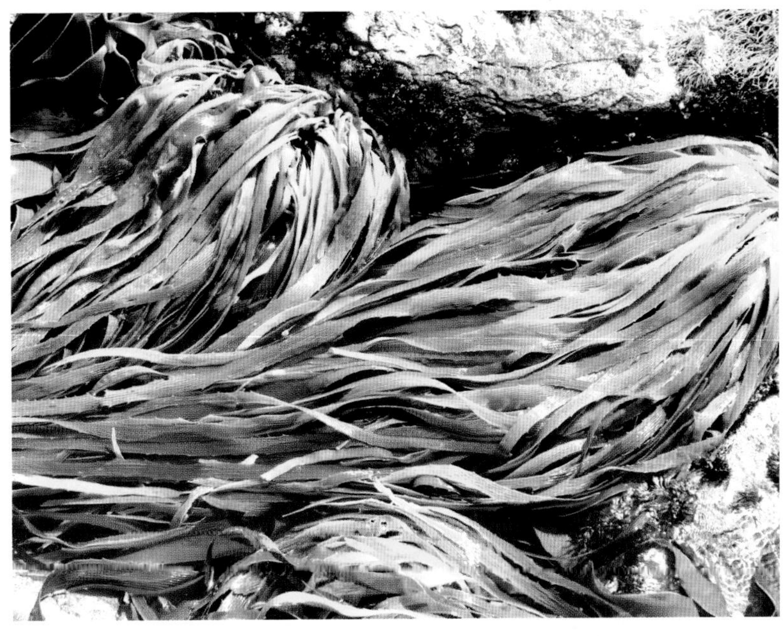

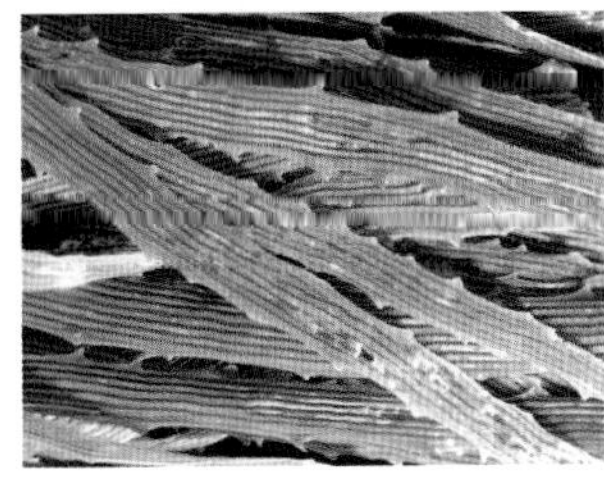

Lessonia corrugata (Fortescue Bay, Tas; photo M.D. Guiry).

(right) *Lessonia corrugata* - detail of thallus (Blackmans Bay, Tas).

Notheia

Notheia includes only the single species *Notheia anomala*, an unusual plant that is found growing almost exclusively on thalli of *Hormosira banksii* (rarely on *Xiphophora*). Plants can be very common wherever *Hormosira* occurs, and the distribution of the species is virtually identical to that of its host. Thalli of *Notheia anomala* are typically small and profusely branched, although the majority of lateral branches are in fact new plants that have grown from within the reproductive cavities of the parent. Structurally the thallus is largely pseudoparenchymatous, with hairs tufts that arise from within the reproductive cavities.

Notheia anomala

TYPE LOCALITY: Bay of Islands, New Zealand.
DISTRIBUTION: From King George Sound, Western Australia, around southern mainland Australia and Tasmania to Port Stephens, New South Wales. New Zealand.
FURTHER READING: Womersley (1987); Saunders & Kraft (1995); Nelson (2013); Scott (2017).

Notheia anomala (Point Lonsdale, Vic).

(top) *Notheia anomala* – as illustrated by Harvey (1862, pl. 213).

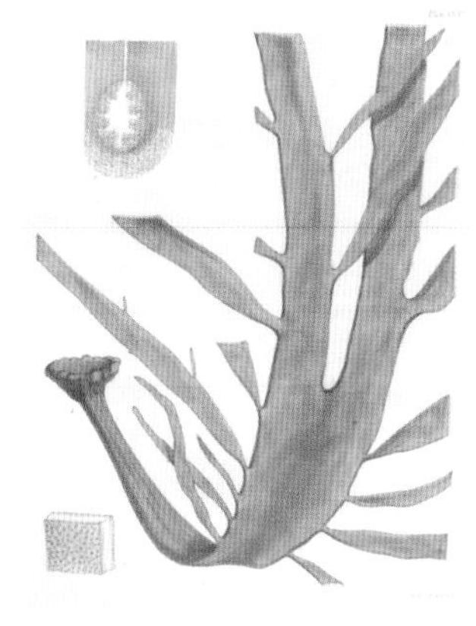

Durvillaea

A genus of nine species found in colder waters of the southern hemisphere, *Durvillaea* is known in Australia by two species, *Durvillaea potatorum* and the recently described *Durvillaea amatheiae* (not shown). *Durvillaea* is a distinctive and easily recognised genus that is also known as 'bull kelp', the name reflecting its massive leathery blades and large discoid holdfast. Plants are common in the upper sublittoral region and often form a fringe at the lower intertidal habitat, especially where there is substantial wave action. In Tasmania, drift thalli of *Durvillaea potatorum* are collected and processed for their alginates - one of very few algal-based industries in Australia.

Durvillaea potatorum

TYPE LOCALITY: South-east Tasmania.
DISTRIBUTION: From Cape Jaffa, South Australia, around south-eastern Australia and the west, south and east coasts of Tasmania to Bermagui, New South Wales.
FURTHER READING: Womersley (1987); Weber et al. (2017); Scott (2017).

Durvillaea potatorum (Lorne, Vic).

(top) *Durvillaea potatorum* – as illustrated by Harvey (1863, pl. 300).

Hormosira

Hormosira is a genus unlikely to be confused with any other. Its single species, *Hormosira banksii*, is a common inhabitant of intertidal regions in southern Australia, where it forms thick, almost monospecific mats. Thalli are composed of chains of subspherical, hollow segments joined by thin constrictions. Reproductive structures and hairs are borne in conceptacles that are scattered over the thallus. *Hormosira banksii* is essentially a cold-water species.

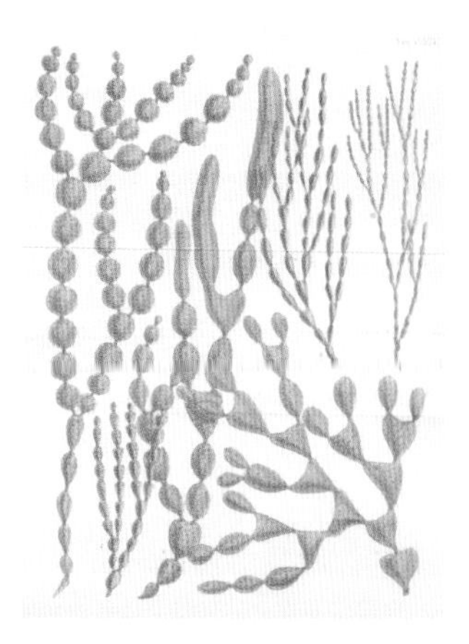

Hormosira banksii

TYPE LOCALITY: Australia.
DISTRIBUTION: From Walpole, Western Australia, around southern Australia and Tasmania to Arrawarra Headland, New South Wales. Lord Howe Island. Norfolk Island. New Zealand.
FURTHER READING: Womersley (1987); Nelson (2013); Scott (2017).

Hormosira banksii (Lorne, Vic).

(top) *Hormosira banksii* – as illustrated by Harvey (1860, pl. 135).

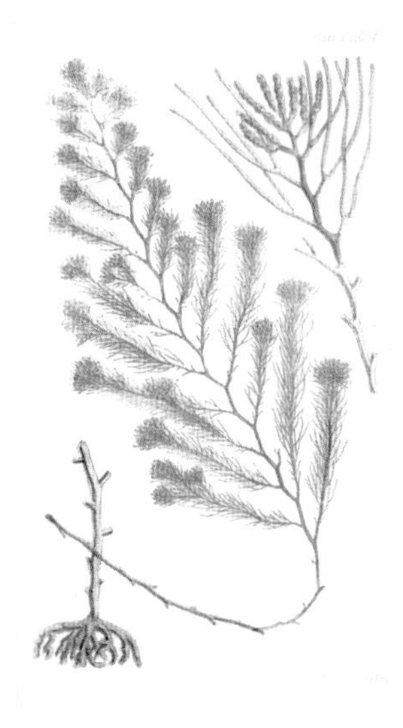

Acrocarpia

Only two species of *Acrocarpia* are known, with *Acrocarpia robusta* found in the colder waters of south-western Australia, where it grows in shallow water on exposed coasts. Thalli have several axes arising from a claw-like holdfast. Lateral branches are arranged spirally or in three ranks, but are often secondarily lost from lower axes, leaving a denuded stem. The lateral branches are clothed with smaller, once-divided branches known as ramuli. Receptacles are terminal in upper ramuli.

Acrocarpia robusta

TYPE LOCALITY: Israelite Bay, Western Australia.
DISTRIBUTION: From Cape Naturaliste to Israelite Bay, Western Australia.
FURTHER READING: Womersley (1987).

(top) *Acrocarpia paniculata* - as illustrated by Harvey (1863, pl. 247, as *Cystophora paniculata*).

Acrocarpia robusta (Albany, WA).

Caulocystis

Only two species of *Caulocystis* are known, both endemic to the southern half of Australia and commonly (but not exclusively) found near the low-tide level. The genus is characterised by its straight, radially branched main axes, upon which vesicles are borne directly. The lateral branches are often worn away from the lower axis, leaving a bare stem with small scars. *Caulocystis uvifera* grows to around 40 cm in length and can be recognised by its nearly spherical vesicles, while those of *Caulocystis cephalornithos*, the second species, are more elongate.

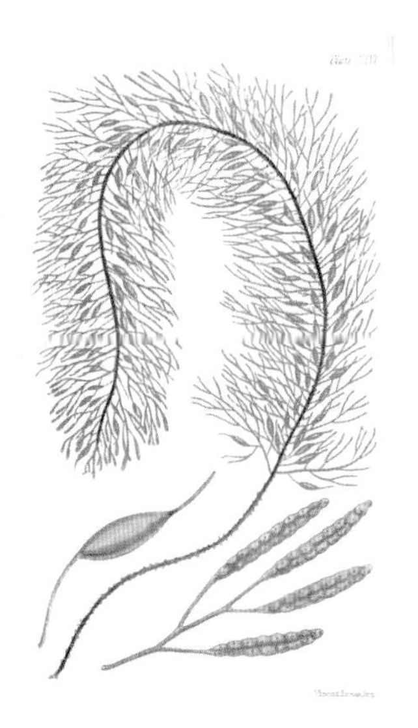

Caulocystis cephalornithos

TYPE LOCALITY: Cape Van Diemen, south-eastern Tasmania.
DISTRIBUTION: From Cape Naturaliste, Western Australia, around south-eastern Australia and Tasmania to Bondi, New South Wales.

Caulocystis uvifera

TYPE LOCALITY: Shark Bay, Western Australia.
DISTRIBUTION: From Shark Bay, Western Australia, around southern Australia and Tasmania to Coogee, New South Wales. Norfolk Island.
FURTHER READING: Womersley (1987); Scott (2017).

Caulocystis cephalornithos (Point Lonsdale, Vic).

(top) *Caulocystis cephalornithos* – as illustrated by Harvey (1859a, pl. 116, as *Cystophora cephalornithos*).

Caulocystis uvifera (Houtman Abrolhos, WA).

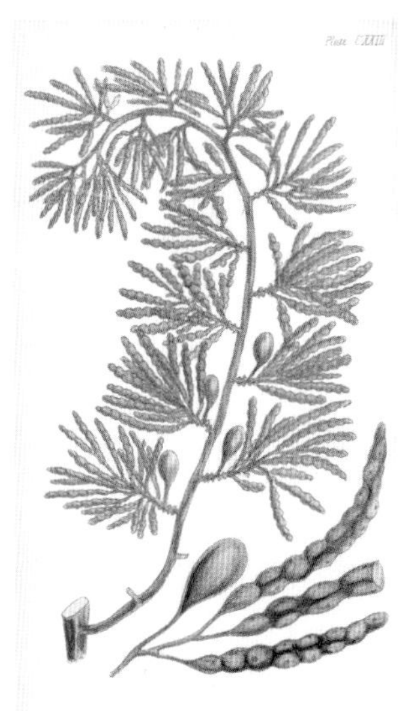

Cystophora

Many of the larger brown algae found in the colder waters along the southern Australian coast belong to the genus *Cystophora*, with some 23 species known from the region. A characteristic feature of the genus is the alternately biased growth at the apices which results in axes that are distinctly flexuous. Reproductive structures are borne in cavities known as conceptacles, which occur at the ends of some lateral branches.

Cystophora grevillei

Cystophora grevillei has terete axes, generally with vesicles.

TYPE LOCALITY: Western Australia, probably near Fremantle.
DISTRIBUTION: From Dongara, Western Australia, around southern mainland Australia to Wilsons Promontory, Victoria, and around Tasmania.

(top) *Cystophora torulosa* – as illustrated by Harvey (1860, pl. 123).

Cystophora grevillei (Rottnest Island, WA).

Cystophora harveyi

A species restricted to the south-west of Western Australia. It is recognisable by its lateral axes being attached only to the central part of the main axis

TYPE LOCALITY: Cape Naturaliste, Western Australia.
DISTRIBUTION: From South Bunker Bay, Geographe Bay, to William Bay, Walpole, Western Australia.

Cystophora monilifera

A closely branched species characterised by its tristichous laterals.

TYPE LOCALITY: Western Australia.
DISTRIBUTION: From Nickol Bay, Western Australia, around southern mainland Australia and northern Tasmania to Long Bay, New South Wales.

Cystophora moniliformis

Cystophora moniliformis is recognisable by its flattened main axis with lateral branches arising from the edges.

TYPE LOCALITY: South coast of Australia.
DISTRIBUTION: From Cape Naturaliste, Western Australia, around southern mainland Australia and Tasmania to Port Stephens, New South Wales.

Cystophora pectinata

Cystophora pectinata is recognisable by its flattened and generally complanate lateral axes. It is similar to *Cystophora racemosa* but differs in lacking vesicles and being more closely branched.

TYPE LOCALITY: Southern Australia.
DISTRIBUTION: From Watermans Bay, Perth, Western Australia, around southern mainland Australia to Gulf St Vincent and Kangaroo Island, South Australia, and Walkerville, Victoria.

Cystophora racemosa

Cystophora racemosa is openly branched with markedly flattened receptacles.

TYPE LOCALITY: Cape Riche, Western Australia.
DISTRIBUTION: From Geographe Bay, Western Australia, around southern mainland Australia to Kangaroo Island, South Australia, and Queenscliff, Victoria.

Cystophora torulosa

A distinctive species in the intertidal on south-eastern Australian coasts, characterised by dense clusters of smooth, terete receptacles.

TYPE LOCALITY: Kent Group Islands, Bass Strait, Australia.
DISTRIBUTION: From Apollo Bay to Wilsons Promontory, Victoria, islands in Bass Strait, and around Tasmania. New Zealand.
FURTHER READING: Womersley (1987); Nelson (2013); Scott (2017).

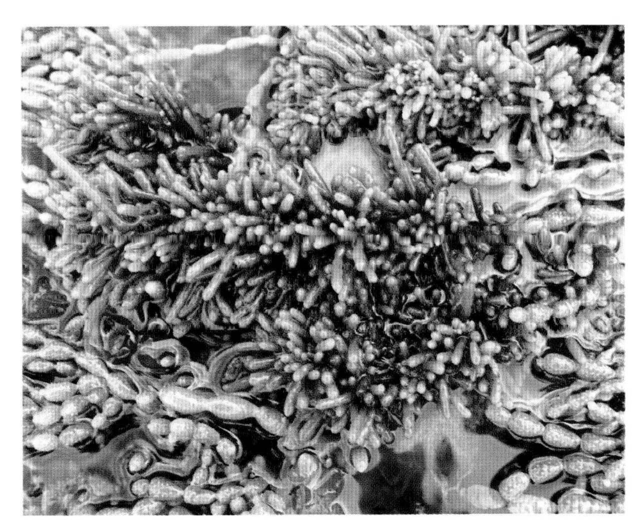

Cystophora harveyi (Gnarabup, WA).

Cystophora monilifera (Augusta, WA).

Cystophora moniliformis (Gnarabup, WA).

Cystophora pectinata (Gnarabup, WA).

Cystophora racemosa (Gnarabup, WA).

Cystophora torulosa (Point Lonsdale, Vic).

Hormophysa

The genus *Hormophysa* includes only a single species, the widely distributed *Hormophysa cuneiformis*. Thalli grow to 40 cm in length and have several branched axes arising from a short stipe. A distinctive feature of the species is the presence of lateral wings, which are borne in threes on the main axes and lateral branches. Thalli can be variable in shape and in some specimens the wings are greatly reduced, leaving only a three-sided axis. Vesicles are borne within the branches.

Hormophysa cuneiformis

TYPE LOCALITY: Suez, Egypt.
DISTRIBUTION: From Augusta, Western Australia, around western and northern Australia to Port Stephens, New South Wales; and isolated records from northern Spencer Gulf, South Australia. Widespread in tropical and subtropical waters of the Indo-Pacific.
FURTHER READING: Womersley (1987, as *Hormophysa triquetra*); Kraft (2009); Dixon & Huisman (2015).

Hormophysa cuneiformis (Rottnest Island, WA).

(above) *Hormophysa cuneiformis* – estuarine form (Shark Bay, WA).

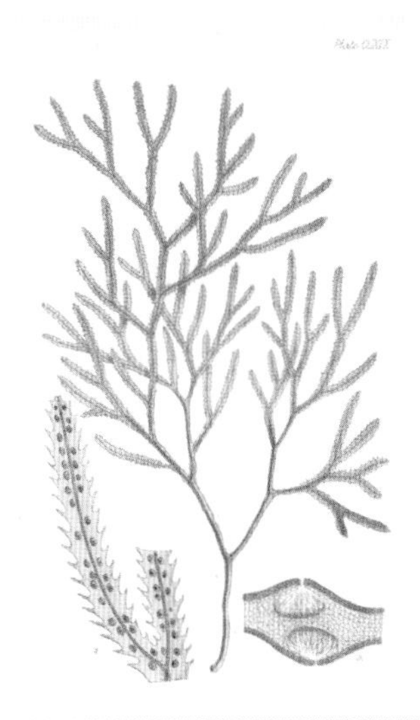

Myriodesma

A genus of eight species found only in Australia, *Myriodesma* is characterised by its subdichotomous, perennial stipe that bears flattened annual fronds. The fronds are pinnately to subdichotomously branched, generally have a prominent midrib and, in most species, have marginal spines. The thallus grows from a single apical cell and is composed of a central medulla and peripheral cortex. Reproductive structures are borne in conceptacles in the annual fronds. *Myriodesma serrulatum* has regularly serrate fronds with an irregular line of conceptacles on each side of the midrib. *Myriodesma quercifolium* is a large species with marginal spines.

Myriodesma serrulatum

TYPE LOCALITY: Australia.
DISTRIBUTION: From Port Denison to Cape Riche, Western Australia.

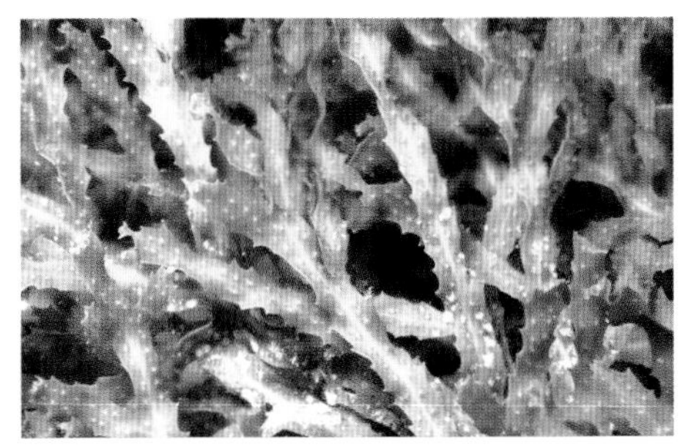

Myriodesma tuberosum

TYPE LOCALITY: Eucla, Western Australia.
DISTRIBUTION: From the Perth region, Western Australia, around south-western Australia to Port Elliot, South Australia.
FURTHER READING: Nizamuddin & Womersley (1967); Womersley (1987); Scott (2017).

Myriodesma quercifolium

TYPE LOCALITY: Southern Australia.
DISTRIBUTION: Geraldton, Western Australia, to Port Elliot, South Australia.

Myriodesma serrulatum (Cape Bouvard, WA).

(top) *Myriodesma serrulatum* – as illustrated by Harvey (1862, pl. 219, as *Myriodesma serrulata*).

Myriodesma tuberosum (Two Rocks, WA).

Myriodesma quercifolium (Cape Peron, WA).

Phyllotricha

Phyllotricha was first described as a genus in 1854, but for most of the subsequent years it has been treated as a subgenus or section of the widespread *Sargassum*. Recently the genus was resurrected as part of major revisions of *Sargassum* based on DNA analyses, which have also included recognition of *Sargassopsis*. In *Phyllotricha* the basal laterals are usually branched, most commonly in a pinnate arrangement. Primary branches may be terete, compressed or angular but never triquetrous. Four species are currently included in the genus, all of which occur in Australia.

Phyllotricha decipiens

TYPE LOCALITY: Port Dalrymple, Tamar Estuary, Tasmania.
DISTRIBUTION: Southern Australia from the Perth region of Western Australia to Tasmania.

Phyllotricha varians

TYPE LOCALITY: Western Australia (probably near Fremantle)
DISTRIBUTION: From northern Tasmania along southern Australia to the Perth region of Western Australia (possibly north to Shark Bay).
FURTHER READING: Womersley (1987, as *Sargassum*); Dixon et al. (2012).

(top) *Phyllotricha sonderi* - as illustrated by Harvey (1863, pl. 243, as *Sargassum sonderi*).

Phyllotricha decipiens (Kangaroo Island, SA; photo M.D. Guiry).

Phyllotricha varians (Gnarabup, WA).

Platythalia

This genus includes two robust species found in south-western Australia. Plants of *Platythalia* are characterised by their stoloniferous holdfast, pinnate branching and reproductive structures borne in normal vegetative branches. Structurally the thallus has a medulla of elongate cells and a small-celled cortex. *Platythalia quercifolia* can be distinguished by its unusual side branches with strongly spinous margins. *Platythalia angustifolia* is separated from *Platythalia quercifolia* by its narrower side branches that lack marginal spines. Both species are relatively common in areas of high wave action.

Platythalia quercifolia

TYPE LOCALITY: South coast, Australia.
DISTRIBUTION: From Geraldton south to the Recherche Archipelago, Western Australia.
FURTHER READING: Womersley (1987).

Platythalia angustifolia

TYPE LOCALITY: Fremantle, Western Australia.
DISTRIBUTION: From Geraldton south to Cape Riche, Western Australia.

Platythalia angustifolia (Cowaramup, WA).

Platythalia quercifolia (Rottnest Island, WA).

(top) *Platythalia quercifolia* – as illustrated by Harvey (1858, pl. 43, as *Carpoglossum quercifolium*).

Sargassopsis

The genus *Sargassopsis* was previously regarded as a synonym of *Sargassum*, but was resurrected for *Sargassum decurrens* after DNA sequence data showed it to be polyphyletic with the rest of *Sargassum* (see opposite page). Thalli of *Sargassopsis*, like those of *Sargassum*, have numerous primary branches arising from a short stipe; however, the lower laterals are branched and tapering, instead of leaf-like. Spherical vesicles are generally present, and reproductive structures are borne in receptacles that arise in the axils of lateral branches. *Sargassopsis decurrens* is a distinctive species, readily recognised by its flattened, winged primary branches.

Sargassopsis decurrens

TYPE LOCALITY: Northern Australia.
DISTRIBUTION: From Rottnest Island, Western Australia, around northern Australia to Keppel Bay, Queensland. New Caledonia.
FURTHER READING: Draisma et al. (2010); Dixon et al. (2012); Dixon & Huisman (2015); Scott (2017).

Sargassopsis decurrens (Cassini Island, WA).

(top) *Sargassopsis decurrens* – as illustrated by Harvey (1860, pl. 145, as *Sargassum decurrens*).

Sargassum

The genus *Sargassum* is ubiquitous in Australian waters. It is perhaps the most common (certainly the most conspicuous) alga occurring in tropical regions. In colder waters, that role is generally taken by the kelps and other large brown algae, but *Sargassum* is nevertheless still present. Numerous species have been described, but accurate identification is often reliant on fertile material and an understanding of the ecological and seasonal variation of the taxon. Thus, many species are poorly defined and the application of correct names is a difficult task, particularly in tropical regions. The species depicted here are provided as examples of the genus and are not comprehensive for Australia.

All species of *Sargassum* have numerous primary branches arising from a short stipe and most have unbranched, leaf-like laterals. The primary branches can be terete (*Sargassum linearifolium*), three-sided (*Sargassum lacerifolium*), or flattened (*Sargassum ligulatum, Sargassum oligocystum*). Spherical vesicles are generally present. Reproductive structures are borne in specialised fertile branches known as receptacles – these can be found in the axils of lateral branches, but occur only at certain times of the year. The morphology of the receptacles is important in distinguishing species. *Sargassum rasta* is an unusual species that has very dense, reduced lateral branches, giving the axes the appearance of dreadlocks.

Sargassum aquifolium

TYPE LOCALITY: Sunda Strait, Indonesia.
DISTRIBUTION: Widespread in the tropical Indo-Pacific.

Sargassum aquifolium (Long Reef, WA).

(top) *Sargassum lacerifolium* – as illustrated by Harvey (1862, pl. 208).

Sargassum lacerifolium

TYPE LOCALITY: Port Dalrymple, Tasmania.

DISTRIBUTION: Known from a few collections from Rottnest Island and Cape Peron, Western Australia. Otherwise from Pearson Island, South Australia, around south-eastern mainland Australia and Tasmania to Pebbly Beach, north of Batemans Bay, New South Wales.

Sargassum ligulatum

TYPE LOCALITY: Western Australia.

DISTRIBUTION: West coast of Western Australia.

Sargassum linearifolium

TYPE LOCALITY: West coast of Australia (probably King George Sound, Western Australia).

DISTRIBUTION: From the Houtman Abrolhos, Western Australia, around southern mainland Australia (possibly northern Tasmania) to New South Wales.

Sargassum oligocystum

TYPE LOCALITY: Lampung Bay, Sumatra, Indonesia.

DISTRIBUTION: Widespread in the tropical Indo-Pacific.

Sargassum lacerifolium (Cape Peron, WA).

Sargassum ligulatum (Houtman Abrolhos, WA).

Sargassum linearifolium (Cape Peron, WA).

Sargassum oligocystum (Heron Island, Qld).

Sargassum podacanthum

TYPE LOCALITY: Western Australia.
DISTRIBUTION: Cape Peron, Western Australia, to Port Noarlunga, South Australia.

Sargassum polycystum

TYPE LOCALITY: Sunda Strait, Indonesia.
DISTRIBUTION: Known from north-western Australia and Queensland. Canary Islands. Africa. Tropical Indo-Pacific as far east as Tonga.

Sargassum rasta

TYPE LOCALITY: Adele Island, Kimberley, Western Australia.
DISTRIBUTION: Widespread in north-western Australia.

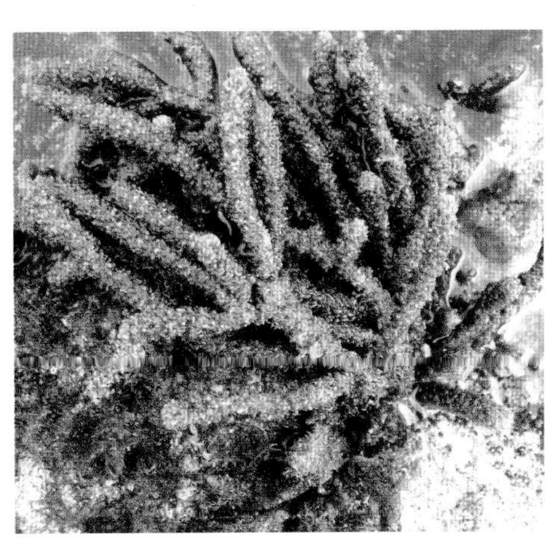

Sargassum tristichum

TYPE LOCALITY: Western Australia.
DISTRIBUTION: From Rottnest Island, Western Australia, to Port Noarlunga, South Australia.
FURTHER READING: Womersley (1987); Kraft (2009); Dixon et al. (2012); Dixon & Huisman (2015); Scott (2017).

Sargassum podacanthum (Cape Peron, WA).

Sargassum polycystum (Heron Island, Qld).

Sargassum rasta (Barrow Island, WA).

Sargassum tristichum (Cape Peron, WA).

Scaberia

The genus *Scaberia* includes only the single species *Scaberia agardhii*, a relatively common species that is usually found in areas of moderate to strong water movement. Thalli are irregularly divided and the upper branches are covered in a distinctive warty layer formed from reduced lateral branches.

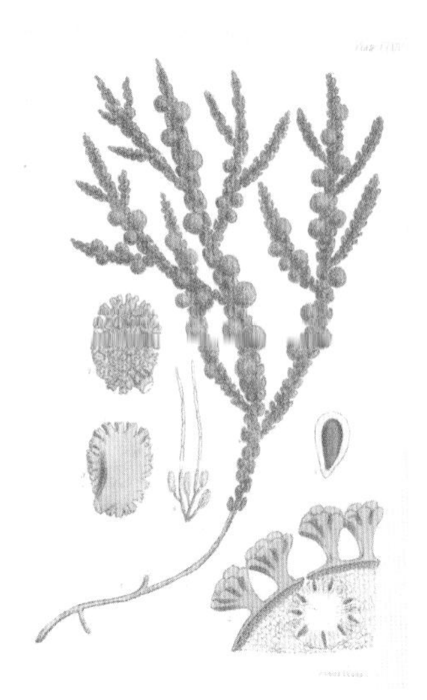

Scaberia agardhii

TYPE LOCALITY: Swan River Estuary, Western Australia.
DISTRIBUTION: Houtman Abrolhos, Western Australia, around southern mainland Australia and northern Tasmania to Collaroy and Lord Howe Island, New South Wales.
FURTHER READING: Womersley (1987); Scott (2017).

Scaberia agardhii (Cape Peron, WA).

(top) *Scaberia agardhii* – as illustrated by Harvey (1860, pl. 164).

Sirophysalis

A genus of two species, *Sirophysalis* is represented in Australia by the common *Sirophysalis trinodis*. Thalli have several main axes that are radially branched. Lateral axes are generally terete and can be branched or unbranched. In warmer waters, flattened, leaf-like branches arise near the base of the thallus, but these do not occur in specimens from southern and south-western Australia. Vesicles are generally formed within the lateral branches – they can be single or in short series of up to four vesicles. *Sirophysalis trinodis* was previously included in *Cystoseira*, but was moved following a DNA sequence study.

Sirophysalis trinodis

TYPE LOCALITY: Tor, Sinai Peninsula, Egypt.
DISTRIBUTION: Widely distributed in warmer Australian waters, with one form extending to Victor Harbor, South Australia. Tropical and subtropical Indo-Pacific.
FURTHER READING: Dixon & Huisman (2015).

Sirophysalis trinodis (Long Reef, WA).

(top) *Sirophysalis trinodis* – as illustrated by Harvey (1860, pl. 139, as *Cystophyllum muricatum*).

Turbinaria

A distinctive genus mostly restricted to the tropics, although *Turbinaria gracilis* can be found as far south as Hamelin Bay in south-western Australia. Thalli are tough and cartilaginous, with upright axes bearing the unusual turbinate laterals characteristic of the genus. The margins of these inflated branches are often bordered by fleshy wings or teeth – in *Turbinaria ornata*, a second ring of teeth arises on the upper surface, slightly offset from the margin. *Turbinaria ornata* is very common on reef flats in northern Australia.

Turbinaria gracilis

TYPE LOCALITY: Western Australia.
DISTRIBUTION: From Hamelin Bay, Western Australia, around northern Australia to Queensland.

Turbinaria ornata

TYPE LOCALITY: Not known.
DISTRIBUTION: From Coral Bay, Western Australia, around northern Australia to the southern Great Barrier Reef, Queensland. Widespread in tropical seas.
FURTHER READING: Kraft (2009); Dixon & Huisman (2015).

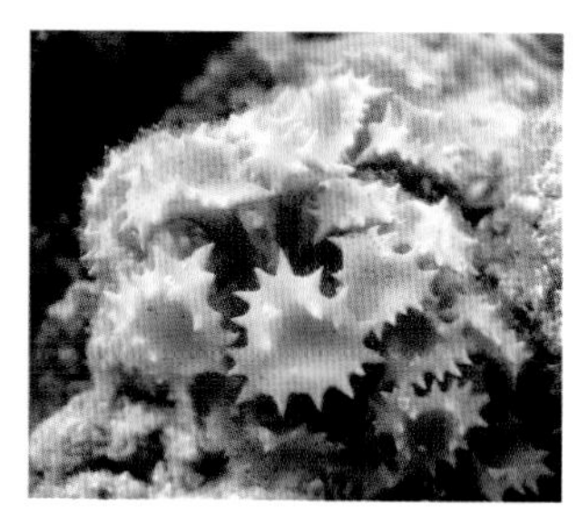

Turbinaria gracilis (Houtman Abrolhos, WA).

(top) *Turbinaria gracilis* – as illustrated by Harvey (1860, pl. 131).

Turbinaria ornata – dense stand on an intertidal platform (Seringapatam Reef, WA).

(left) *Turbinaria ornata* (Scott Reef, WA).

Phyllospora

The genus *Phyllospora* is restricted to the colder waters of south-eastern Australia, where it is common in the upper sublittoral region on rough-water coasts. It includes only a single species, *Phyllospora comosa*, also known as 'cray weed'. Thalli grow to about 3 m in length and have a flattened main axis that bears numerous, closely set, lateral branches. *Phyllospora* differs from the closely related *Scytothalia* (see p. 334) in producing vesicles at the bases of some lateral branches. Structurally the thallus has a central medulla composed of elongate cells and hyphae, and a parenchymatous cortex. Reproductive structures are borne in the lateral branches.

Phyllospora comosa

TYPE LOCALITY: South-east Tasmania.
DISTRIBUTION: From Robe, South Australia, around south-eastern mainland Australia and Tasmania to Port Macquarie, New South Wales.
FURTHER READING: Womersley (1987); Scott (2017).

Phyllospora comosa (Lorne, Vic).

(top) *Phyllospora comosa* – as illustrated by Harvey (1860, pl. 153).

Scytothalia

Scytothalia includes only the single species *Scytothalia dorycarpa*, a robust species generally found on rough-water coasts. Thalli grow to 2 m in length and have flattened axes that are slightly flexuous and alternately branched. Structurally the thallus has a central medulla, composed of elongate cells and hyphae, and a peripheral cortex of highly pigmented cells. Reproductive structures are borne in specialised short, lateral branches, known as receptacles, that are clustered on the margins of the thallus. *Scytothalia dorycarpa* is a distinctive species that can be recognised by its habit and the clusters of lateral receptacles.

Scytothalia dorycarpa

TYPE LOCALITY: King George Sound, Western Australia.
DISTRIBUTION: From Geraldton, Western Australia, around southern mainland Australia to Point Lonsdale, Victoria, and the north coast of Tasmania.
FURTHER READING: Womersley (1987); Scott (2017).

Scytothalia dorycarpa (Rottnest Island, WA).

(top) *Scytothalia dorycarpa* – as illustrated by Harvey (1858, pl. 9).

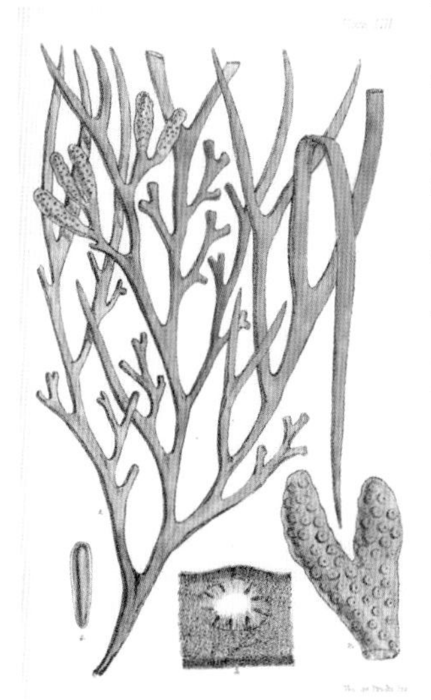

Xiphophora

A genus of two species, *Xiphophora chondrophylla* and *Xiphophora gladiata*, both endemic to southern Australia and New Zealand. Thalli of *Xiphophora* are dichotomously branched with compressed axes that lie more or less in one plane. Vesicles are absent. The two species differ in their habit and features of their reproductive structures (receptacles), with *Xiphophora chondrophylla* having shorter branch intervals (1–2 cm) and several times divided receptacles, while branch intervals in *Xiphophora gladiata* are 10–30 cm and receptacles are simple or only one or two times divided.

Xiphophora gladiata

TYPE LOCALITY: South-east Tasmania.
DISTRIBUTION: Around Tasmania (rare on the north coast) and Western Port, Victoria.
FURTHER READING: Womersley (1987); Nelson (2013); Scott (2017).

Xiphophora chondrophylla

TYPE LOCALITY: Port Dalrymple (Tamar Estuary mouth), Tasmania.
DISTRIBUTION: From Point Labatt, Searcy Bay, South Australia, to Walkerville, Victoria, Bass Strait islands and the north coast of Tasmania. New Zealand.

Xiphophora chondrophylla (Lorne, Vic).

Xiphophora gladiata (Fortescue Bay, Tas; photo M.D. Guiry).

(top) *Xiphophora gladiata* – as illustrated by Harvey (1858, pl. 53, as *Fucodium gladiatum*).

Division Chlorophyta

THE GREEN ALGAE

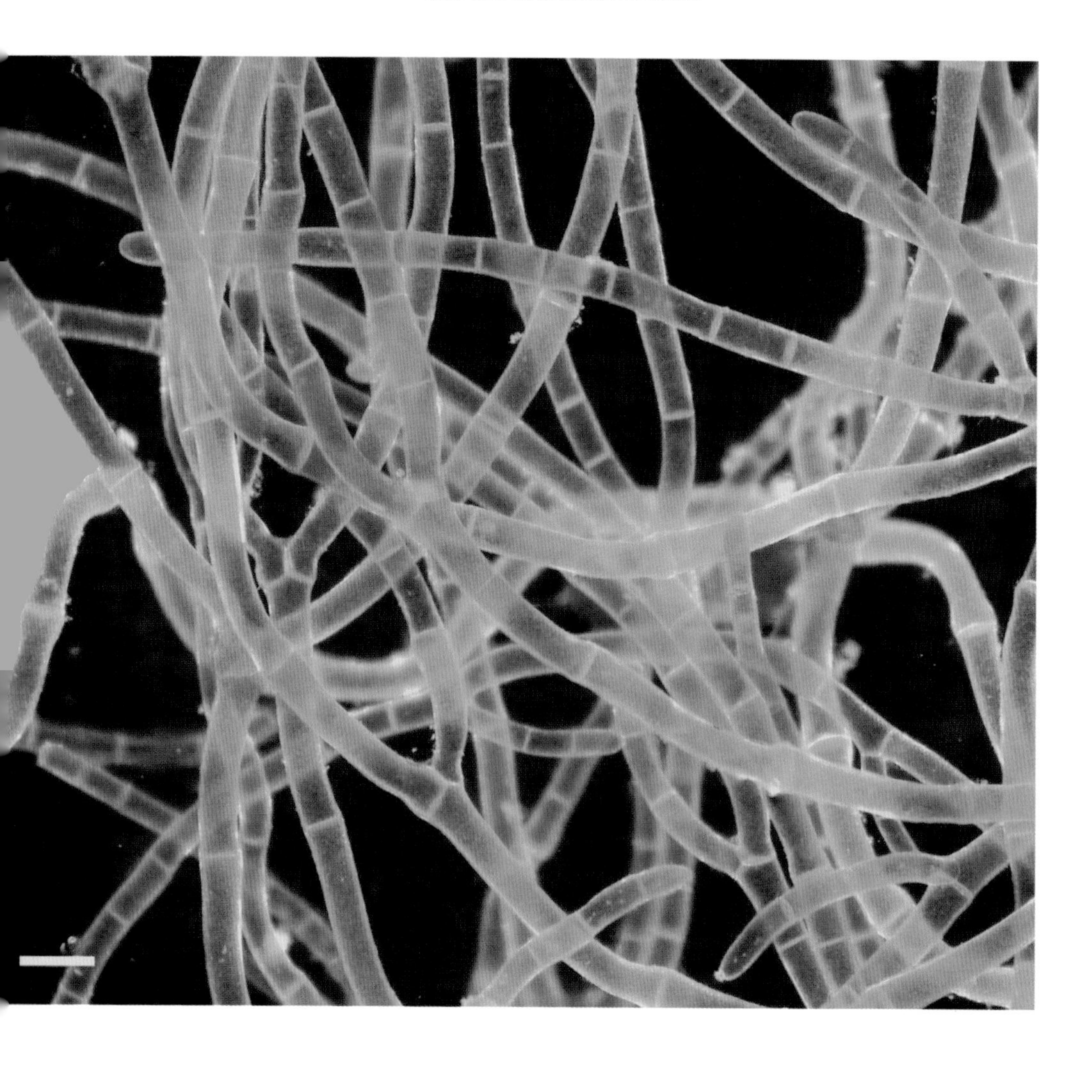

Cladophora valonioides – detail of filaments. Scale = 500 µm.

The division Chlorophyta (from the Greek* chloros *meaning 'green' and* phyton *meaning 'plant') enjoys a special place in evolutionary history, for it is thought that all terrestrial plants evolved from the green algae.

This belief is based on a number of characteristics shared by the two groups. The green colouration of the Chlorophyta is due to the presence of chlorophyll *a* and *b*, and the absence of any masking accessory pigments. In most green algae, a motile, unicellular phase is present during some stage of the life history. These motile phases bear two or more flagella that are similar in structure, a condition known as 'isokont'.

Green algae can be found in most habitats, from marine to freshwater to terrestrial, and can have unicellular, multicellular, or coenocytic thalli (with a large, multinucleate cell without cross-walls comprising the entire thallus). Some 8,000 species are known, of which approximately 2,000 are thought to occur in Australia (from all habitats).

Recognition of the green algae is generally straightforward – their colour is usually grass-green or a slightly greyish hue of the same colour. They never have a yellowish tinge as can be found in some red or brown algae.

Palmoclathrus

Palmoclathrus includes only *Palmoclathrus stipitatus*, a very rare species known from deep-water habitats in southern Australia. Thalli are dark green, with terete stipes bearing expanded blades terminally. The blades have perforations and are structurally composed of dispersed cells in a mucilaginous matrix. A similar structure is also found in *Palmophyllum* (see opposite page), but that genus only forms simple crusts and does not have the morphological complexity of *Palmoclathrus*.

Palmoclathrus stipitatus

TYPE LOCALITY: Waldegrave Island, South Australia.
DISTRIBUTION: Known from the Recherche Archipelago, Western Australia, to Portland Bay, Victoria.
FURTHER READING: Womersley (1971), (1984).

Palmoclathrus stipitatus (Waldegrave Island, SA; photo K.L. Branden).

Palmophyllum

A genus of two species, with *Palmophyllum umbracola* known from isolated locations in Australia. Plants form dark green, irregularly contoured crusts that are firmly attached to the substratum, generally in deeper water, but the species can also be found on shaded undercut reef walls in the shallows. The thallus can be smooth or warty and has an unusual structure with numerous isolated cells embedded in a clear gelatinous matrix. The cells are spherical and very small, generally less than 10 µm in diameter.

Palmophyllum umbracola

TYPE LOCALITY: L'Esperance Rock, Kermadec Islands, New Zealand.
DISTRIBUTION: Known from Rottnest Island and Cape Peron, Western Australia, south-eastern Australia, Lord Howe Island, the southern Great Barrier Reef. New Zealand.
FURTHER READING: Nelson & Ryan (1986); Kraft (2007); Nelson (2013).

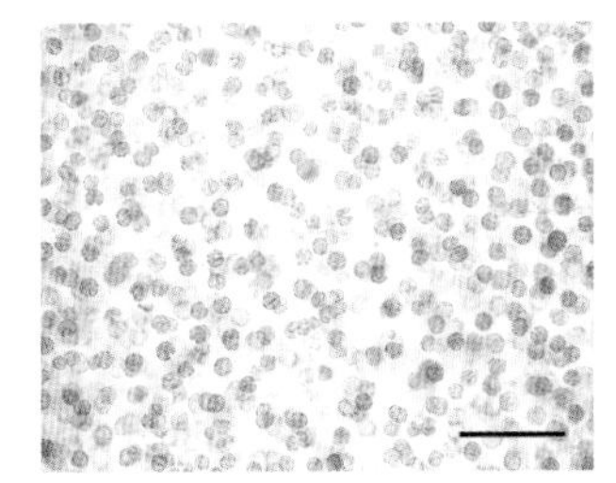

Palmophyllum umbracola (Cape Peron, WA).

(above) *Palmophyllum umbracola* – section of thallus (Cape Peron, WA). Scale = 50 µm.

Uronema

Thalli of *Uronema* are microscopic and consist of narrow, unbranched green filaments anchored by a small gelatinous pad at the base. The genus includes primarily freshwater species, with *Uronema marinum* the sole marine representative. Plants often occur as epiphytes on other algae, but because of their small size – generally less than 1 mm in length – they tend to be overlooked. Other genera of unbranched filamentous green algae, such as *Chaetomorpha* (see p. 352), are considerably larger and their cells contain multiple chloroplasts, whereas cells of *Uronema* have only a single chloroplast.

Uronema marinum

TYPE LOCALITY: Kellidie Bay, Coffin Bay, South Australia.
DISTRIBUTION: Occurs in southern and western Australia, Lord Howe Island, the Great Barrier Reef. Micronesia. Hawaiian Islands.
FURTHER READING: Womersley (1984); Kraft (2007); Huisman (2015).

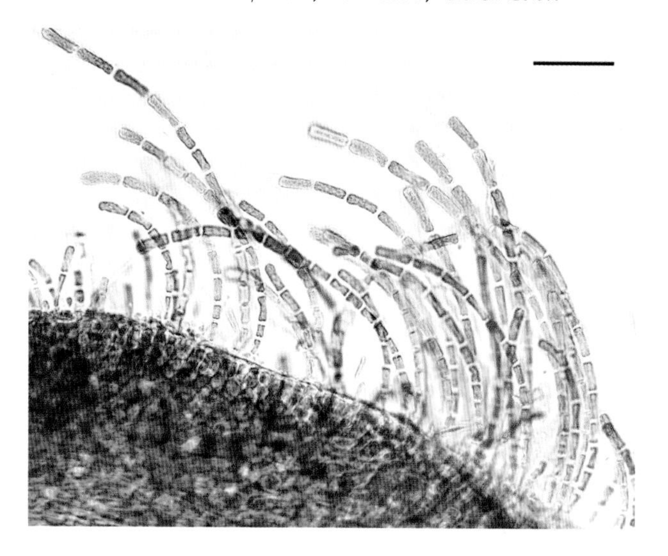

Uronema marinum (Ningaloo Reef, WA). Scale = 50 µm.

Gayralia

Gayralia includes two species, of which the widespread *Gayralia oxysperma* occurs in Australia. Thalli are green and membranous and, in many respects, resemble blade-like species of *Ulva* (see p. 344), but *Gayralia* can be distinguished by its blades being only a single cell layer thick, while those of *Ulva* have two layers. In surface view, the cells are often arranged in discrete packets of two or four, these separated by thick layers of extracellular material. *Gayralia oxysperma* commonly occurs as an epiphyte on mangrove pneumatophores.

Gayralia oxysperma

TYPE LOCALITY: Baltic Sea.
DISTRIBUTION: Widely distributed in temperate to tropical regions of the world.
FURTHER READING: Womersley (1984, as *Ulvaria oxysperma*); Huisman et al. (2007); Huisman (2015).

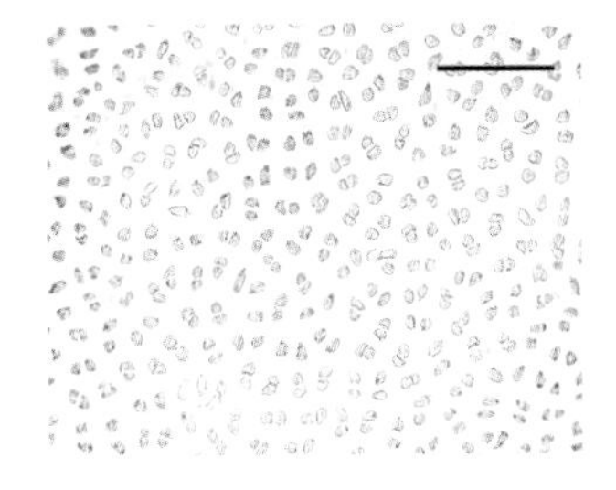

Gayralia oxysperma – growing on pneumatophores of *Avicennia marina* (Shark Bay, WA).

(above) *Gayralia oxysperma* – surface view of thallus (Shark Bay, WA). Scale = 50 µm.

Ulva

Species of the genus *Ulva* occur throughout the world and are common in most regions. In the past, the genus was restricted to species composed of flattened blades that are two cells thick, but recent studies have shown that tubular species, previously classified in the separate genus *Enteromorpha*, should also be included in *Ulva*. Thalli of blade-forming species can be divided into branches (*Ulva stenophylloides*) or remain as a single frond (*Ulva fasciata* and *Ulva lactuca*), although there is also considerable habit variation within single species. Tube-forming species such as *Ulva flexuosa* often occur on rocks in the high intertidal. Separation of species is generally based on thallus form and microscopic features, but this can be unreliable and DNA sequencing is now the preferred method. *Ulva* is one of the few algae that are known by a common name, in this case 'sea lettuce'. This name reflects both the appearance of the alga and its use as a food in many countries.

Ulva compressa

TYPE LOCALITY: Probably Bognor, Sussex, England.
DISTRIBUTION: Widespread in most seas.

Ulva fasciata

TYPE LOCALITY: Alexandria, Egypt.
DISTRIBUTION: Widespread in most seas.

Ulva flexuosa

TYPE LOCALITY: Duino (near Trieste), Adriatic Sea.
DISTRIBUTION: Widespread in most seas.

Ulva lactuca

TYPE LOCALITY: West coast of Sweden.
DISTRIBUTION: Widespread in most oceans, mostly in colder waters.

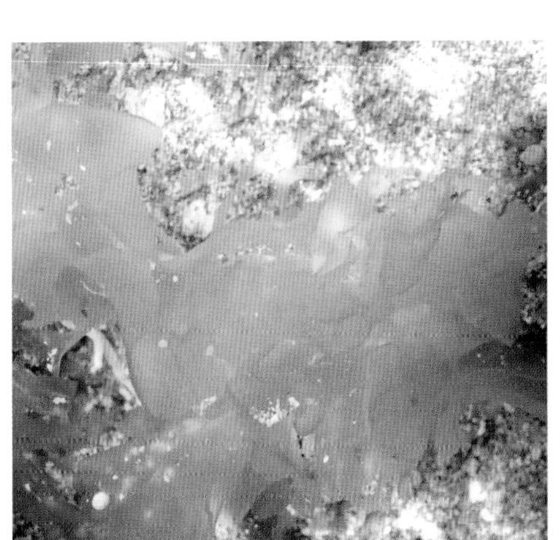

Ulva stenophylloides

TYPE LOCALITY: Lighthouse Reef, Point Lonsdale, Victoria.
DISTRIBUTION: Cape Peron, Western Australia, probably around southern Australia to Point Lonsdale, Victoria.
FURTHER READING: Hayden et al. (2003); Kraft (2007); Kraft et al. (2010); Kirkendale et al. (2013); Huisman (2015); Scott (2017).

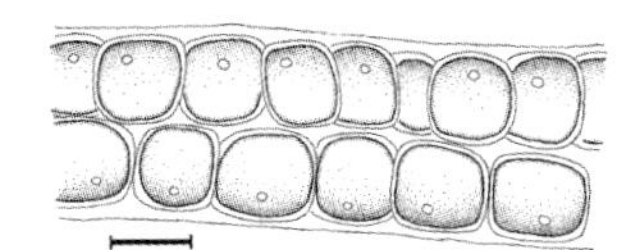

Ulva compressa (Point Roadknight, Vic).

(top) *Ulva compressa* – detail of thallus (Point Roadknight, Vic). Scale = 100 µm.

Ulva fasciata (Cape Peron, WA).

Ulva flexuosa (Maret Islands, WA).

Ulva stenophylloides (Cape Peron, WA).

Ulva lactuca (Barrow Island, WA).

(right) *Ulva lactuca* – section of thallus (Houtman Abrolhos, WA). Scale = 20 µm.

Ulvella

Ulvella forms microscopic plants that grow on the surfaces of a variety of algal and other hosts. Thalli are composed of tufted or prostrate radiating filaments that coalesce laterally to form a flattened, irregularly shaped or circular disc, which mostly remains one cell thick but occasionally produces extra layers towards the centre. In some countries *Ulvella lens* is cultivated and fed to abalone.

Ulvella lens

TYPE LOCALITY: Brest, Finistère, France.
DISTRIBUTION: Widespread in most seas.
FURTHER READING: Kraft (2007); Nielsen et al. (2013); Huisman (2015).

Ulvella lens – epiphytic on *Pterocladiella caerulescens* (Ningaloo Reef, WA). Scale = 25 µm.

Anadyomene

Anadyomene forms blade-like thalli that are composed of polychotomously branched filaments with numerous lateral branches that anastomose with adjacent filaments. Plants are seen at their best when viewed under a microscope or magnifying glass, when the intricate patterns of the filaments become clear. Several species are recorded from Australia. *Anadyomene plicata* is commonly found in intertidal regions in tropical Australia, where its crisp, bright-green thalli stand erect even during periods of exposure.

Anadyomene plicata

TYPE LOCALITY: Rauki, Waigeo Island, Maluku, Indonesia.
DISTRIBUTION: From the Houtman Abrolhos, Western Australia, across northern Australia. Indonesia. Solomon Islands. Philippines.
FURTHER READING: Huisman & Leliaert (2015).

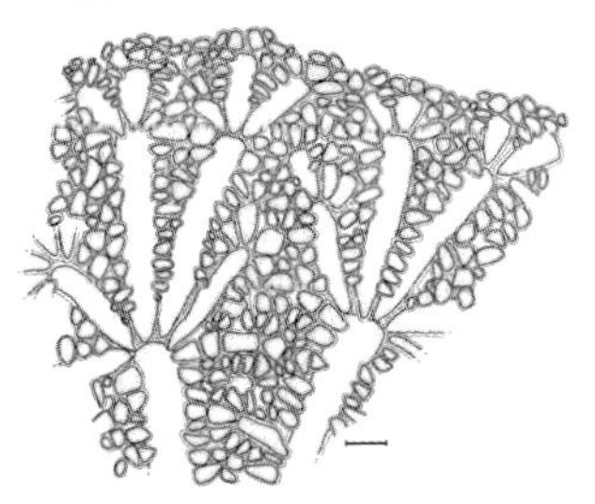

Anadyomene plicata (North Turtle Island, WA).

(above) *Anadyomene plicata* – detail of thallus (North Turtle Island, WA). Scale = 100 µm.

(right) *Anadyomene plicata* – detail of thallus (Barrow Island, WA). Scale = 100 µm.

Microdictyon

The blade-like thallus of *Microdictyon* is formed of anastomosing branched filaments that are arranged in one plane. Plants bear some resemblance to *Ulva*, but are readily distinguished by the reticulate thallus. *Microdictyon umbilicatum* is widespread in most Australian seas.

Microdictyon umbilicatum

TYPE LOCALITY: New South Wales (possibly Port Jackson).
DISTRIBUTION: Probably Australia-wide and widespread in the Indian and Pacific Oceans.
FURTHER READING: Kraft (2007); Huisman & Leliaert (2015).

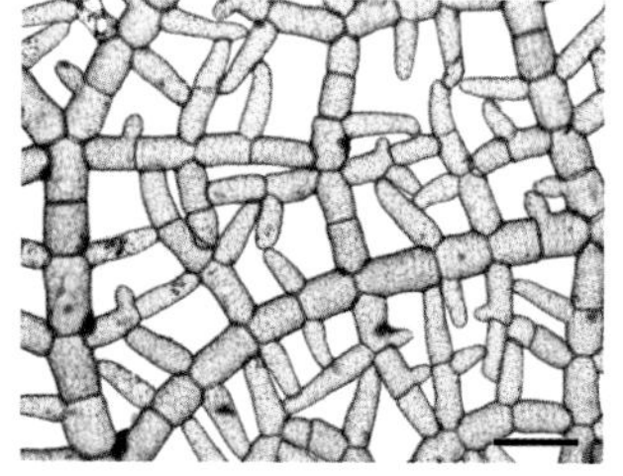

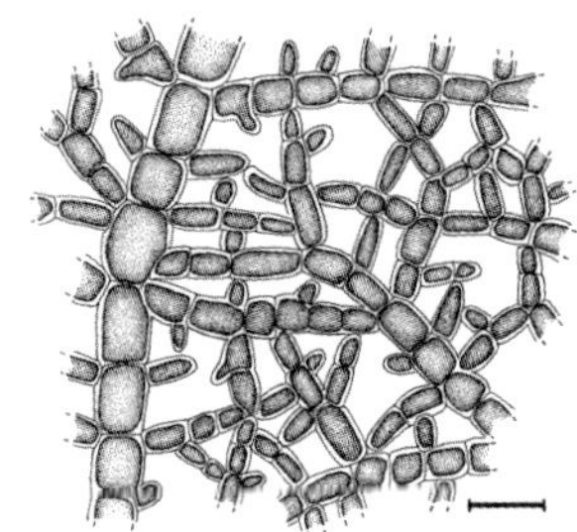

(top) *Microdictyon umbilicatum* – as illustrated by Harvey (1858: pl. 50, as *Microdictyon agardhianum*).

Microdictyon umbilicatum (Masthead Island, Qld).

(above) *Microdictyon umbilicatum* – detail of thallus (Rottnest Island, WA). Scale = 250 µm.

(left) *Microdictyon umbilicatum* – detail of thallus (Rottnest Island, WA). Scale = 200 µm.

Boodlea

Thalli of *Boodlea* have the appearance of an amorphous green sponge. They are composed of filaments that branch in all directions, with the branches becoming secondarily attached to one another. Reproduction is probably by fragmentation, as portions of the thallus are easily dislodged and can re-attach. *Boodlea composita* is common in tropical regions and can be a conspicuous element of intertidal reefs. *Boodlea vanbosseae* is found on several offshore atolls and islands, where it forms low dense mats on hard substrata.

Boodlea composita

TYPE LOCALITY: Mauritius.
DISTRIBUTION: From the Houtman Abrolhos, Western Australia, around northern Australia to Queensland and Lord Howe Island, New South Wales. Warmer waters of the Indo-Pacific.

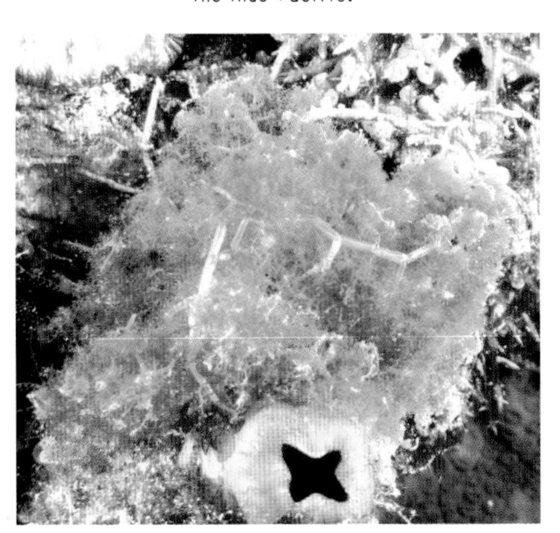

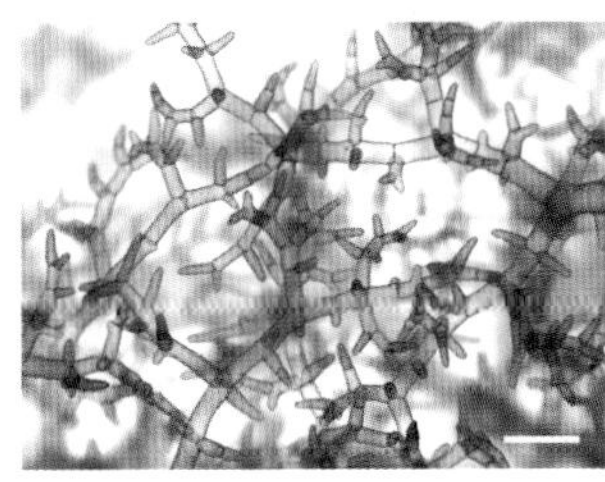

Boodlea vanbosseae

TYPE LOCALITY: Lucipara Island, Indonesia.
DISTRIBUTION: North-western Australia and south-eastern Queensland. Tropical Indo-Pacific.
FURTHER READING: Kraft (2007); Leliaert et al. (2007); Huisman & Leliaert (2015).

Boodlea composita (Sykes Reef, Qld).

(above) *Boodlea composita* – detail of thallus (Coral Bay, WA). Scale = 500 µm.

Boodlea vanbosseae (Scott Reef, WA).

Cladophoropsis

Thalli of *Cladophoropsis* are composed of entangled filaments that form spreading mats or cushions. The genus differs from the closely related *Cladophora* in the absence of cellular cross-walls at the bases of lateral branches, which therefore have cytoplasmic continuity with the bearing cell. In addition, rhizoids often form from the lower ends of cells. Several species of *Cladophoropsis* are known in Australian seas. *Cladophoropsis vaucheriiformis* differs from other species in the genus in forming a regular association with a sponge symbiont. It occurs as mats or as erect forms with finger-like processes.

Cladophoropsis vaucheriiformis

TYPE LOCALITY: Mauritius.
DISTRIBUTION: From the North West Cape region, Western Australia, around northern Australia to Lord Howe Island, New South Wales. Widespread in warmer waters.
FURTHER READING: Huisman & Leliaert (2015).

Cladophoropsis vaucheriiformis (Broomfield Reef, Qld)

(above) *Cladophoropsis vaucheriiformis* (Cassini Island, WA).

Struvea

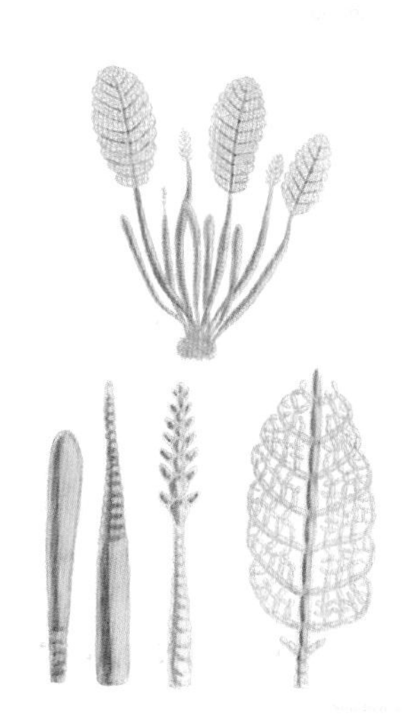

Thalli of *Struvea* have an elongate stipe bearing a flattened, reticulate blade derived from the secondary attachment of lateral branches. *Struvea plumosa* is one of the more distinctive and attractive members of the Australian marine flora. Individual thalli occur, but the species is more often encountered growing in clusters. Stipes are clavate or elongate and have regular constrictions. Blade development is seasonal and stipes are often found without any lateral branches. These can be easily confused with the stipes of *Apjohnia laetevirens* (which displays a similar seasonal growth), and without mature blades there is very little to distinguish between the two species. *Struvea thoracica* is a deep-water species known only from the southern Great Barrier Reef.

Struvea plumosa

TYPE LOCALITY: Western Australia.
DISTRIBUTION: From Barrow Island, Western Australia, around southern Australia to Encounter Bay, South Australia.

Struvea thoracica

TYPE LOCALITY: Wistari Channel, Capricorn Group, southern Great Barrier Reef, Queensland, Australia.
DISTRIBUTION: Known only from the vicinity of the type locality.
FURTHER READING: Kraft & Wynne (1996); Kraft (2007); Huisman (2015).

Struvea plumosa (Cape Peron, WA).

(top) *Struvea plumosa* – as illustrated by Harvey (1858, pl. 32).

Struvea thoracica (Heron Island, Qld).

Chaetomorpha

Plants of *Chaetomorpha* are similar in appearance to *Cladophora* (see p. 354) but differ in being entirely unbranched, instead typically composed of long, often entangled, filaments. Separation of species is based on habit and the diameter and length/breadth ratios of cells within the filaments. Two of the species included here, *Chaetomorpha aerea* and *Chaetomorpha linum*, are distinguished based on the presence of a basal attachment in *Chaetomorpha aerea* versus loose-lying in *Chaetomorpha linum*, but may be forms of a single species. *Chaetomorpha coliformis* is an attractive species with bead-like cells.

Chaetomorpha aerea

TYPE LOCALITY: Cromer, Norfolk, England.
DISTRIBUTION: Widespread in warm and temperate waters. This species also occurs in freshwater.

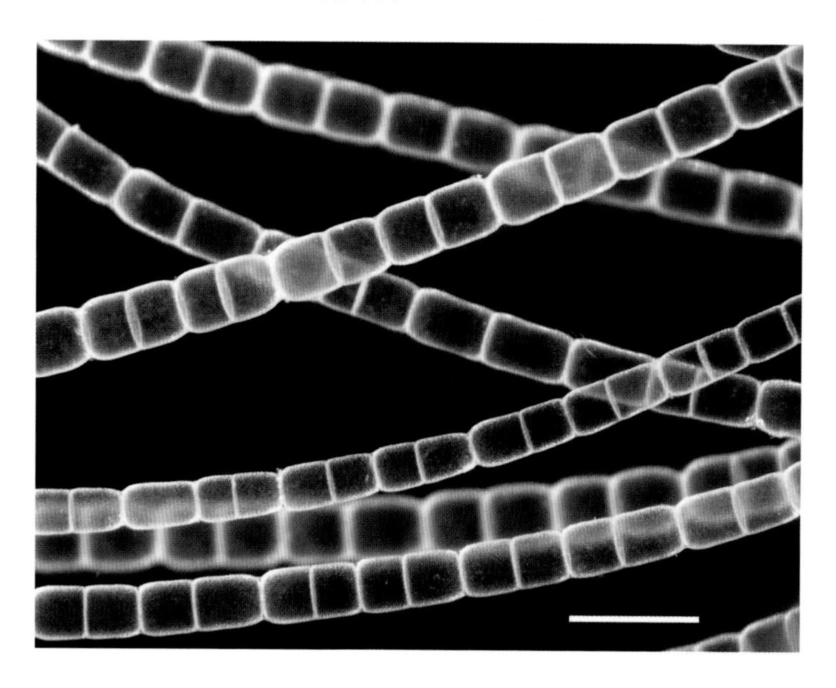

Chaetomorpha coliformis

TYPE LOCALITY: Probably near Hobart, Tasmania.
DISTRIBUTION: From Venus Bay, Eyre Peninsula, South Australia, to Walkerville, Victoria, and around Tasmania. New Zealand. South America.

Chaetomorpha linum

TYPE LOCALITY: Nakskov Fjord, Lolland, Denmark.
DISTRIBUTION: Widespread in warm and temperate waters.
FURTHER READING: Womersley (1984); Kraft (2007); Huisman & Leliaert (2015); Scott (2017).

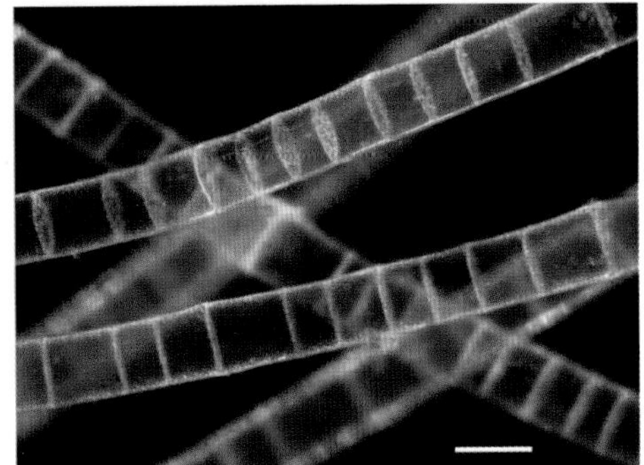

Chaetomorpha aerea – detail of filaments (Cape Peron, WA). Scale = 1 mm.

Chaetomorpha coliformis (Fortescue Bay, Tas; photo M.D. Guiry).

Chaetomorpha linum (Walpole, WA).

(above right) *Chaetomorpha linum* – detail of filaments (Swan Estuary, WA). Scale = 400 µm.

Cladophora

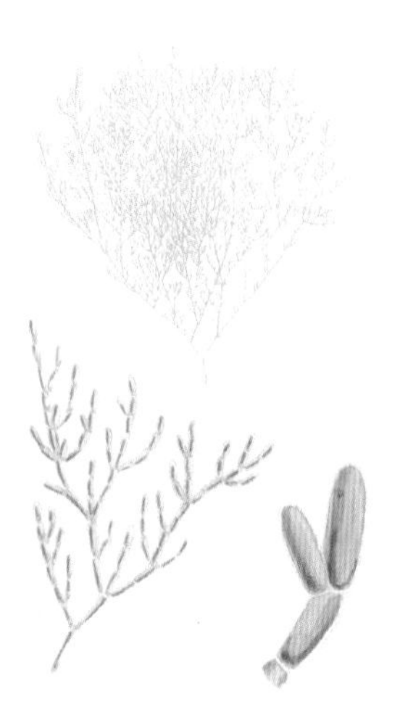

A genus of many widespread species, *Cladophora* includes branched, filamentous plants with medium to large multinucleate cells. A distinctive feature is the presence of complete cellular cross-walls forming discrete cells. In other superficially similar genera (e.g. *Bryopsis*) cross-walls are not formed and the cell contents are continuous. Species of *Cladophora* occupy a variety of habitats, ranging from individual thalli attached to rock in coastal regions, to dense, free-floating mats in estuaries. The genus is also well represented in freshwater lakes and streams. Many species occur in Australia and accurate identification beyond genus level is best attempted with fresh material and knowledge of its growth habit.

Cladophora coelothrix

Cladophora coelothrix is a tufted, dark-green species, often common in shallow water on rough coasts.

TYPE LOCALITY: Livorno, Italy.
DISTRIBUTION: A cosmopolitan species.

Cladophora fuliginosa

A coarse species that forms clumps of arching filaments, generally epilithic in the intertidal and subtidal, especially on reef flats.

TYPE LOCALITY: Bahamas.
DISTRIBUTION: Widespread in tropical seas.

Cladophora vagabunda

Branching in *Cladophora vagabunda* is generally secund, with recurved branchlets.

TYPE LOCALITY: A salt marsh on the island of Selsey, Sussex, England.
DISTRIBUTION: A cosmopolitan species.

Cladophora valonioides

Cladophora valonioides can form dense blooms during summer, coating other algae such as *Sargassum*.

TYPE LOCALITY: Western Australia.
DISTRIBUTION: From Champion Bay, Western Australia, around south-western Australia to Guichen Bay, South Australia.
FURTHER READING: Kraft (2007); Huisman & Leliaert (2015); Scott (2017).

Cladophora vagabunda (Cape Peron, WA).

(above) *Cladophora vagabunda* - branch detail (Cape Peron, WA).

Cladophora valonioides (Cape Peron, WA).

(opposite top) *Cladophora valonioides* - as illustrated by Harvey (1859a, pl. 78).

Cladophora coelothrix (Shoalwater Bay, WA).

Cladophora fuliginosa (Little Turtle Island, WA).

Lychaete

Lychaete includes several species previously included in the genus *Cladophora*, but shown by DNA sequence analyses to represent a separate genus. Most species in *Lychaete* have large, coarse thalli, with growth occurring near branch apices and often with distinct stipes. However, some species, such as *Lychaete herpestica*, differ in forming cushion-like clumps composed of arching filaments. The species is recognisable by its unusual morphology at branching points and non-septate lateral branches.

Lychaete herpestica

TYPE LOCALITY: Bay of Islands, New Zealand.

DISTRIBUTION: Known from the Houtman Abrolhos, Western Australia, around northern Australia to Queensland. Also in the Indo-Pacific, including Sri Lanka, Japan, Lord Howe Island and New Zealand.

FURTHER READING: Kraft (2007, as *Cladophora herpestica*); Huisman & Leliaert (2015, as *Cladophora herpestica*); Boedeker et al. (2016, as *Acrocladus herpesticus*); Wynne (2017).

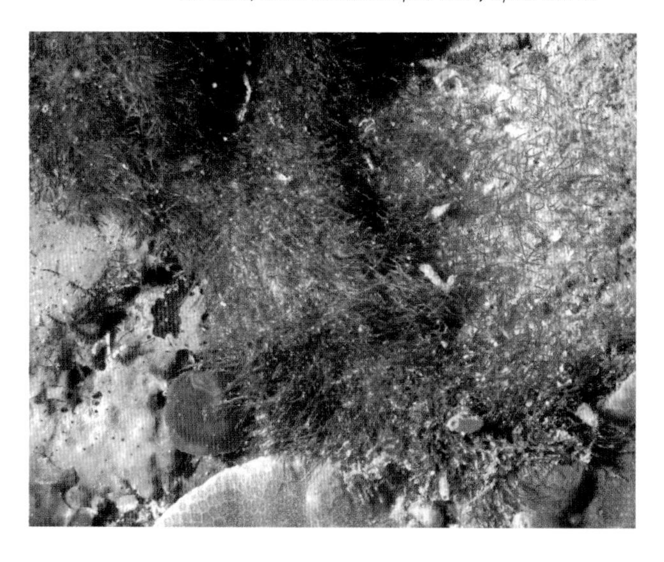

Lychaete herpestica (Rowley Shoals, WA).

Dictyosphaeria

Dictyosphaeria includes species with clusters of cells aggregated into various forms, but with a common feature of undergoing an unusual form of cell division known as segregative division. In this type of division, the cell contents of the parent cell form numerous spheres of cytoplasm well before the cell walls of the new cells are laid down. This condition is not obvious in cells that are not dividing and can be difficult to observe. Species of *Dictyosphaeria* can usually be recognised by their habit – the tropical species (such as *Dictyosphaeria cavernosa* and *Dictyosphaeria versluysii*) form very firm, irregularly shaped but generally globose thalli that are usually found in shallow waters. They differ in their colour and structure (dark green and becoming hollow in *Dictyosphaeria cavernosa*, grey-green and solid in *Dictyosphaeria versluysii*). *Dictyosphaeria sericea* is unusual in this normally tropical genus in forming a blade-like thallus and also in inhabiting the cooler waters of southern Australia.

Dictyosphaeria cavernosa

TYPE LOCALITY: Al-Qunfudhab, Saudi Arabia. Mokha, Yemen.
DISTRIBUTION: From the Houtman Abrolhos, Western Australia, around northern Australia to Lord Howe Island, New South Wales. Widespread in tropical and subtropical seas.

Dictyosphaeria cavernosa
(Heron Island, Qld).

Dictyosphaeria sericea

TYPE LOCALITY: Western Australia.
DISTRIBUTION: From Rottnest Island, Western Australia, around southern mainland Australia to Walkerville, Victoria, and northern Tasmania.

Dictyosphaeria versluysii

TYPE LOCALITY: Malaysian Archipelago.
DISTRIBUTION: From Rottnest Island (rarely), Western Australia, around northern Australia to the southern Great Barrier Reef, Queensland. Widespread in tropical seas.
FURTHER READING: Womersley (1984); Kraft (2007); Huisman (2015).

Dictyosphaeria sericea (Rottnest Island, WA).

Dictyosphaeria versluysii (Cassini Island, WA).

Valoniopsis

A genus of two species, with *Valoniopsis pachynema* being widely distributed in tropical and subtropical seas, including those of Australia. Plants are a medium to dark-green colour and composed of cylindrical segments that are closely aggregated and mostly oriented perpendicular to the substratum, forming firm, cushion-like clumps. The habit of *Valoniopsis pachynema* is similar to that of *Cladophora fuliginosa*, and the two species can occupy similar habitats. The former is characterised by compact cushions with mostly erect filaments and branches arising in clusters, while the latter has narrower, curved filaments with branches arising singly and reattaching at the apices.

Valoniopsis pachynema

TYPE LOCALITY: Benkulen [Bengkulu] and Pulau Tikus, near Bengkulu, Sumatra, Indonesia.

DISTRIBUTION: From Direction Island, Western Australia, around northern Australia to Jervis Bay, New South Wales. Widespread in tropical and subtropical seas.

FURTHER READING: Kraft (2007); Huisman & Leliaert (2015).

Valoniopsis pachynema (Barrow Island, WA).

Apjohnia

Apjohnia is a small genus comprising only two species worldwide, of which *Apjohnia laetevirens* can be found in Australia. *Apjohnia laetevirens* has an essentially southern Australian distribution. Plants can grow to a height of 25 cm and are often found in shallow waters. They are composed of firm filaments and can be easily recognised because of the regular annular constrictions formed at the base of each branch. Immature plants, which consist of only a single, club-shaped cell, are often found.

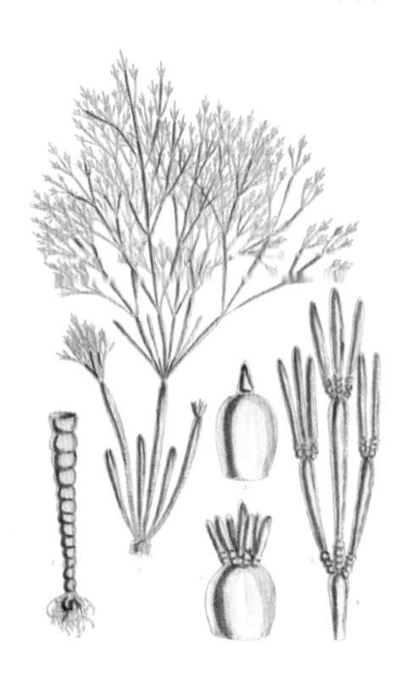

Apjohnia laetevirens

TYPE LOCALITY: Phillip Island, Victoria.
DISTRIBUTION: From Green Head, Western Australia, around southern mainland Australia and King and Deal Islands, Tasmania, to Collaroy, New South Wales.
FURTHER READING: Womersley (1984).

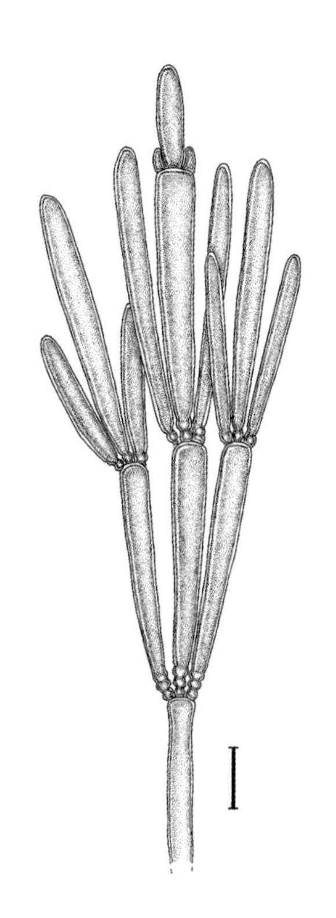

Apjohnia laetevirens (Port Lincoln, SA; photo G.J. Edgar).

(top) *Apjohnia laetevirens* – as illustrated by Harvey (1858, pl. 5)

(above) *Apjohnia laetevirens* – detail of branches with annular constrictions (Rottnest Island, WA). Scale = 2 mm.

Boergesenia

A distinctive genus, recognisable by its large, club-shaped vesicles that occur singly or in clusters. Three species are known, of which *Boergesenia forbesii* is locally common in tropical Australia, where it can be found in both the intertidal and shallow subtidal habitats, often growing in rosette clusters. A much rarer species is *Boergesenia magna*, which is known only from deep-water habitats. It is larger, and occurs as solitary individuals.

Boergesenia forbesii

TYPE LOCALITIES: Ryukyo-retto, Japan. Sri Lanka.
DISTRIBUTION: From the North West Cape region, Western Australia, around northern Australia to the Solitary Islands, New South Wales. Widespread in the Indo-Pacific.

Boergesenia magna

TYPE LOCALITY: Wistari Reef, Capricorn Group, southern Great Barrier Reef, Queensland.
DISTRIBUTION: Known only from the southern Great Barrier Reef, Queensland.
FURTHER READING: Kraft (2007); Huisman & Leliaert (2015).

Boergesenia forbesii (Little Turtle Island, WA).

Boergesenia magna (Heron Island, Qld).

Siphonocladus

Thalli of *Siphonocladus* form tufts of cylindrical or club-shaped axes, which secondarily divide into numerous segments and become radially branched by a process known as 'segregative division'. In this process, the cell contents of the parent filament form numerous small spheres that eventually distend, become contiguous, and lay down cell walls. Lateral branches are formed by protrusions of the outer wall. One species, *Siphonocladus tropicus*, occurs in Australia.

Siphonocladus tropicus

TYPE LOCALITIES: Moule, Basse-Terre, Marie Galante, Grand-Bourg, Guadeloupe.
DISTRIBUTION: Widespread in most warm seas.
FURTHER READING: Huisman et al. (2007); Huisman & Leliaert (2015).

Siphonocladus tropicus (Dampier Archipelago, WA).

Valonia

Thalli of *Valonia* are composed of large vesicular cells that are generally interwoven to form a hemispherical dome or an irregular cushion. Adjacent vesicles are joined by small, tenacular cells that form secondary connections. New vesicles are borne from small lenticular cells that arise on the parent vesicle. Several species of *Valonia* are known, mostly from tropical and warmer seas. *Valonia macrophysa* and *Valonia fastigiata* grow in shallow coral habitats, where they form thick, firm cushions that firmly adhere to the substratum. *Valonia ventricosa* differs in being composed of a single large, generally spherical cell. It is a common and distinctive plant found in tropical and subtropical regions worldwide. Living thalli have a distinctive refractive sheen that can make them conspicuous underwater, although they are equally likely to be overgrown with, and obscured by, epiphytic coralline or filamentous algae. The species has been given the common name of 'sailor's eyeballs'.

Valonia fastigiata

TYPE LOCALITIES: Sri Lanka; Tonga.
DISTRIBUTION: Widespread in the tropical Indo-Pacific.

Valonia fastigiata (Seringapatam Reef, WA).

Valonia macrophysa

TYPE LOCALITY: Hvar Island, Yugoslavia.
DISTRIBUTION: From the Houtman Abrolhos, Western Australia, around northern Australia to the Coffs Harbour region, New South Wales. Widespread in warmer seas.

Valonia ventricosa

TYPE LOCALITY: Guadeloupe, West Indies.
DISTRIBUTION: From Rottnest Island, Western Australia, around northern Australia to Queensland. Widespread in warmer seas.
FURTHER READING: Kraft (2007); Huisman & Leliaert (2015).

Valonia macrophysa (Houtman Abrolhos, WA).

Valonia ventricosa (Ashmore Reef).

Bryopsis

Thalli of *Bryopsis* are composed of branched filaments, in all cases without complete cellular cross-walls so that the cytoplasm of the entire plant is continuous. In some species, the side-branches remain short and are all of similar length, giving the plant the appearance of a slightly ruffled feather (e.g. *Bryopsis pennata and Bryopsis plumosa*). In others, such as *Bryopsis minor*, the side-branches are of similar length to the main axes and no distinct axes can be readily recognised. *Bryopsis foliosa* has mostly radial branching, although occasional opposite branches also occur.

Bryopsis foliosa

TYPE LOCALITY: Western Australia, probably near Fremantle.

DISTRIBUTION: Whitfords Beach to Hamelin Bay, Western Australia.

Bryopsis minor

TYPE LOCALITY: American River Inlet, Kangaroo Island, South Australia.

DISTRIBUTION: Known from the Houtman Abrolhos, Western Australia, and from Kangaroo Island and Port MacDonnell, South Australia.

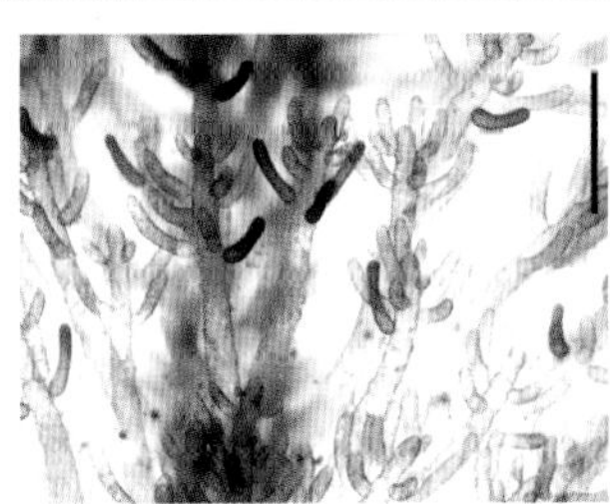

Bryopsis foliosa (Cape Peron, WA).

(above) *Bryopsis foliosa* – detail of thallus (Cape Peron, WA). Scale = 1 mm.

Bryopsis minor (Houtman Abrolhos, WA).

(far right) A sacoglossan sea slug *Caliphylla* sp., which feeds on *Bryopsis* and incorporates the chloroplasts into its leaf-like appendages.

Bryopsis pennata

TYPE LOCALITY: Antilles, West Indies.
DISTRIBUTION: Widespread in tropical seas.

Bryopsis plumosa

TYPE LOCALITY: Exmouth, England.
DISTRIBUTION: Widespread in temperate seas.
FURTHER READING: Womersley (1984); Kraft (2007); Huisman (2015); Scott (2017).

Bryopsis pennata (Little Turtle Island, WA).

Bryopsis plumosa (Carnac Island, WA).

Pseudobryopsis

Thalli of *Pseudobryopsis* are superficially similar to those of *Bryopsis* in having a central axis with numerous lateral filaments without cellular cross-walls. The genera differ, however, when fertile. Gametes in *Pseudobryopsis* are borne in specialised gametangia that arise on the lateral filaments, while in *Bryopsis* the filaments themselves act as gametangia. Only a single species, *Pseudobryopsis hainanensis*, is known from Australia.

Pseudobryopsis hainanensis

TYPE LOCALITY: Kuan-nen, Wenchang, Hainan Dao, China.
DISTRIBUTION: From the Perth region to the Houtman Abrolhos, Western Australia, and Lord Howe Island and northern New South Wales. China. South Africa.
FURTHER READING: Henne & Schnetter (1999); Kraft (2000), (2007); Huisman (2015).

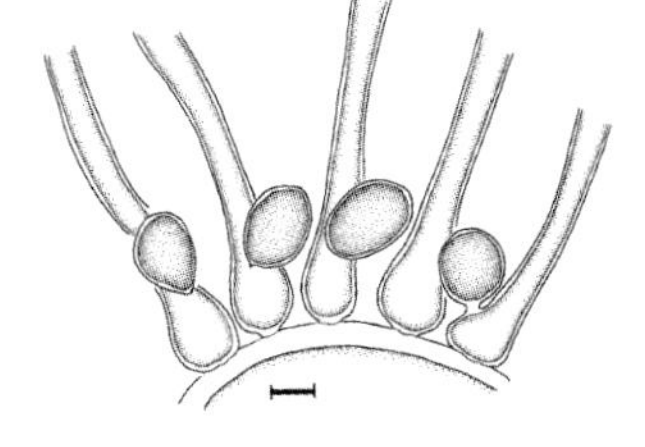

Pseudobryopsis hainanensis (Cape Peron, WA).

(above) *Pseudobryopsis hainanensis* – partial section of thallus, showing wall of primary axis and the bases of lateral filaments with ellipsoid gametangia (Map Reef, WA). Scale = 50 µm.

Pseudoderbesia

A rarely encountered genus, with only two known species, *Pseudoderbesia* was unknown in Australia until recently, when the new species *Pseudoderbesia eckloniae* was discovered growing epiphytically on degrading branches of the kelp *Ecklonia radiata* at Cape Peron in Western Australia. Plants are grass green and form a velvety layer that spreads laterally for up to 10 cm, with a stoloniferous base and dichotomously branched upright axes. The filaments are siphonous (lacking transverse cell walls) and taper upwardly, with slight constrictions at the branch bases and are filled with numerous chloroplasts, each with a single pyrenoid.

Pseudoderbesia eckloniae

TYPE LOCALITY: Cape Peron, Western Australia.
DISTRIBUTION: Known only from Cape Peron.
FURTHER READING: Huisman & Verbruggen (2020).

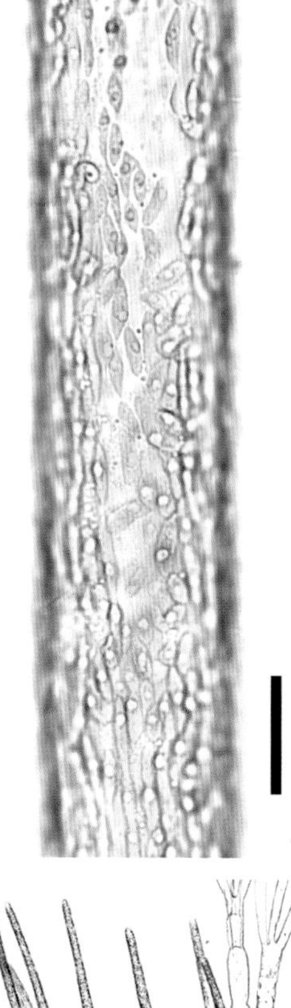

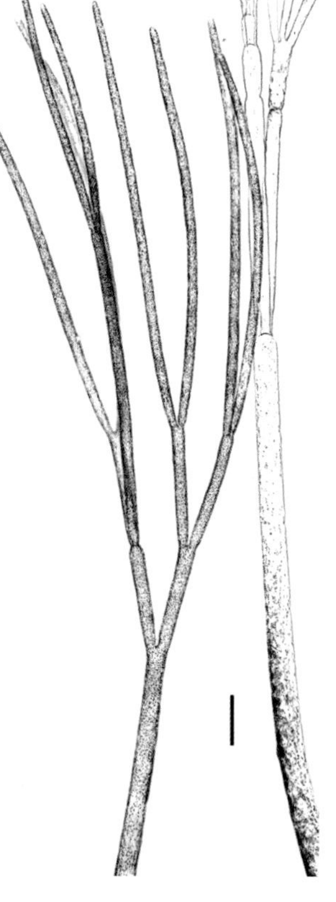

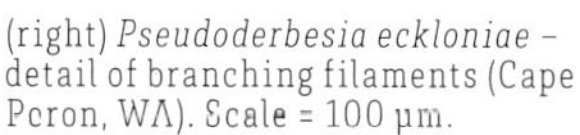

(right) *Pseudoderbesia eckloniae* – detail of branching filaments (Cape Peron, WA). Scale = 100 µm.

Pseudoderbesia eckloniae – epiphytic on *Ecklonia radiata* (Cape Peron, WA).

(above) *Pseudoderbesia eckloniae* – detail of chloroplasts (Cape Peron, WA). Scale = 25 µm.

Caulerpa

Species of *Caulerpa* are common throughout tropical regions of the world. Australia is unusual in having a large number of species also occurring in colder waters, those of southern Australia. Species of the genus can be easily recognised by their creeping stolon with numerous upright axes of various forms, depending on the species. In many species, the upright axes bear lateral branches, known as ramuli. Plants can be grass-green to grey-green in colour. The distinctive features of the individual species are given under the specific headings.

Caulerpa articulata

A very rare species, in Australia only known from a few collections from Rottnest Island, Western Australia. It has numerous elongate vesicles in opposite rows on the upright axes.

TYPE LOCALITY: East coast of New Zealand, probably from the Wairarapa coast.
DISTRIBUTION: In Australia, known only from Rottnest Island. New Zealand.

Caulerpa biserrulata

One of a small number of species in which the upright axes are leaf-like, in this case with small marginal spines.

TYPE LOCALITY: Cape York, Queensland, Australia.
DISTRIBUTION: Widespread in the tropical Indo-Pacific.

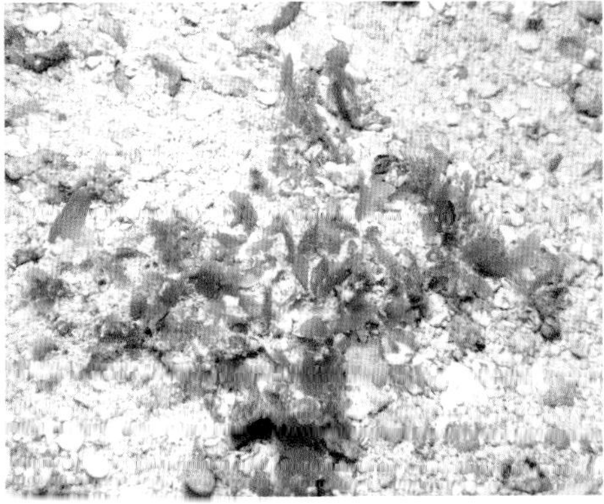

Caulerpa articulata (Rottnest Island, WA).
Caulerpa biserrulata (Heron Island, Qld).
(top) *Caulerpa scalpelliformis* - as illustrated by Harvey (1858, pl. 17).

Caulerpa brachypus

Caulerpa brachypus has strap-like upright axes with tiny marginal teeth.

TYPE LOCALITY: Tanega-shima, Kagoshima Prefecture, Japan.
DISTRIBUTION: Widely distributed in the tropical Pacific and Indian Oceans as well as the Caribbean Sea.

Caulerpa brownii

A common species usually found on rocks just below low tide level. It can be recognised by the curved branchlets found on the upright axes.

TYPE LOCALITY: Kent Island, Bass Strait, Australia.
DISTRIBUTION: From Whitfords Beach, Western Australia, around southern mainland Australia and Tasmania to Walkerville, Victoria. Lord Howe Island. New Zealand.

Caulerpa cactoides

An easily recognised species with its large, club-shaped vesicles. A distinctive feature is the annulations found at the base of the upright axes.

TYPE LOCALITY: Southern coast of Australia.
DISTRIBUTION: From the Houtman Abrolhos, Western Australia, around southern mainland Australia and Tasmania to the Richmond River mouth, New South Wales.

Caulerpa chemnitzia

A species characterised by ramuli that broaden gradually to flat-topped apices.

TYPE LOCALITY: Malabar Coast, India.
DISTRIBUTION: Widespread in tropical seas.

Caulerpa brachypus (Heron Island, Qld).
Caulerpa brownii (Gnarabup, WA).
Caulerpa cactoides (Cape Peron, WA).
Caulerpa chemnitzia (Lizard Island, Qld).

Caulerpa constricta

An unusual species in which the upright axes are constricted and do not bear lateral ramuli.

TYPE LOCALITY: North of Beacon Island, Wallabi Group, Houtman Abrolhos, Western Australia.

DISTRIBUTION: Known from the Geraldton area (Port Denison and Houtman Abrolhos) and North West Cape, Western Australia.

Caulerpa coppejansii

A tropical species similar to *Caulerpa sedoides*, but with slightly larger ramuli.

TYPE LOCALITY: Wistari Channel, Heron Island, Queensland.

DISTRIBUTION: Great Barrier Reef, Queensland. Fiji. Papua New Guinea.

Caulerpa corynephora

Somewhat similar to *Caulerpa cactoides* in appearance, *Caulerpa corynephora* differs by usually being smaller, with ramuli mostly borne on successive segments, as opposed to often being separated by a naked segment.

TYPE LOCALITY: Tudu Island, also known as Warrior Islet, Torres Strait, Queensland.

DISTRIBUTION: Known from scattered localities in northern Australia. South to Albany in Western Australia.

Caulerpa crispata

Thalli of this taxon have upright axes bearing unbranched ramuli. In Western Australian specimens, the erect axes have clustered branches several centimetres above the base.

TYPE LOCALITY: Port Phillip Heads, Victoria.

DISTRIBUTION: From the Perth region, Western Australia, around southern mainland Australia to Waratah Bay, Victoria, and Tasmania.

Caulerpa constricta (Houtman Abrolhos, WA).

Caulerpa coppejansii (Wistari Channel, Heron Island, Qld).

Caulerpa corynephora (James Price Point, WA).

Caulerpa crispata (Cape Peron, WA).

Caulerpa cupressoides var. *flabellata*

This variety has flattened upright axes with marginal spines; it is similar to some forms of *Caulerpa serrulata*, but with a slightly upward orientation of spines.

TYPE LOCALITY: St Jan (St John) and St Thomas, Virgin Islands.
DISTRIBUTION: Widespread in tropical seas.

Caulerpa cupressoides var. *mamillosa*

This variety is characterised by its distinctly inflated ramuli, particularly near the base of the erect axes.

TYPE LOCALITIES: Agalega Islands, south-west Indian Ocean. Mangareva (Island), Îles Gambier, French Polynesia.
DISTRIBUTION: Widespread in tropical seas.

Caulerpa cylindracea

Upright axes bear vesiculate ramuli that are club shaped and unconstricted at the base.

TYPE LOCALITY: Western Australia.
DISTRIBUTION: In Western Australia, found south to Esperance. Papua New Guinea. New Caledonia. Invasive in the Mediterranean Sea and southern Australia.

Caulerpa fergusonii

Upright branches in *Caulerpa fergusonii* have opposite lateral vesicles that are mostly circular and slightly compressed.

TYPE LOCALITY: Tuticorin, India.
DISTRIBUTION: India. Sri Lanka. Indonesia. Malaysia. Papua New Guinea. Philippines. Japan. Fiji. In Western Australia, it is known from the Kimberley as far south as Canal Rocks in the south-west.

Caulerpa cupressoides var. *flabellata* (Heron Island, Qld).

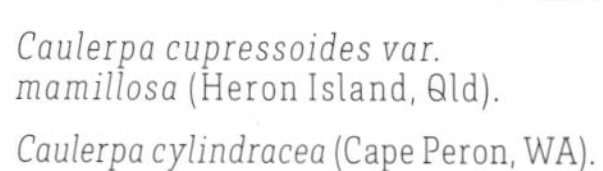
Caulerpa cupressoides var. *mamillosa* (Heron Island, Qld).

Caulerpa cylindracea (Cape Peron, WA).

Caulerpa fergusonii (Rottnest Island, WA).

Caulerpa filicoides

Caulerpa filicoides has relatively small thalli, and erect branches that bear terminal fern-like rosettes. Thallus height to 3.5 cm, spreading laterally.

TYPE LOCALITY: Miyako-zima and Naha, Ryukyu, Japan.

DISTRIBUTION: Widespread in the tropical Pacific. Not recorded from the Indian Ocean coast of Australia other than at Ashmore Reef.

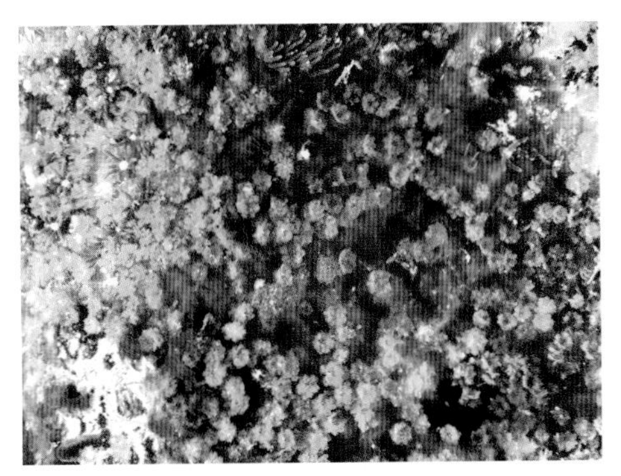

Caulerpa hedleyi

An uncommon species that has feather-like upright axes.

TYPE LOCALITY: Off Kangaroo Island, South Australia.

DISTRIBUTION: Known from Rottnest Island, Western Australia, and Pearson Island, Investigator Strait and the Isles of St Francis, South Australia.

Caulerpa flexilis

In *Caulerpa flexilis* the upright branches produce secondary branches that are arranged opposite one another. These in turn are densely covered with smaller, spiny ramuli that are bifurcate at the base. *Caulerpa flexilis* is largely restricted to the colder waters of southern Australia.

TYPE LOCALITY: Esperance, Western Australia.

DISTRIBUTION: From Geraldton, Western Australia, around southern mainland Australia and Tasmania to Collaroy, New South Wales. New Zealand.

Caulerpa heterophylla

Caulerpa heterophylla has upright axes with short conical spines and distichous lateral branches. The latter are covered with short ramuli. The species is similar to *Caulerpa flexilis*, but is less densely branched, and the short ramuli are undivided as opposed to bifurcate.

TYPE LOCALITY: Dyer Island, adjacent to Rottnest Island, Western Australia.

DISTRIBUTION: Known from Rottnest Island and near Albany, Western Australia.

Caulerpa filicoides (Ashmore Reef).
Caulerpa flexilis (Cape Peron, WA).

Caulerpa hedleyi (Rottnest Island, WA).
Caulerpa heterophylla (Rottnest Island, WA).

Caulerpa lamourouxii

Upright axes in *Caulerpa lamourouxii* are fleshy and generally compressed. Some specimens (as in the one pictured) lack side-branches altogether.

TYPE LOCALITY: Red Sea.
DISTRIBUTION: From the Houtman Abrolhos, Western Australia, around northern Australia to Queensland. Widespread in tropical and warmer seas.

Caulerpa lentillifera

A mostly tropical species with upright axes that are densely covered by subspherical vesicles that are generally arranged in longitudinal rows. The vesicles are borne on distinctly constricted stalks.

TYPE LOCALITY: Coast of Eritrea.
DISTRIBUTION: From the Houtman Abrolhos, Western Australia, around northern Australia to Queensland. Widespread in tropical seas.

Caulerpa nummularia

A low-growing species with disk-shaped branches.

TYPE LOCALITY: Friendly Islands.
DISTRIBUTION: Widespread in the Indo-Pacific.

Caulerpa obscura

Upright axes of *Caulerpa obscura* bear numerous laterals that are densely clothed with unbranched ramuli.

TYPE LOCALITY: Western Australia.
DISTRIBUTION: From the Houtman Abrolhos, Western Australia, around southern mainland Australia to Walkerville, Victoria, and Tasmania.

Caulerpa lamourouxii (Barrow Island, WA).
Caulerpa lentillifera (Long Reef, WA).
Caulerpa nummularia (Heron Island, Qld).
Caulerpa obscura (Cape Peron, WA).

Caulerpa perplexa

Caulerpa perplexa is a delicate plant, with flattened, upright axes that are somewhat leaf-like in shape.

TYPE LOCALITY: Swirl Reef, Rottnest Island, Australia.
DISTRIBUTION: From Cape Naturaliste north to the Dampier Archipelago, Western Australia.

Caulerpa racemosa

The upright axes of *Caulerpa racemosa* produce many small, vesiculate side-branches. Several varieties of *Caulerpa racemosa* were previously recognised, with most now regarded as independent species following DNA sequence analyses. *Caulerpa racemosa* is found in tropical regions worldwide and is generally restricted to warmer waters in Australia.

TYPE LOCALITY: Suez, Egypt.
DISTRIBUTION: Widely distributed in tropical seas.

Caulerpa scalpelliformis

Displaying yet another of the myriad of forms presented by the genus, *Caulerpa scalpelliformis* has flattened upright axes with two rows of side-branches. Two forms are found in Australia: the robust type in colder waters and *Caulerpa scalpelliformis* variety *denticulata* in tropical regions.

TYPE LOCALITY: Southern coast of Australia.
DISTRIBUTION: From Whitfords Beach, Western Australia, around southern mainland Australia and Tasmania to Jervis Bay, New South Wales.

Caulerpa sedoides

Caulerpa sedoides has vesiculate ramuli and bears a strong likeness to *Caulerpa racemosa*. It differs in its ramuli being constricted near the base.

TYPE LOCALITY: Kent Islands, Bass Strait, Australia.
DISTRIBUTION: From Port Denison, Western Australia, around southern mainland Australia and Tasmania to Bowen, Queensland. New Zealand.

Caulerpa racemosa (Ashmore Reef).

Caulerpa perplexa (Rottnest Island, WA).

Caulerpa scalpelliformis (Cape Peron, WA).

Caulerpa sedoides (Rottnest Island, WA).

Caulerpa serrulata

Caulerpa serrulata is more commonly found in tropical seas, but occurs as far south as the Houtman Abrolhos on the coast of Western Australia. It is generally a grey-green colour, with spiny margins on the flattened, upright branches.

TYPE LOCALITY: Mokha, Yemen.
DISTRIBUTION: From the Houtman Abrolhos, Western Australia, around northern Australia to Queensland. Widely distributed in tropical seas.

Caulerpa sertularioides

A common species in tropical regions, *Caulerpa sertularioides* has upright axes with narrow, terete ramuli borne in two (occasionally three) opposite rows, giving the thallus the appearance of a coarse feather.

TYPE LOCALITY: Tropical America.
DISTRIBUTION: From the North West Cape region, Western Australia, around northern Australia to Queensland. Widely distributed in tropical seas.

Caulerpa simpliciuscula

A colder-water species that has many ovoid to clavate vesicles tightly packed around the upright axes.

TYPE LOCALITY: Kent Island, Bass Strait.
DISTRIBUTION: From the Houtman Abrolhos, Western Australia, around southern mainland Australia and Tasmania to Walkerville, Victoria.

Caulerpa taxifolia var. *taxifolia*

A tropical species with upright axes bearing pinnate lateral ramuli that are compressed, upwardly directed and closely set.

TYPE LOCALITY: St Croix, Virgin Islands.
DISTRIBUTION: From the Montebello Islands, Western Australia, around northern Australia to Queensland. Widespread in tropical seas.

Caulerpa serrulata (Mavis Reef, WA).
Caulerpa sertularioides (Coral Bay, WA).

Caulerpa simpliciuscula (Cape Peron, WA).

Caulerpa taxifolia var. taxifolia (Cassini Island, WA).

Caulerpa taxifolia var. *distichophylla*

Variety *distichophylla* is more slender than typical *Caulerpa taxifolia*.

TYPE LOCALITY: Western Australia, probably near Fremantle.
DISTRIBUTION: From Dongara to King George Sound, Western Australia.

Caulerpa trifaria

A distinctive species with terete ramuli borne in three distinct rows on the upright axes.

TYPE LOCALITY: Port Phillip Heads, Victoria.
DISTRIBUTION: From Cottesloe, Western Australia, around southern mainland Australia and Tasmania to Western Port, Victoria.

Caulerpa verticillata

This small species has slender erect branches that are delicate and filamentous, with lateral branches arising in whorls.

TYPE LOCALITY: West Indies.
DISTRIBUTION: Widely distributed in warmer waters of the Indian and Pacific Oceans and the Caribbean Sea.

Caulerpa vesiculifera

Caulerpa vesiculifera has upright axes surrounded by ovoid vesicles that have a pedicel below the constriction.

TYPE LOCALITY: Western Port, Victoria
DISTRIBUTION: From Shark Bay, Western Australia, to Phillip Island, Victoria, and the north coast of Tasmania.

Caulerpa taxifolia var. *distichophylla* (Cape Peron, WA).

Caulerpa trifaria (Cape Peron, WA).

Caulerpa verticillata (Ashmore Reef).

Caulerpa vesiculifera (Cape Peron, WA).

Caulerpa webbiana

A small species, restricted to warmer waters, *Caulerpa webbiana* forms a dense, low covering over hard substrata. It has mostly close, short, often imbricate ramuli that are repeatedly forked, pointed at the tips, and arranged in whorls.

TYPE LOCALITY: Arrecife, Isla Lanzarote, Canary Islands.
DISTRIBUTION: From the Houtman Abrolhos, Western Australia, around northern Australia to Lord Howe Island, New South Wales. Widespread in tropical and warmer seas.

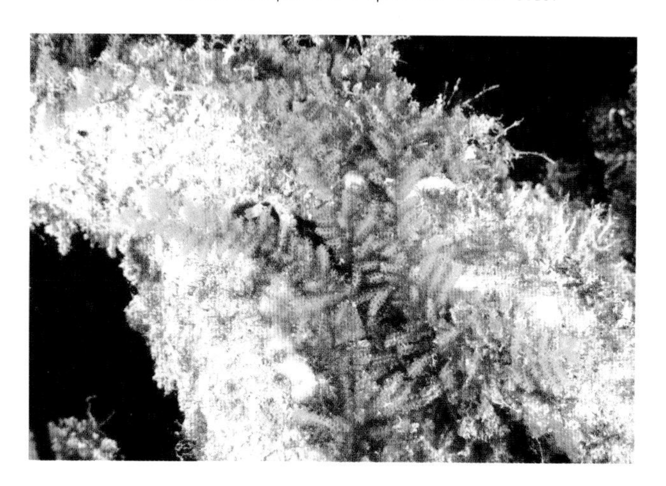

Caulerpa webbiana f. *disticha*

Form *disticha* has lateral branches arising in a single plane.

TYPE LOCALITY: Canary Islands.
DISTRIBUTION: Widespread in tropical seas.
FURTHER READING: Womersley (1984); Adams (1994); Kraft (2007); Belton et al. (2014, 2015, 2019).

Caulerpa webbiana (Heron Island, Qld).

Caulerpa webbiana f. *disticha* (Sykes Reef, Qld).

Codium

Species of *Codium* come in a wide variety of forms, from prostrate, amorphous crusts to upright, dichotomously branched thalli. Although their external appearance does not suggest a close relationship, an examination of their internal structures reveals a surprisingly similar construction. All are composed of branched filaments without cellular cross-walls. At the periphery of the thallus, these filaments form swollen projections, known as utricles, that are regularly aligned and form the cortex. Hairs are often borne on the walls of the utricles and project beyond the cortex. Reproductive bodies are formed in specialised structures known as gametangia that are borne laterally on the utricles. Numerous species occur in Australia; those included here are representative of the various morphological forms that the genus can exhibit.

Codium arabicum

The thallus is convoluted and irregularly shaped, growing prostrate on rock surfaces, with margins occasionally unattached. Similar to *Codium spongiosum*, but with firmer consistency and shorter utricles (less than 1 mm in length).

TYPE LOCALITY: El-Tor, Sinai Peninsula, Red Sea, Egypt.
DISTRIBUTION: Widespread in the Indo-Pacific.

Codium duthieae

Thalli of *Codium duthieae* are large and generally regularly dichotomously branched. A distinctive feature of the species is the slightly compressed region at each point of branching.

TYPE LOCALITY: Strandfontein, South Africa.
DISTRIBUTION: From Champion Bay, Western Australia, around southern Australia to Walkerville, Victoria, and the north coast of Tasmania. South Africa.

Codium arabicum (Cassini Island, WA).

Codium duthieae (Cape Peron, WA).

(top) *Codium spongiosum* – as illustrated by Harvey (1858, pl. 15).

Codium dwarkense

A common, relatively small, dichotomously branched species in the tropical Indian Ocean.

TYPE LOCALITY: Dwarka, India.
DISTRIBUTION: Tropical Indian Ocean.

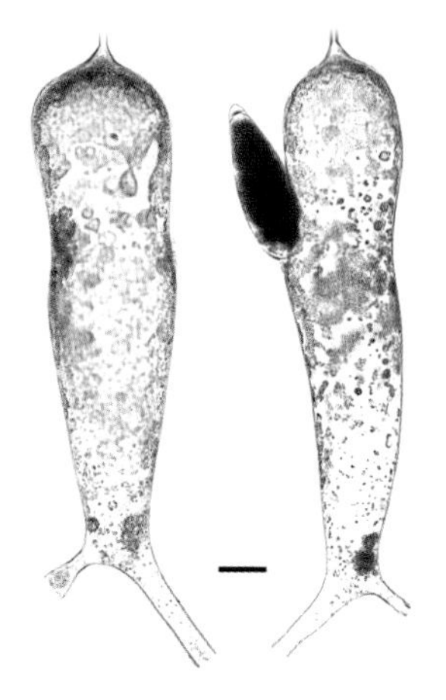

Codium fragile

A medium to large dichotomously branched species that has been introduced to numerous locations worldwide, *Codium fragile* has utricles that have a terminal spine and a slight narrowing at mid-level.

TYPE LOCALITY: Japan.
DISTRIBUTION: Widespread in temperate seas.

Codium galeatum

One of the larger dichotomously branched species, typically a very dark green colour.

TYPE LOCALITY: Port Phillip, Victoria.
DISTRIBUTION: From Champion Bay, Western Australia, around southern Australia and Tasmania to Ballina, New South Wales.

(top) *Codium fragile* - detail of utricles with terminal spine and lateral gametangium (Albany, WA). Scale = 100 µm.

Codium dwarkense (Barrow Island, WA).

Codium fragile (Jervis Bay, NSW; photo R. Hilliard).

Codium galeatum (Perth, WA; photo J. Phillips).

Codium geppiorum

A decumbent species with dichotomously divided, terete axes, often forming secondary attachments with the reef surface and other branches.

TYPE LOCALITY: Kai Islands and Celebes, Indonesia.
DISTRIBUTION: From the Houtman Abrolhos, Western Australia, around northern Australia to Queensland. Widespread in the warmer waters of the Indo-Pacific.

Codium gongylocephalum

The name refers to the unusual form of utricles and reproductive structures, in which the upper surface is ringed by four to eight blunt knobs. Found only in deep water in areas subjected to strong currents.

TYPE LOCALITY: Coral Gardens, Heron Island, Capricorn Group, southern Great Barrier Reef, Queensland.
DISTRIBUTION: Endemic to Heron Island and Wistari Reef on the southern Great Barrier Reef, Queensland.

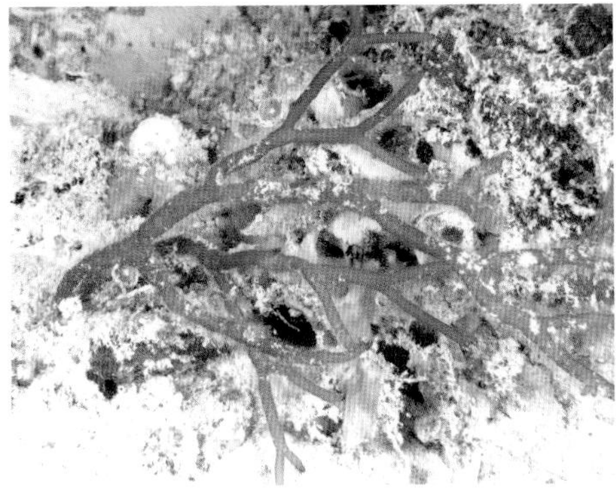

Codium harveyi

A light-green, dichotomously branched species with narrow branches.

TYPE LOCALITY: Vivonne Bay, Kangaroo Island, South Australia.
DISTRIBUTION: From Shark Bay, Western Australia, around southern Australia and Tasmania, to Lake Macquarie, New South Wales. New Zealand.

Codium laminarioides

One of the more unusual species, with a flat, undivided blade-like thallus.

TYPE LOCALITY: Rottnest Island, Western Australia.
DISTRIBUTION: From Port Denison, Western Australia, to Pearson Island, South Australia.

Codium geppiorum (Heron Island, Qld).
Codium gongylocephalum (Heron Island, Qld).
Codium harveyi (Rous Head, WA).
Codium laminarioides (Rottnest Island, WA).

Codium muelleri

A bright green species, distinguished by its terete branches and short utricles with a dome-shaped thickening on the inner surface of the outer wall.

TYPE LOCALITY: Lefevre Peninsula, South Australia.
DISTRIBUTION: Dongara, Western Australia to Gabo Island, Victoria, and Tasmania.

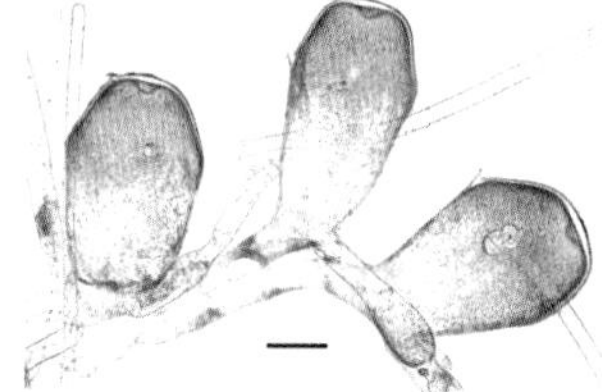

Codium lucasii

Codium lucasii forms a thick, spreading, crust-like thallus.

TYPE LOCALITY: Bondi, New South Wales.
DISTRIBUTION: From Ningaloo Reef, Western Australia, around southern and eastern mainland Australia to Redcliffe, Queensland, and the north coast of Tasmania.

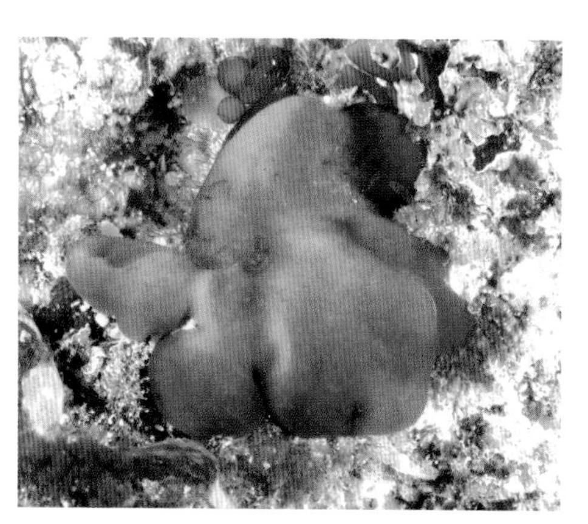

Codium saccatum

Codium saccatum is unusual in having a hollow thallus, the only species with this structure in the tropical Indo-Pacific.

TYPE LOCALITY: Futae, Amakusa-shoto, Kumamoto Prefecture, Japan.
DISTRIBUTION: In Australia known from New South Wales and Queensland. Japan. Korea. Taiwan. Several Pacific islands.

Codium spongiosum

Codium spongiosum forms large, sponge-like thalli that are firm when living but quickly degenerate once detached. Large numbers of this species are commonly found washed up on tropical shores.

TYPE LOCALITY: King George Sound, Western Australia.
DISTRIBUTION: Throughout Australia. Widespread in warmer seas.

Opposite page:
Codium lucasii (Rottnest Island, WA).
Codium muelleri (Cape Peron, WA).
Codium muelleri – detail of utricles (Cape Peron, WA). Scale = 100 µm.
Codium saccatum (Heron Island, Qld).

Codium spongiosum (Heron Island, Qld).
Codium sursum (Heron Island, Qld).
Codium tenue (Walpole, WA).

Codium sursum

A light-green, dichotomously branched, erect species. The species name refers to the thallus reaching upwardly towards light from its deep-water attachment.

TYPE LOCALITY: Old Gulch, Lord Howe Island, New South Wales.
DISTRIBUTION: Lord Howe Island, Wistari Reef, central and southern Great Barrier Reef, Queensland. Fiji.

Codium tenue

In Australia, *Codium tenue* is only known from silty estuarine habitats in the south-west, similar to the locations in which the species grows in South Africa.

TYPE LOCALITY: Eastern Cape Province, South Africa.
DISTRIBUTION: Known reliably only from South Africa and south-west Australia.
FURTHER READING: Kraft (2000), (2007); Huisman et al. (2015); McDonald et al. (2015); Silva & Chacana (2015); Scott (2017).

Pedobesia

Pedobesia is known in Australia by the single species *Pedobesia clavaeformis*. Thalli are composed of clusters of undivided, clavate branches arising from a filamentous base. They grow to around 5 cm in height and are generally a dark-green colour. Lateral sporangia are often present on the upper portion of the branches, as can be seen in the photograph.

Pedobesia clavaeformis

TYPE LOCALITY: Western Port, Victoria.
DISTRIBUTION: From the Perth region, Western Australia, around southern mainland Australia to Cape Patterson, Victoria, and Tasmania. Northern New Zealand.
FURTHER READING: MacRaild & Womersley (1974); Womersley (1984); Nelson (2013).

Pedobesia clavaeformis (Map Reef, WA).

Avrainvillea

Thalli of *Avrainvillea* have branched or unbranched stipes bearing terminal fan-shaped, usually spongy blades. They are often found in sand patches on reef flats, anchored by a large fibrous holdfast that binds sand grains. The blades are uncalcified and are composed of aggregated, dichotomously branched siphons that lack lateral branchlets. Some species of *Avrainvillea* can appear similar to *Udotea* (see p. 400), but their fronds are not calcified and tend to have a spongy texture. *Avrainvillea amadelpha* typically has clustered blades and the species forms extensive mats in the Hawaiian Islands, where it is regarded as a pest. *Avrainvillea calathina* is characterised by a bowl-like appearance resulting from curvature of the blade, whereas *Avrainvillea erecta* is more upright with kidney-shaped blades. *Avrainvillea clavatiramea* is a cold-water species found in southern Australia.

Avrainvillea amadelpha

TYPE LOCALITY: Agalega Islands, north-east of Madagascar.

DISTRIBUTION: Widespread in the tropical Indo-Pacific. Introduced in the Mediterranean.

Avrainvillea calathina

TYPE LOCALITY: Old Gulch, Lord Howe Island.

DISTRIBUTION: Known from Lord Howe Island and the southern Great Barrier Reef. Also reported from Iran.

Avrainvillea amadelpha (Scott Reef, WA).

Avrainvillea calathina (Heron Island, Qld).

Avrainvillea clavatiramea

TYPE LOCALITY: Port Phillip, Victoria.
DISTRIBUTION: From Rottnest Island, Western Australia, to Port Phillip, Victoria.

Avrainvillea erecta

TYPE LOCALITY: Philippines.
DISTRIBUTION: Widespread in the Indian and western Pacific Oceans.
FURTHER READING: Kraft & Olsen-Stojkovich (1985); Kraft (2007); Huisman (2015); Verlaque et al. (2017).

Avrainvillea clavatiramea (Canal Rocks, WA; photo G.J. Edgar).

Avrainvillea erecta (Champagney Island, WA).

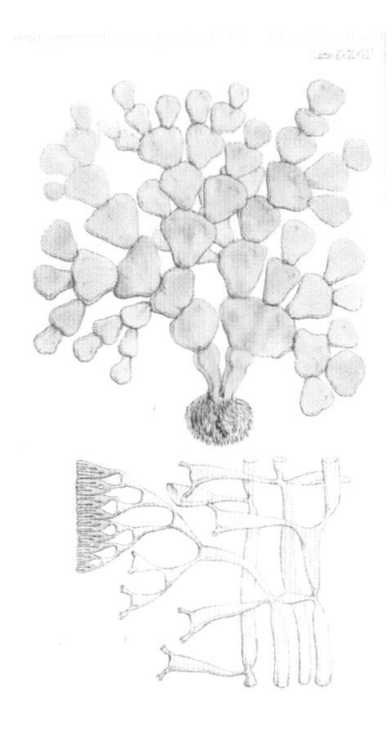

Halimeda

Thalli of *Halimeda* are segmented and conspicuously calcified. Segments can be variously shaped, depending on the species. Those of *Halimeda versatilis* are flattened and triangular to almost discoid, while in *Halimeda cylindracea* the segments are cylindrical, as the specific epithet suggests. Structurally the thallus is composed of elongate filaments without cellular cross-walls. At the cortex, these are expanded and divided to form one to several layers of utricles. *Halimeda* reproduces by forming small spherical gametangia on the surface of the thallus. Once the gametes are released, the parent plant dies, leaving only the calcified skeleton. Calcium carbonate from *Halimeda* is a major component of sediments in tropical coral reefs and is often present in greater quantities than sediments derived from the corals. *Halimeda* is at its most diverse in tropical regions, where many species are known, with *Halimeda versatilis* being the only species found in colder seas.

Halimeda cylindracea

A sand-dwelling species with a massive submerged holdfast and cylindrical segments.

TYPE LOCALITY: Nosy-Bé, Madagascar.
DISTRIBUTION: From Coral Bay, Western Australia, around northern Australia to the Great Barrier Reef, Queensland. Tropical Indo-Pacific.

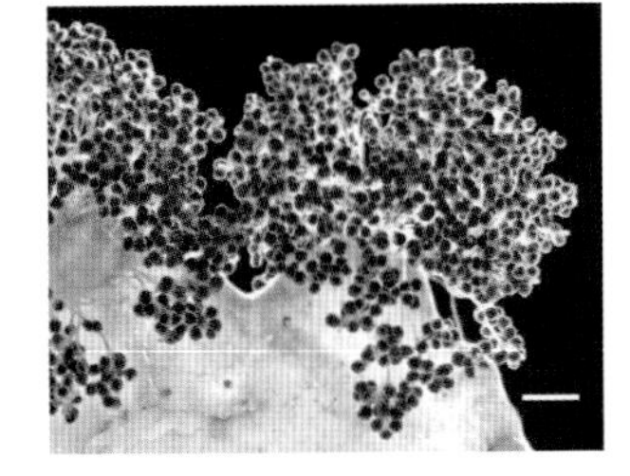

(top) *Halimeda versatilis* – as illustrated by Harvey (1863, as *Halimeda macroloba*).

(above) *Halimeda versatilis* – reproductive thallus with gametangia. These reproductive structures are typical of the genus. (Cape Peron, WA). Scale = 1 mm.

Halimeda cylindracea (Heron Island, Qld).

Halimeda discoidea

A common species, often with a variety of segment shapes in a single thallus.

TYPE LOCALITY: Unknown.
DISTRIBUTION: Widespread in the tropical Indo-Pacific.

Halimeda distorta

Distinguished by a sprawling thallus, with successive segments often not aligned.

TYPE LOCALITY: Atoll of Ant, Ponape, Caroline Islands, Micronesia.
DISTRIBUTION: Widespread in the tropical Indo-Pacific.

Halimeda fragilis

A bushy species, generally with intermingling branches.

TYPE LOCALITY: Elugelab Island, Eniwetok Atoll, Marshall Islands.
DISTRIBUTION: Widespread in the tropical Indo-Pacific.

Halimeda gigas

Segments of *Halimeda gigas* are relatively large and often shiny.

TYPE LOCALITY: Eniwetok Atoll, Marshall Islands.
DISTRIBUTION: Widespread in the tropical Indo-Pacific.

Halimeda discoidea (Scott Reef, WA).

Halimeda distorta (Scott Reef, WA).

Halimeda fragilis (Heron Island, Qld).

Halimeda gigas (Long Reef, WA).

Halimeda macroloba

Another sand-dwelling species (see also *Halimeda cylindracea*, p. 387) with large, oval- to kidney-shaped segments.

TYPE LOCALITY: Red Sea.
DISTRIBUTION: Widespread in the tropical Indo-Pacific.

Halimeda macrophysa

The large diameter utricles of *Halimeda macrophysa* give the thallus surface a speckled appearance.

TYPE LOCALITY: Matuku, Fiji islands, South Pacific.
DISTRIBUTION: Widespread in the tropical Indo-West Pacific.

Halimeda micronesica

Distinguished by its often-hemispherical thallus and unusual fan-shaped basal segment, which supports numerous segments from its margin.

TYPE LOCALITY: Ant Atoll, Pohnpei, Caroline Islands.
DISTRIBUTION: Widespread in the tropical Indo-West Pacific.

Halimeda minima

A species with relatively small segments. Plants of *Halimeda minima* can be upright or pendant from reef walls.

TYPE LOCALITY: Bikini Lagoon, Bikini Atoll, Marshall Islands.
DISTRIBUTION: Widespread in the tropical Indo-West Pacific.

Halimeda macroloba (Cassini Island, WA).

Halimeda macrophysa (Heron Island, Qld).

Halimeda micronesica (Heron Island, Qld).

Halimeda minima (Lizard Island, Qld).

Halimeda opuntia

A species that forms numerous attachments to the substratum and is often found filling gaps between coral on intertidal reefs

TYPE LOCALITY: Jamaica.
DISTRIBUTION: Pantropical.

Halimeda velasquezii

Another species with small segments, often found in shallow rock pools.

TYPE LOCALITY: Santa Ana, Cagayan Province, Luzon, Philippines.
DISTRIBUTION: Widespread in the tropical Indo-West Pacific.

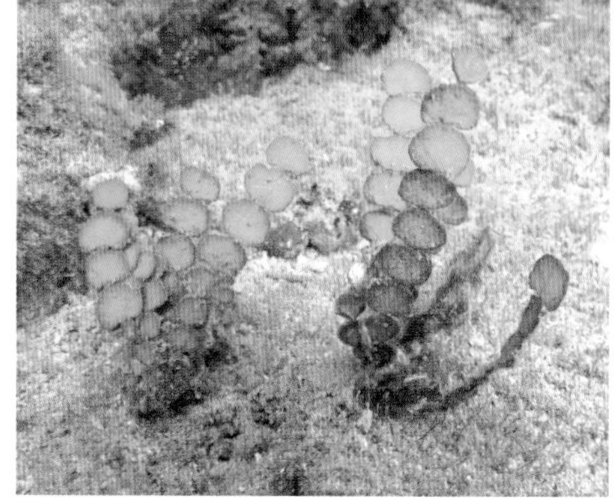

Halimeda versatilis

A colder-water species, the only *Halimeda* found in southern Australia.

TYPE LOCALITY: Cape Riche, Western Australia.
DISTRIBUTION: South-western Australia.

Halimeda xishaensis

A distinctive species, recognisable by its intertidal habitat and large segments with undulate margins.

TYPE LOCALITY: Chenhangdao, Xisha Island, Guangdong Province, China.
DISTRIBUTION: Widespread in the tropical Indo-West Pacific.
FURTHER READING: Hillis-Colinvaux (1980); Noble & Kraft (2007); Huisman & Verbruggen (2015a); Cremen et al. (2016).

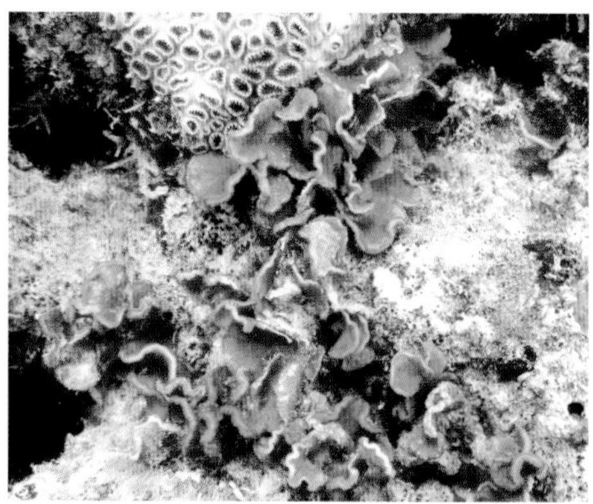

Halimeda opuntia (Cassini Island, WA).

Halimeda velasquezii (James Price Point, WA).

Halimeda versatilis (Cape Peron, WA).

Halimeda xishaensis (Cassini Island, WA).

Pseudocodium

Pseudocodium, as the name suggests, appears superficially like a species of *Codium* (see p. 379). Structurally, however, the genus most closely resembles an uncalcified *Halimeda* (see p. 387), as it has a central core of siphons that give rise to a layer of closely packed utricles forming the thallus surface. *Pseudocodium devriesii* is an African species with a disjunct occurrence in the vicinity of Perth, suggesting it might be introduced. It has dichotomously divided, terete branches. *Pseudocodium* sp., a recent discovery from near Heron Island (Queensland), is an undescribed species that has slightly flattened branches.

Pseudocodium devriesii

TYPE LOCALITY: Isipingo Beach, near Durban, Kwazulu-Natal, South Africa.

DISTRIBUTION: In Australia, known only from the Perth region, Western Australia. South Africa. Southern Mozambique. Madagascar. Oman.

Pseudocodium sp.

DISTRIBUTION: Known only from the vicinity of Heron Island, Queensland, where it grows in rubble or sand in deep water in areas of strong currents.

FURTHER READING: Womersley (1984); De Clerck et al. (2008).

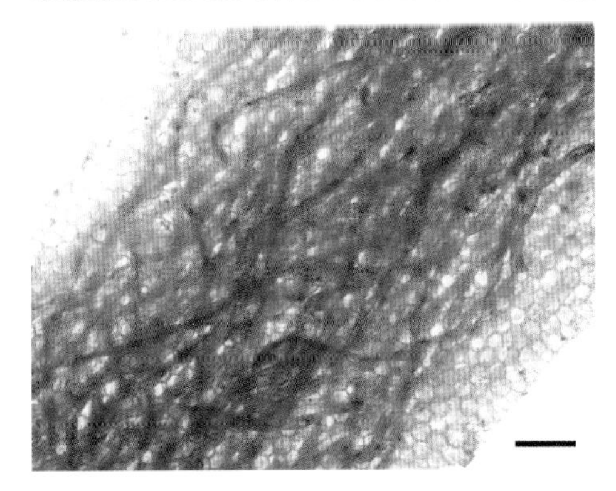

Pseudocodium devriesii (Perth, WA).

Pseudocodium sp. (Heron Island, Qld).

(right) *Pseudocodium* sp. – closer view of thallus (Heron Island, Qld). Scale = 250 µm.

Rhipilia / Kraftalia complex

The genus *Kraftalia* was described following a DNA study and comprises species previously included in *Rhipilia* and *Rhipiliopsis*. All three genera include several species in Australia, mostly in tropical seas. Thalli are uncalcified and predominantly fan-shaped. Structurally they are composed of branched filaments that lack cellular cross-walls (although constrictions do occur). *Kraftalia crassa* often occurs in clusters, the closely abutting thalli occasionally fusing at their margins. *Rhipilia diaphana* has larger, spongy fronds and is common on the Great Barrier Reef. *Kraftalia gracilis* (see opposite page) is a relatively small species, with blades composed of a single layer of distinctively sinuous, laterally attached siphons.

Kraftalia crassa

TYPE LOCALITY: Heron Island, Capricorn Group, southern Great Barrier Reef, Queensland.
DISTRIBUTION: Tropical Australia. Philippines.

Rhipilia diaphana

TYPE LOCALITY: Bikini Atoll, Marshall Islands.
DISTRIBUTION: Widespread in tropical seas.
FURTHER READING: Millar & Kraft (2001); Kraft (2007); Huisman & Verbruggen (2015b); Lagourgue & Payri (2021).

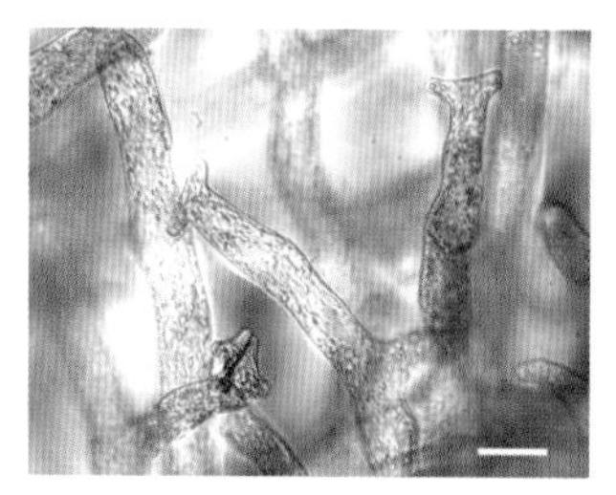

Kraftalia crassa (Heron Island, Qld).

Rhipilia diaphana (Sykes Reef, Qld).

(above) *Rhipilia diaphana* detail showing tenacula with finger-like projections (Sykes Reef, Qld). Scale = 50 µm.

Rhipiliopsis

Rhipiliopsis includes a large number of species distributed mainly in tropical and subtropical seas. Thalli have a corticated monosiphonous stalk that bears a circular or fan-shaped blade. The blade is composed of one to several layers of filaments that adhere laterally to one another, forming a spongy texture in those species with multiple layers. All filaments lack cellular cross-walls. Unlike several closely related genera of similar form (e.g. *Udotea*), *Rhipiliopsis* is never calcified. *Rhipiliopsis multiplex* is one of the more distinctive species, as it produces an extensive prostrate system from which the upright axes are formed. A further distinguishing feature is the production of multiple blades in series on a single central stalk. In contrast, *Rhipiliopsis peltata* has a single blade. *Rhipiliopsis multiplex* was once common on vertical rock walls and undercuts at Rottnest Island, Western Australia, generally in a partially shaded position, but has not been observed recently. Blades in *Rhipiliopsis millarii* are coarse and mesh-like and the siphons have short lateral papillae with spines.

Kraftalia gracilis

TYPE LOCALITY: Heron Island, southern Great Barrier Reef, Queensland.
DISTRIBUTION: Capricorn Group, Great Barrier Reef.

Rhipiliopsis millarii

TYPE LOCALITY: Wistari Reef, southern Great Barrier Reef, Queensland.
DISTRIBUTION: Known from the Great Barrier Reef.

Kraftalia gracilis (Heron Island, Qld).

(right) *Kraftalia gracilis* – detail of sinuous filaments (Heron Island, Qld). Scale = 25 µm.

Rhipiliopsis millarii (Lizard Island, Qld).

Rhipiliopsis multiplex

TYPE LOCALITY: Fish Hook Bay, Rottnest Island, Western Australia.
DISTRIBUTION: Known only from Rottnest Island.

Rhipiliopsis peltata

TYPE LOCALITY: Port Phillip Heads, Victoria.
DISTRIBUTION: From Map Reef, Western Australia, around southern mainland Australia to Inverloch, Victoria.
FURTHER READING: Kraft (1986a); Kraft (2007); Huisman & Verbruggen (2015b).

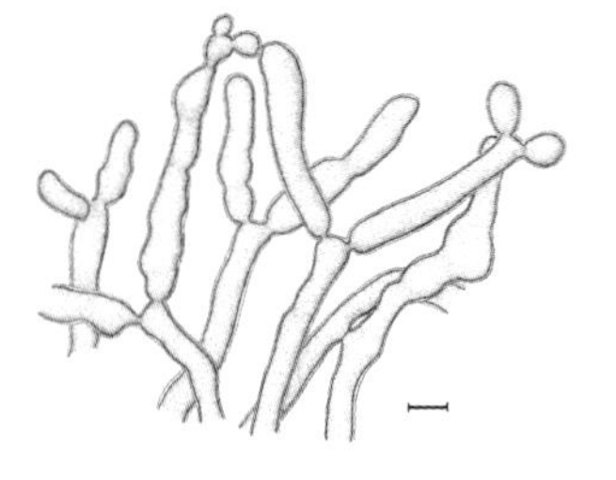

Rhipiliopsis multiplex (Rottnest Island, WA).

Rhipiliopsis peltata (Canal Rocks, WA; photo G.J. Edgar).

(above) *Rhipiliopsis peltata* – detail of filaments at thallus apex (Map Reef, WA). Scale = 30 µm.

Callipsygma

Callipsygma includes the single species *Callipsygma wilsonis*, named for John Bracebridge Wilson (1828–1895), who collected numerous algal specimens from the vicinity of Port Phillip Heads, including the type specimen of *Callipsygma wilsonis* in 1881. Plants are erect and grow to about 35 cm tall; they consist of coarse axes that taper upwardly to younger branches that are somewhat fan-shaped. The upper branches are composed of dichotomously branched siphons that can become attached laterally to adjacent siphons. *Callipsygma* is unusual in that the siphons are partitioned into segments by cross-walls that are incomplete, leaving a small gap in the centre. It is usually found in deep water or in shaded locations.

Callipsygma wilsonis

TYPE LOCALITY: Sorrento, Victoria.
DISTRIBUTION: Rottnest Island, Western Australia, and from Kangaroo Island, South Australia, to Sorrento, Victoria, several islands in Bass Strait, and Musselroe Bay, Tasmania.
FURTHER READING: Womersley (1984); Scott (2017).

Callipsygma wilsonis (Rottnest Island, WA).

(above) *Callipsygma wilsonis* – closer view of thallus (Rottnest Island, WA). Scale = 2 mm.

Chlorodesmis

Chlorodesmis is represented in Australia by four species, of which *Chlorodesmis fastigiata* is included here. The species is common in the tropical waters of the Great Barrier Reef, where its bright green colour stands in marked contrast with its surroundings. It is yet to be recorded from the west coast of Australia. Thalli of *Chlorodesmis* are repeatedly dichotomously branched, with filaments that lack cellular cross-walls but are regularly constricted. In *Chlorodesmis fastigiata* these constrictions are above the dichotomies and are at slightly different heights. A closely related species, *Chlorodesmis major*, has constrictions at equal distances above the dichotomies. *Chlorodesmis fastigiata* is also known by the common name 'turtle weed'.

Chlorodesmis fastigiata

TYPE LOCALITY: Mariana Islands.
DISTRIBUTION: Tropical waters of eastern Australia. Indo-Pacific.
FURTHER READING: Ducker (1967), (1969); Kraft (2007).

Chlorodesmis fastigiata (Lizard Island, Qld).

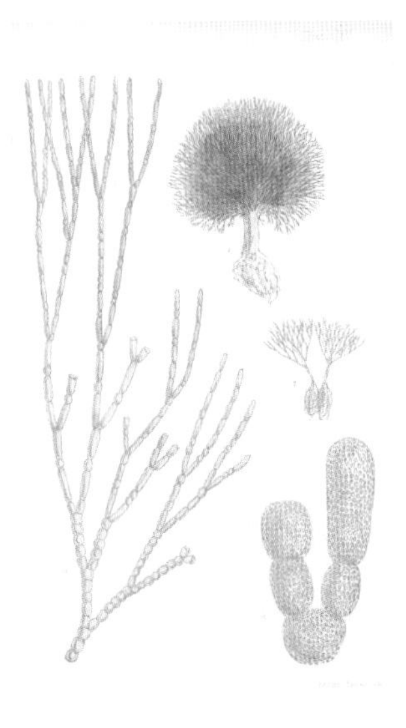

Penicillus

Penicillus is a distinctive genus, with its calcified stalk bearing a cluster of divided branches at the apex. The pioneering Irish botanist William Henry Harvey encountered *Penicillus* at Rottnest Island and later wrote that it 'may be compared to a miniature tree, or to a shaving brush' (Harvey, 1855). His description, although quaint, perfectly characterises the species. *Penicillus* has not been found at Rottnest in recent times, but is a common inhabitant of shallow, sandy habitats in warmer seas. Two species are known from tropical regions of Australia, but *Penicillus nodulosus* is by far the more commonly encountered.

Penicillus nodulosus

TYPE LOCALITY: Shark Bay, Western Australia.
DISTRIBUTION: From Rottnest Island, Western Australia (rarely), around northern Australia to Queensland. Warmer waters of the Indo-Pacific.
FURTHER READING: Kraft (2007); Huisman (2015).

Penicillus nodulosus (Heron Island, Qld).

(top) *Penicillus nodulosus* – as illustrated by Harvey (1858, pl. 22, as *Penicillus arbuscula*).

(above) *Penicillus nodulosus* – unusual growth form with cup-shaped thallus (Lizard Island, Qld).

Rhipidosiphon

Rhipidosiphon includes three tropical species, of which *Rhipidosiphon javensis* is found in Australia. Plants are small, generally less than 3 cm tall, and have a slender, monosiphonous stalk that bears a terminal fan-shaped blade one-cell-layer thick. The siphons of the blade lack any form of lateral branches and are held together by calcium carbonate. *Rhipidosiphon* is somewhat similar to *Udotea* (see p. 400) but differs in the stalk being composed of a single siphon, whereas stalks of *Udotea* are thicker with numerous siphons.

Rhipidosiphon javensis

TYPE LOCALITY: Leiden Island (Nyamuk-besar), near Jakarta, Java, Indonesia.
DISTRIBUTION: Widespread in the tropical Indo-Pacific.
FURTHER READING: Littler & Littler (1990); Kraft (2007); Coppejans et al. (2011); Huisman (2015).

Rhipidosiphon javensis (Lizard Island, Qld).

Tydemania

Tydemania was named for Gustaaf Frederik Tydeman, the captain of H.M.S. *Siboga* during the Dutch biological and hydrographic expedition to Indonesia from March 1899 to February 1900. Included among the scientists on board was Anna Weber-van Bosse, who subsequently described the algae collected during the expedition in a series of landmark publications. *Tydemania* includes only the single species, but it can appear is several different forms. The photo shows the more common and distinctive form, which looks very much like a chain of dark-green pompoms.

Tydemania expeditionis

TYPE LOCALITY: Indonesia.
DISTRIBUTION: Widespread in the tropical Indo-Pacific. In Australia, known primarily from the Great Barrier Reef, but also found at Ashmore Reef.
FURTHER READING: Coppejans et al. (2001); Littler & Littler (2003).

Tydemania expeditionis (Ashmore Reef).

Udotea

Species of *Udotea* are common in tropical seas and several occur in Australia. Plants often grow on sandy substrata, and produce a massive, fibrous holdfast to anchor themselves. Thalli are calcified and have a narrow stipe and a broad, flattened upper blade. Structurally the blade is composed of numerous branched siphons that either are naked or bear lateral appendages of various forms, depending on the species. These appendages serve to bind adjacent siphons. *Udotea flabellum* is very common and is one of the few species of the genus found in tropical regions worldwide. The four species included here can be separated on features of their blades, whether they have a single layer of siphons (*Udotea glaucescens*) or several (all other species), and the presence (*Udotea argentea* and *Udotea flabellum*) or absence (*Udotea orientalis*) of lateral appendages on the siphons. *Udotea argentea* differs from *Udotea flabellum* in producing unbranched lateral appendages.

Udotea argentea

TYPE LOCALITY: Suez, Egypt.
DISTRIBUTION: From the North West Cape region, Western Australia, around northern Australia to Queensland. Tropical Indo-Pacific.

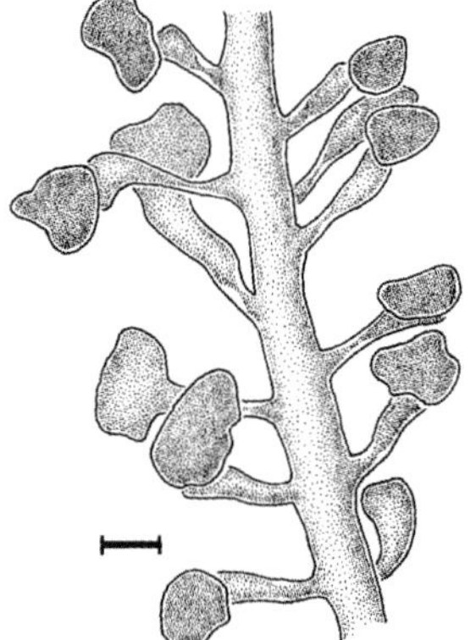

Udotea flabellum

TYPE LOCALITY: West Indies.
DISTRIBUTION: From the North West Cape region, Western Australia, around northern Australia to Queensland. Widespread in tropical seas.

Udotea glaucescens

TYPE LOCALITY: Tonga.
DISTRIBUTION: From Shark Bay, Western Australia, around northern Australia to Thursday Island, Queensland. Widespread in the tropical Indo-Pacific.

Udotea orientalis

TYPE LOCALITY: Kambaragi Bay, Tanahjampea, South Sulawesi, Indonesia.
DISTRIBUTION: From the Dampier Archipelago, Western Australia, around northern Australia to Gladstone, Queensland. Widespread in the tropical Indo-Pacific.
FURTHER READING: Kraft (2007); Huisman (2015).

Udotea argentea (Montebello Islands, WA).

(left) *Udotea argentea* – detail of filaments (Montebello Islands, WA). Scale = 50 µm.

Udotea flabellum (Heron Island, Qld).

Udotea glaucescens (Scott Reef, WA).

Udotea orientalis (Hamilton Island, Qld).

Bornetella

Three species of *Bornetella* are recorded from Australia. Thalli are small and spherical or club-shaped, and generally occur in clusters on rock in shallow water. They are often shiny and bright green in colour, but are equally likely to have a reddish-brown tinge. *Bornetella oligospora* is similar in appearance to *Bornetella nitida*, with the two species differing mainly in reproductive features (*Bornetella oligospora* having four or more sporangia per lateral branch, each with two to nine spores, whereas *Bornetella nitida* has two sporangia per lateral branch, each with eight to 26 spores). The third species found in Australia, *Bornetella sphaerica*, has, as the name suggests, a spherical thallus.

Bornetella nitida

TYPE LOCALITY: Tonga.
DISTRIBUTION: Tropical Indo-Pacific, but in Australia known only from the east coast.

Bornetella oligospora

TYPE LOCALITIES: Indonesia: Macassar [Ujung Pandang], Celebes, and Ban, Flores.
DISTRIBUTION: From the North West Cape region, Western Australia, around northern Australia to the Great Barrier Reef, Queensland.

Bornetella sphaerica

TYPE LOCALITY: Sorong, Irian Jaya, Indonesia.
DISTRIBUTION: Tropical Indo-Pacific.
FURTHER READING: Berger & Kaever (1992); Kraft (2007); Huisman (2015).

Bornetella nitida (Heron Island, Qld).

Bornetella oligospora (Little Turtle Island, WA).

Bornetella sphaerica (Long Reef, WA).

Neomeris

Neomeris is generally found growing in clusters on rock in shallow tropical waters. The genus is similar in form and growth habit to *Bornetella oligospora*, but differs in being more heavily calcified (calcification is recognisable as the white lower half of the thallus). Three species are known in Australia: *Neomeris vanbosseae*, *Neomeris annulata* (not shown) and *Neomeris bilimbata*. *Neomeris vanbosseae* and *Neomeris bilimbata* are characterised by the particulate nature of the calcified region. In *Neomeris annulata* the segments of the calcified region are united into transverse rows.

Neomeris bilimbata

TYPE LOCALITY: Itu Aba, Tizard Bank, South China Sea.

DISTRIBUTION: Widespread in the tropical Indo-Pacific.

Neomeris vanbosseae

TYPE LOCALITY: Sikka, Flores, Indonesia.

DISTRIBUTION: From the North West Cape region, Western Australia, around northern Australia to the southern Great Barrier Reef, Queensland. Tropical Indo-Pacific.

FURTHER READING: Berger & Kaever (1992); Kraft (2007); Huisman (2015).

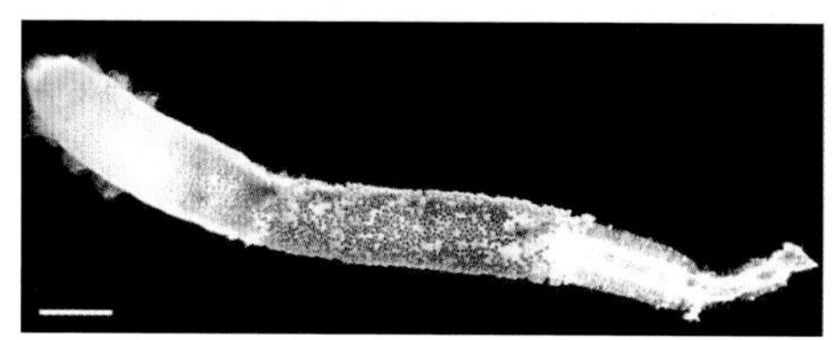

Neomeris bilimbata (Heron Island, Qld).

(left) *Neomeris bilimbata* – detail of decalcified plant (Cassini Island, WA). Scale = 1 mm.

Neomeris vanbosseae (Dampier Archipelago, WA).

Acetabularia

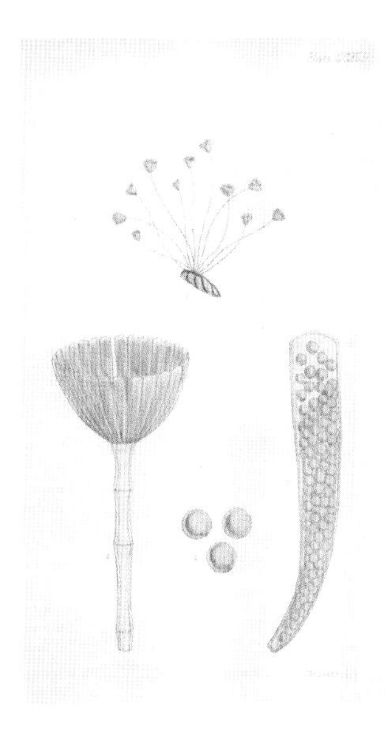

Individual thalli of *Acetabularia* have a central stalk bearing a cluster of gametangial rays that form the characteristic cup at the apex. The stalks are lightly calcified. When fertile, the gametangial rays can be seen to contain a number of cysts that are eventually released. Plants of *Acetabularia caliculus* are often found growing in dense clusters on rocks or dead shells that are partly buried in sand. In *Acetabularia caliculus*, the gametangial rays are attached laterally, whereas in *Acetabularia peniculus* they remain free.

Acetabularia peniculus

TYPE LOCALITY: King George Sound, Western Australia.
DISTRIBUTION: From Shark Bay, Western Australia, around southern Australia to Lord Howe Island, New South Wales, and Tasmania. New Caledonia.
FURTHER READING: Womersley (1984); Kraft (2007); Huisman (2015); Scott (2017).

Acetabularia caliculus

TYPE LOCALITY: Shark Bay, Western Australia.
DISTRIBUTION: Throughout Australia. Widely distributed in warmer seas.

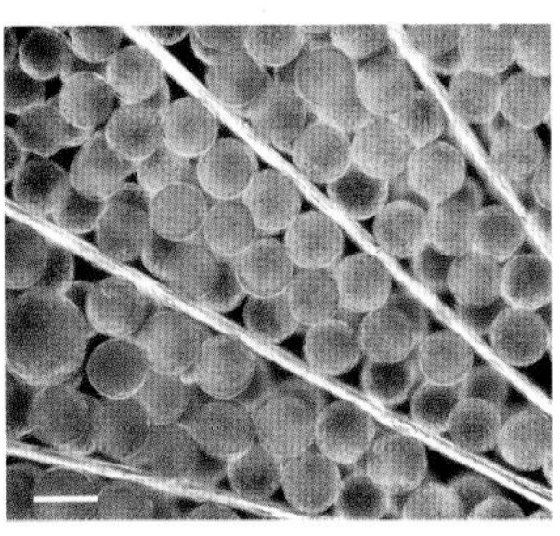

(top) *Acetabularia caliculus* – as illustrated by Harvey (1863, pl. 249).

Acetabularia caliculus (Cape Peron, WA).

(above) *Acetabularia caliculus* – detail of gametangial rays (Cape Peron, WA). Scale = 200 µm.

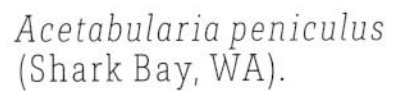

Acetabularia peniculus (Shark Bay, WA).

(above) *Acetabularia peniculus* – closer view (Shark Bay, WA).

Parvocaulis

Parvocaulis is similar to *Acetabularia* (see p. 405), differing in having an annulate (rather than smooth) stalk and lacking an inferior corona (a ring of colourless hairs below the gametangial rays). Two species occur in tropical Australia and they can be distinguished by their gametangial rays, fused laterally (*Parvocaulis parvulus*) or remaining free (*Parvocaulis exiguus*, not illustrated). Both species are diminutive, at most reaching 5 mm in height, and frequently grow on dead coral, but they are often overlooked because of their small size.

Parvocaulis parvulus

TYPE LOCALITY: Makasar, Sulawesi, Indonesia.
DISTRIBUTION: Widespread in tropical and subtropical seas.
FURTHER READING: Kraft (2007); Huisman (2015).

Parvocaulis parvulus (Scott Reef, WA). Scale = 500 µm.

Division Cyanobacteria

THE BLUE-GREEN ALGAE

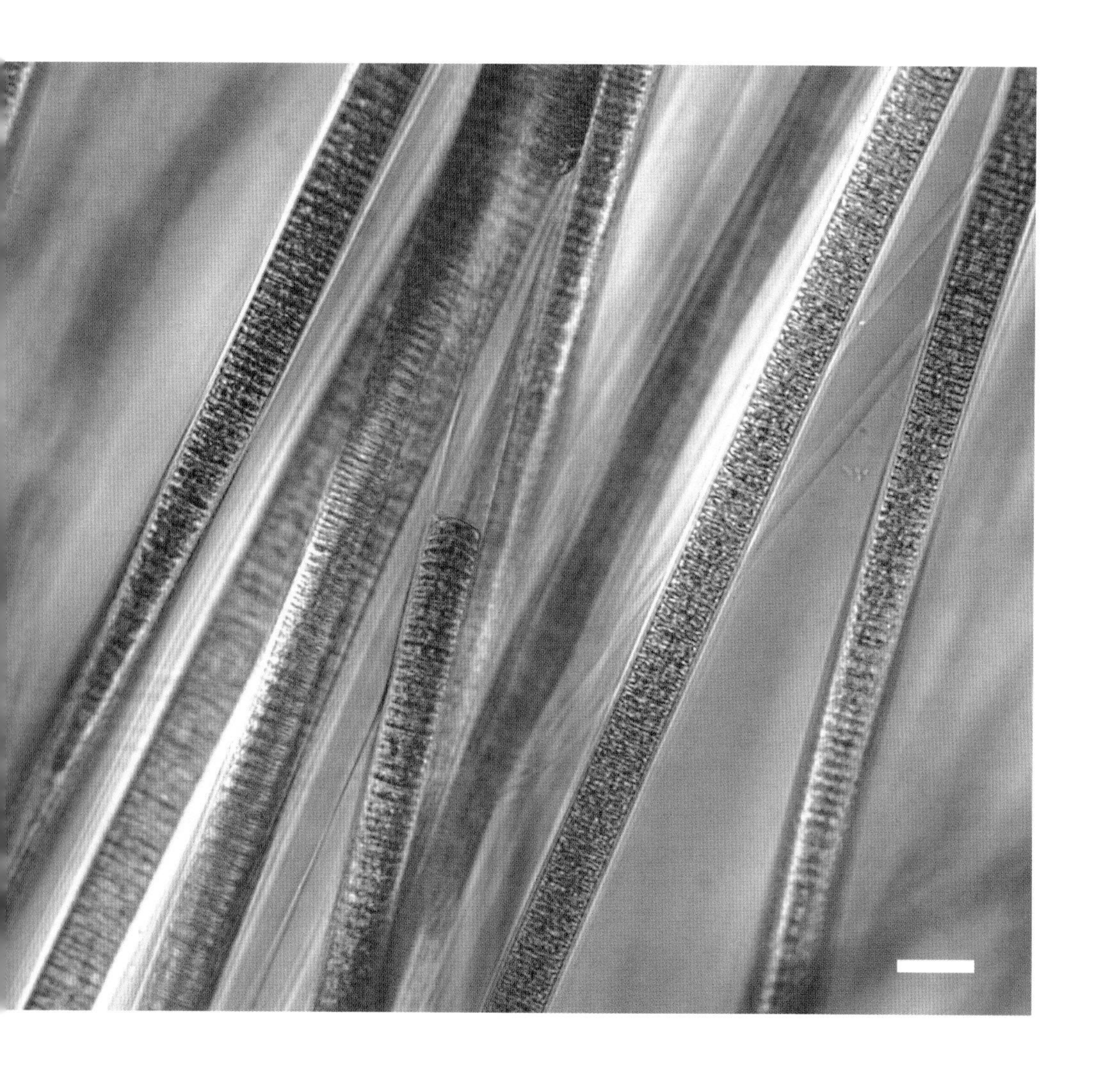

Lyngbya semiplena – detail of filaments. Scale = 25 µm.

Strictly speaking, the division Cyanobacteria (from the Greek *cyanos* meaning 'dark blue') includes photosynthetic bacteria and should not really be included in a book on marine plants. Nevertheless, the blue-green algae occasionally appear plant-like and are often encountered in the marine ecosystem. Members of the Cyanobacteria contain chlorophyll *a* and several accessory pigments. Numerous species occur, but most are small and inconspicuous and unlikely to be noticed.

Recognition of the blue-green algae is generally straightforward, as they are typically blue-green to purple in colour and are simply constructed.

Hydrocoryne

Hydrocoryne is represented in Australia by the widespread *Hydrocoryne enteromorphoides*, which forms wispy green tufts growing on sand or rock or epiphytically on other algae. The filaments are unbranched and composed of rounded, generally pale blue-green cells with occasional heterocytes. The species can be common on intertidal reefs.

Hydrocoryne enteromorphoides

TYPE LOCALITY: Guadeloupe, Caribbean, West Indies.
DISTRIBUTION: Widespread.
FURTHER READING: Umezaki & Watanabe (1994); Huisman et al. (2007, as *Hormothamnion enteromorphoides*); Komárek (2013).

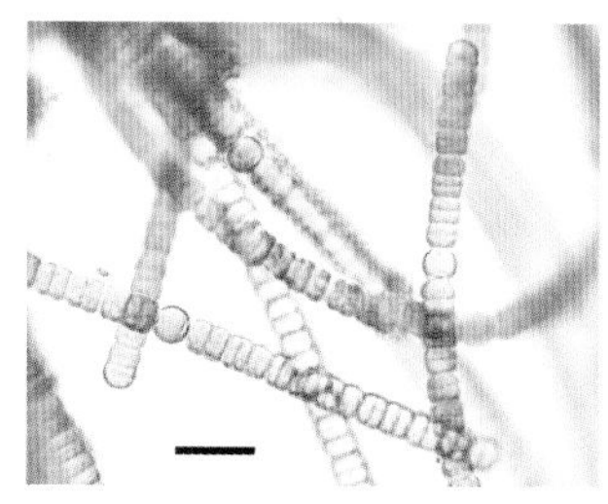

Hydrocoryne enteromorphoides (Cape Peron, WA).

(above) *Hydrocoryne enteromorphoides* (Lizard Island, Qld).

(left) *Hydrocoryne enteromorphoides* – detail of filaments. The spherical cells are heterocytes (Cape Peron, WA). Scale = 25 µm.

Rivularia

Rivularia is a widespread genus that forms hemispherical or mat-like thalli composed of one to several layers of aggregated trichomes in a gelatinous matrix. The trichomes are oriented vertically (when young with a basal heterocyte and tapering to an apical hair), but as they mature several intercalary heterocytes can develop. *Rivularia firma* was the first species of marine algae to be described by Bryan Womersley, who went on to author the six volume series *The Marine Benthic Flora of Southern Australia*.

Rivularia firma

TYPE LOCALITY: Kangaroo Island, South Australia.
DISTRIBUTION: From Two Peoples Bay, Western Australia, around southern mainland Australia and Tasmania to Wilsons Promontory, Victoria.
FURTHER READING: Womersley (1946); Scott (2017).

Rivularia firma
(Point Lonsdale, Vic).

Brachytrichia

Brachytrichia is widely distributed in most seas and includes three species, of which *Brachytrichia quoyi* occurs in Australia. The species forms light to dark-green globular thalli on intertidal rock, these often collapsing when desiccated. The walls of the thallus are composed of occasionally branched filaments in a matrix, with intercalary heterocytes that are spherical or barrel-shaped. *Brachytrichia quoyi* is similar in form to the southern Australian *Rivularia firma* (see p. 411), but in that genus the filaments do not branch.

Brachytrichia quoyi

TYPE LOCALITY: Mariana Islands.
DISTRIBUTION: Widespread in warmer seas.
FURTHER READING: Whitton (2011); Komárek (2013).

Brachytrichia quoyi (James Price Point, WA).

Symploca

Symploca is a filamentous cyanobacterium that is widespread and can occupy a range of habitats. The marine species most commonly encountered is *Symploca hydnoides*, which has a characteristic wick-like appearance due to the filaments aggregating. The filaments are unbranched, with cylindrical cells that are generally longer than broad or equidimensional.

Symploca hydnoides

TYPE LOCALITY: Appin, Argyll, Scotland.
DISTRIBUTION: Widespread, more common in warmer seas.
FURTHER READING: Komárek et al. (2014).

Symploca hydnoides (Sykes Reef, Qld).

Trichodesmium

Trichodesmium is perhaps the most conspicuous of the marine blue-green algae, by virtue of its tendency to form dense aggregations on the surface of the ocean. These are particularly conspicuous during calm conditions and are often misinterpreted as pollution. The genus is strictly planktonic and occurs as clusters of trichomes held together by a sheath. *Trichodesmium erythraeum* is most common in tropical waters, but can also be seen further south on both the east and west coasts of Australia. The species is also known by the common name 'sea sawdust'.

Trichodesmium erythraeum

TYPE LOCALITY: Near Tor, Sinai Peninsula, Egypt.
DISTRIBUTION: Widespread, more common in warmer seas.
FURTHER READING: Komárek & Anagnostidis (2005); Whitton (2011); Komárek et al. (2014).

Trichodesmium erythraeum (near Brue Reef, WA).

Lyngbya

Species of *Lyngbya* are common worldwide and are found in a variety of habitats. Thalli are filamentous and unbranched, their uniseriate trichomes with cylindrical or barrel-shaped cells that are always shorter than wide. Heterocytes do not occur. The colour of thalli can vary considerably, from blue-green to pinkish to dark purple. Some species, such as *Lyngbya majuscula* (not shown), contain biologically active compounds that are potentially carcinogenic, and contact should be avoided.

Lyngbya semiplena

TYPE LOCALITY: Trieste, Italy.
DISTRIBUTION: Widespread, more common in warmer seas.
FURTHER READING: Komárek et al. (2014).

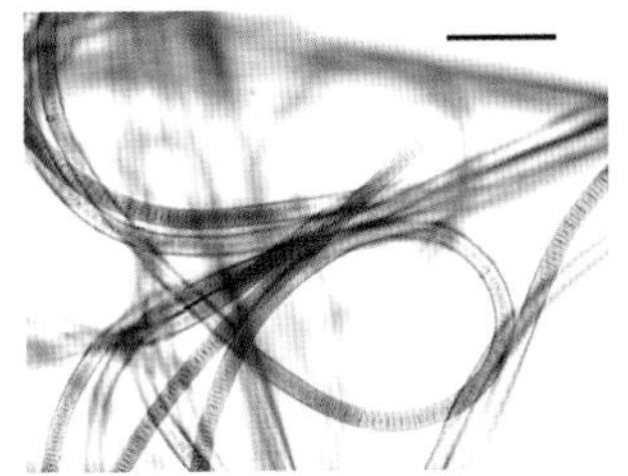

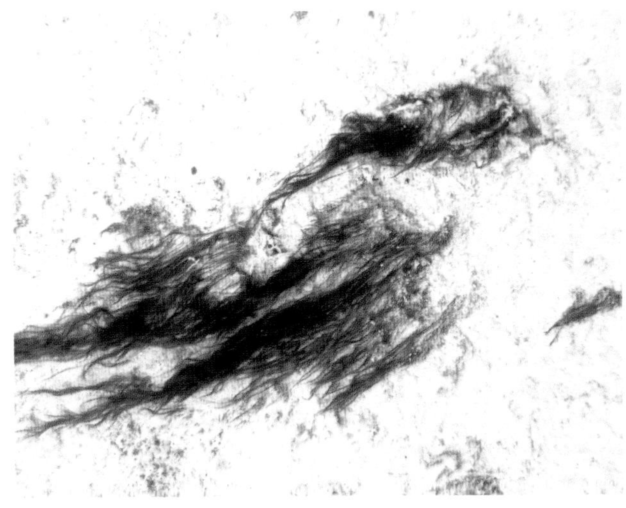

Lyngbya semiplena (Lizard Island, Qld)

(above) *Lyngbya semiplena* – detail of filaments (Cape Peron, WA). Scale = 100 µm.

Spirulina

Spirulina is a filamentous cyanobacterium that can form wispy green mats on the seabed, generally in sand or mud in low energy habitats. The filaments are unbranched and tightly coiled along their whole length, and they often display a rotating movement. Heterocytes are absent. *Spirulina* is edible and is sold in health food stores as a powder or tablet. *Spirulina subsalsa* is a cosmopolitan species that is generally green, but red mats can form under low light conditions.

Spirulina subsalsa

TYPE LOCALITY: Odense Fjord, Denmark.
DISTRIBUTION: Cosmopolitan.
FURTHER READING: McGregor (2007); Komárek et al. (2014).

Spirulina subsalsa (Nornalup Estuary, WA).

Division Magnoliophyta

THE SEAGRASSES

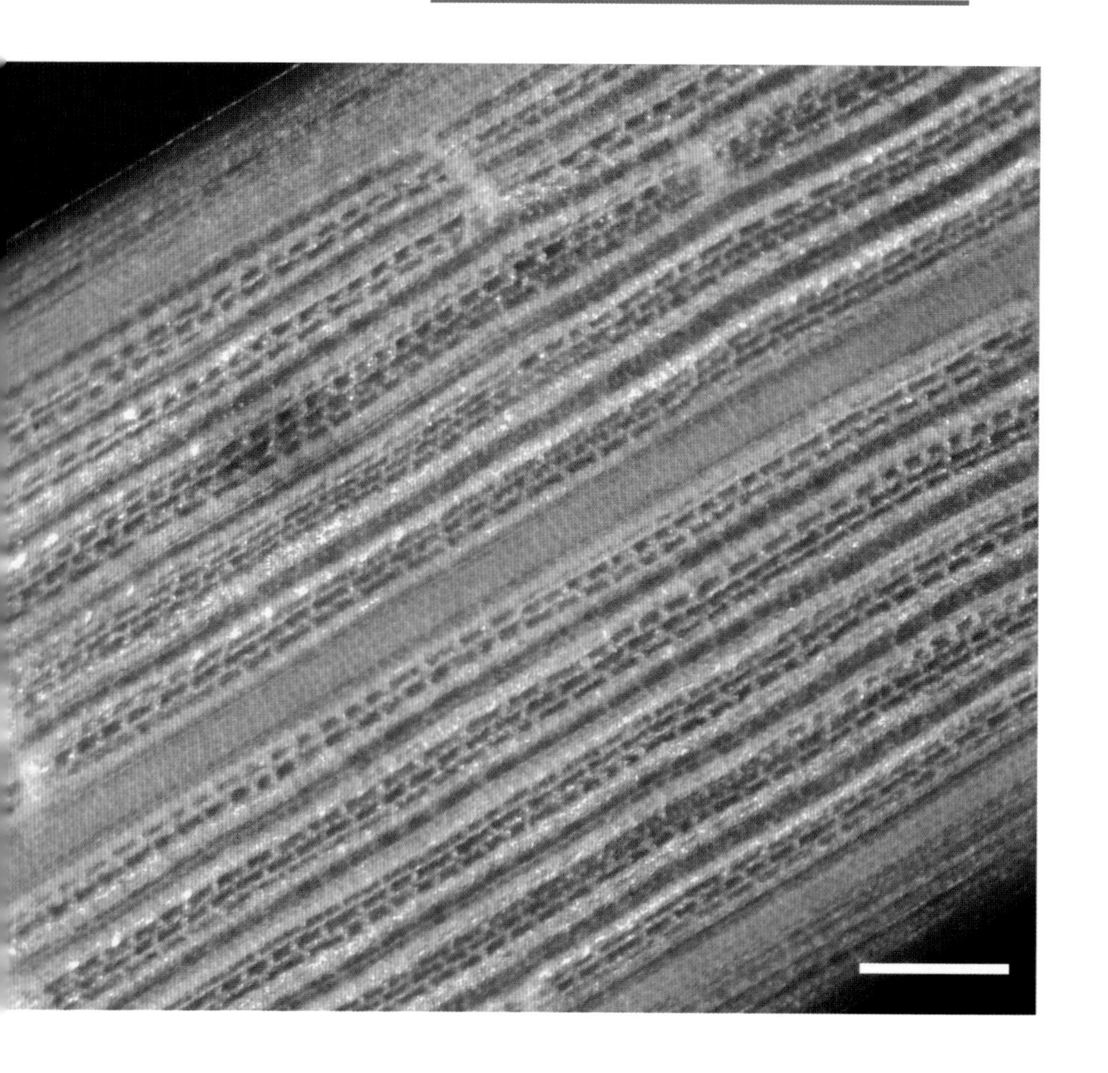

Heterozostera nigricaulis – detail of leaf surface. Scale = 250 µm.

The seagrasses are true flowering plants that have invaded the oceans. As such, they are only remotely related to the algae that make up the majority of this book, and have more in common with their terrestrial ancestors. Seagrasses produce flowers – admittedly inconspicuous versions in comparison with those of some of their flamboyant cousins – and also possess vascular tissue (veins) similar to that found in higher plants. Although represented worldwide by only around 60 species (and in Australia by half that number), seagrasses can often be the dominant element in shallow coastal systems. Shark Bay in Western Australia, for example, has over 4,000 square kilometres of almost monospecific beds of seagrass, with the Wooramel Seagrass Bank being the largest single seagrass bed in the world. Seagrasses can be very important in stabilising sediments and in providing a habitat for juvenile fish and invertebrates.

Seagrasses can usually be recognised by their green colour and propensity for forming dense beds. Some genera are grass-like in appearance; others can be leaf-like, or have woody stems. All have vascular tissue, a feature not found in the algae.

Amphibolis

Amphibolis forms extensive beds in shallow, sandy areas. The genus is readily recognised by the woody, terete stalk that supports the terminal cluster of leaves. Only two species are known, both of which occur in Australian waters. In *Amphibolis antarctica*, the leaf blades are usually proportionally shorter (2.5–10 times as long as they are broad) and are twisted; those of *Amphibolis griffithii* are longer (10–15 times as long as broad) and not twisted. The genus *Thalassodendron* (p. 424) is closely related but has very coarse and stiff rhizomes and the leaf blades have denticulate margins. *Amphibolis* is also known by the common name 'wire-weed'.

Amphibolis antarctica

TYPE LOCALITY: Esperance Bay, Western Australia.
DISTRIBUTION: From Carnarvon, Western Australia, around southern mainland Australia and the north coast of Tasmania to Wilsons Promontory, Victoria.

Amphibolis antarctica (Cape Peron, WA).

Amphibolis griffithii

TYPE LOCALITY: Henley Beach, South Australia.
DISTRIBUTION: From Champion Bay, Western Australia, around southern Australia to Victor Harbor, South Australia.
FURTHER READING: Robertson (1984); Waycott et al. (2014).

Amphibolis griffithii (Cape Peron, WA).

Cymodocea

Cymodocea is another seagrass genus with flat, strap-like leaves. In this case these are often at the tip of a short vertical stem that arises from the horizontal rhizome. Two species are known in Australia, of which the common *Cymodocea serrulata* is included here. Plants often occur in lagoon and reef habitats and have distinctive leaves with tips that are rounded and have serrated edges.

Cymodocea serrulata

TYPE LOCALITY: Tropical Australia.
DISTRIBUTION: From Exmouth Gulf, Western Australia, around northern Australia to the Gold Coast, Queensland. Widespread in the tropical Indo-Pacific.
FURTHER READING: Waycott et al. (2004).

Cymodocea serrulata
(Bali; photo: M. van Keulen).

Halodule

Halodule can be recognised by its narrow, strap-like leaves, with one to three points on each leaf tip and a distinct central vein. Plants can reach up to 25 cm in height, and the width of the leaf can vary considerably, depending on the habitat, but is usually less than 4 mm. *Halodule uninervis* can be common in tropical areas, where it generally grows on sandy or muddy substrata, often mixed with other seagrasses. This species is grazed by dugongs.

Halodule uninervis

TYPE LOCALITY: Mokha, Yemen.
DISTRIBUTION: From Exmouth Gulf, Western Australia, around northern Australia to the Gold Coast, Queensland. Widespread in the tropical Indo-Pacific.
FURTHER READING: Waycott et al. (2004).

Halodule uninervis (Hamilton Island, Qld).

Syringodium

Syringodium is represented in Australia by the single species *Syringodium isoetifolium*, an essentially tropical species that can occasionally be found in subtropical waters. The species is readily recognised by the production of leaves that are circular in cross-section, whereas the majority of other marine seagrasses have flattened leaves. *Syringodium isoetifolium* generally grows in isolated patches on sandy substrata.

Syringodium isoetifolium

TYPE LOCALITY: Unknown.
DISTRIBUTION: From the Perth region, Western Australia, around northern Australia to southern Queensland. Widespread in the tropical Indo-West Pacific.
FURTHER READING: Waycott et al. (2004).

Syringodium isoetifolium (Cape Peron, WA).

Thalassodendron

Thalassodendron includes two species in Australia, with *Thalassodendron ciliatum* widespread in tropical regions and *Thalassodendron pachyrhizum* found only in the south-west of Western Australia. Thalli have a robust rhizome with upright stems that bear apical tufts of flattened, curved leaves. The rhizome can become lignified and very coarse with age. *Thalassodendron pachyrhizum* is a rare species that is generally found growing on rock – it is similar in appearance to *Amphibolis* but can be distinguished by its coarse rhizome and leaf margins with fine teeth.

Thalassodendron ciliatum

TYPE LOCALITY: Al Mukha, Yemen, southern Red Sea.
DISTRIBUTION: Widespread in the tropical Indo-Pacific.

Thalassodendron pachyrhizum

TYPE LOCALITY: Leighton, Western Australia.
DISTRIBUTION: From the Houtman Abrolhos south to Bremer Bay, Western Australia.
FURTHER READING: Robertson (1984); Waycott et al. (2004), (2014).

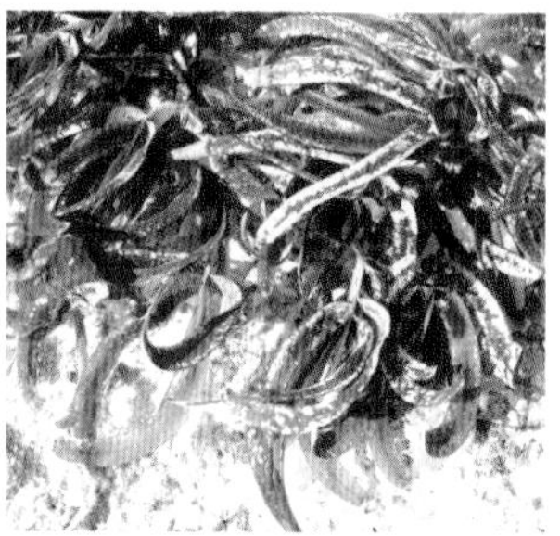

Thalassodendron ciliatum (Hamilton Island, Qld).

(far left) *Thalassodendron ciliatum* – intertidal plant with deep red leaves (Barrow Island, WA).

(left) *Thalassodendron ciliatum* – apex of leaf with marginal teeth (Barrow Island, WA). Scale = 2 mm.

Thalassodendron pachyrhizum (Rottnest Island, WA).

Enhalus

Enhalus is the largest growing of the tropical seagrasses, with the leaves of its single species, *Enhalus acoroides,* reaching up to 2 m in length. The species usually occurs in areas with a fine sand or muddy bottom, often forming dense meadows in the intertidal or shallow subtidal. The leaves of *Enhalus* are long and thick, and can be recognised not only by their size but also by their unusual inrolled margins, which give the impression of a thickened rim. The seeds of *Enhalus* are eaten in several Asian countries, raw, boiled or roasted.

Enhalus acoroides

TYPE LOCALITY: Not known.
DISTRIBUTION: From the Dampier Archipelago, Western Australia, around northern Australia to Magnetic Island, Queensland. Widespread in the tropical Indo-west Pacific.
FURTHER READING: Waycott et al. (2004).

Enhalus acoroides
(Bali; photo: M. van Keulen).

Halophila

Species of the seagrass genus *Halophila* are also known by the common name of 'paddleweed'. Plants are easily recognised by their upright branches that bear leaves in pairs. Like other genera of seagrasses, *Halophila* has horizontal stolons ('runners') that grow just below the surface of the sand. Five species can be found in Australia, of which the most widespread and common is *Halophila ovalis*, which occurs in shallow sand beds where it often colonises the spaces between other seagrass species. It is common in the warmer waters of the Indo-Pacific region but also extends into the colder waters of south-west and south-east Australia. *Halophila australis* has more elongate leaves that taper towards the base. *Halophila decipiens* occurs in tropical regions worldwide and can be distinguished from *Halophila ovalis* by its smaller leaves and the presence of fine serrations on the leaf margins, plus small hairs on both leaf surfaces. *Halophila capricorni* is similar to *Halophila decipiens*, but the hairs occur only on one leaf surface. *Halophila spinulosa* has a vertical stem and numerous leaves.

Halophila australis

TYPE LOCALITY: Queenscliff, Victoria.
DISTRIBUTION: Dongara, Western Australia, around southern mainland Australia and Tasmania to Sydney, New South Wales.

Halophila capricorni

TYPE LOCALITY: 'Steve's Bommie', One Tree Reef, Capricornia Section, Great Barrier Reef.
DISTRIBUTION: Great Barrier Reef and Coral Sea.

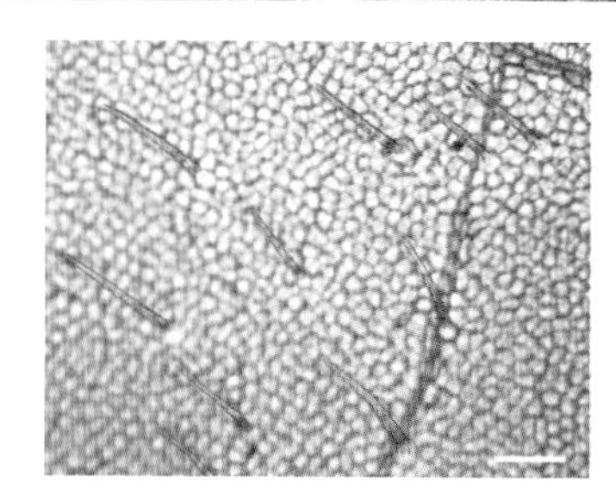

Halophila australis (Cape Peron, WA).

Halophila capricorni (Heron Island, Qld).

(left) *Halophila capricorni* – leaf surface with hairs (Heron Island, Qld). Scale = 250 µm.

Halophila decipiens

TYPE LOCALITY: Off Koh Kahdat, Gulf of Thailand.
DISTRIBUTION: From Cockburn Sound, Western Australia, around western, northern and eastern Australia to Mallacoota Inlet, Victoria. Widespread in tropical and warmer waters of the Indo-Pacific.

Halophila ovalis

TYPE LOCALITY: Queensland.
DISTRIBUTION: From Cowaramup Bay, Western Australia, around western, northern and eastern Australia to the New South Wales-Victoria border. Widespread in tropical and warmer waters of the Indo-Pacific.

Halophila spinulosa

TYPE LOCALITY: Tropical Australia.
DISTRIBUTION: Widespread in the tropical Indo-Pacific.
FURTHER READING: Robertson (1984); Larkum (1995); Waycott et al. (2004), (2014); Scott (2017).

Halophila decipiens (Lizard Island, Qld).

Halophila ovalis (Cape Peron, WA).

Halophila spinulosa (Lizard Island, Qld).

Thalassia

Thalassia includes two species: the generitype *Thalassia testudinum* (not shown), restricted to the tropical Americas, and *Thalassia hemprichii,* found in the tropical Indo-west Pacific. *Thalassia hemprichii* has grass-like upright leaves that are often arching and have red or brown speckles (caused by tannins in the epidermal cells). The species commonly forms meadows on shallow sandy reef platforms and is an important food source for green turtles.

Thalassia hemprichii

TYPE LOCALITY: Mesewa (Massaua), Ethiopia, southern Red Sea.
DISTRIBUTION: From the North West Cape region, Western Australia, around northern Australia to the Whitsunday Islands, Queensland. Widespread in the tropical Indo-west Pacific.
FURTHER READING: Waycott et al. (2004).

Thalassia hemprichii (Scott Reef, WA).

Posidonia

Species of *Posidonia* form dense and extensive meadows in shallow subtidal areas. In most species, the leaf blades are flattened, with the terete to biconvex leaves of *Posidonia ostenfeldii* (not illustrated) the only exception. The remaining species can be separated by the breadth of their leaf blades and the dimensions and appearance of blade epidermal cells. Blades in *Posidonia australis* are generally over 10 mm broad with epidermal cells mostly square in surface view. In *Posidonia angustifolia* and *Posidonia sinuosa*, the blades are narrower and epidermal cells are elongate in surface view – those of *Posidonia sinuosa* have distinctive sinuous margins. Blades in *Posidonia sinuosa* can also be slightly concave-convex. *Posidonia* is known by the common name 'tapeweed'.

Posidonia australis

TYPE LOCALITY: Georgetown, Tasmania.
DISTRIBUTION: From Shark Bay, Western Australia, around southern mainland Australia and northern and eastern Tasmania to Lake Macquarie, New South Wales.

Posidonia australis (Cape Peron, WA).

(top) *Posidonia australis* – fruit (Cape Peron, WA).

(above) *Posidonia australis* – on decorator crab (Cape Peron, WA).

Posidonia angustifolia

TYPE LOCALITY: Four kilometres north-east of Cape Naturaliste, Western Australia.
DISTRIBUTION: From the Houtman Abrolhos, Western Australia, around southern Australia to Port MacDonnell, South Australia.

Posidonia coriacea

TYPE LOCALITY: Georgetown, Tasmania.
DISTRIBUTION: From Coral Bay, Western Australia, around southern mainland Australia and northern and eastern Tasmania to Lake Macquarie, New South Wales.

Posidonia sinuosa

TYPE LOCALITY: Point Atwick, Garden Island, Western Australia.
DISTRIBUTION: From the Houtman Abrolhos, Western Australia, around southern Australia to Kingston, South Australia.
FURTHER READING: Cambridge & Kuo (1979); Robertson (1984); Waycott et al. (2014).

Posidonia angustifolia (Augusta, WA).

Posidonia coriacea (Bateman Bay, WA; photo M. van Keulen).

Posidonia sinuosa (Cape Peron, WA).

(above) *Posidonia sinuosa* – closer view of leaves (Cape Peron, WA).

Heterozostera

Heterozostera includes three species in Australia, all of which were previously included in a broadly defined *Heterozostera tasmanica* (not shown), a species now regarded as restricted to south-eastern Australia. *Heterozostera* is commonly found in shallow coastal waters of temperate Australia. Plants have branched, creeping rhizomes from which upright stems and leaves develop. The leaf blades are linear, generally only 1–2 mm broad, and have three longitudinal veins. *Heterozostera* is also known by the common names of 'garweed' and 'eelgrass'. *Heterozostera nigricaulis* is the most commonly encountered species and can be distinguished by the presence of erect wiry black stems and blades with a rough surface (when viewed under the microscope), whereas in *Heterozostera polychlamys*, black stems are lacking and blade surfaces are smooth. Some authors include the species of *Heterozostera* in a broadly defined *Zostera*.

Heterozostera nigricaulis

TYPE LOCALITY: Kangaroo Island, South Australia.
DISTRIBUTION: From Shark Bay, Western Australia, around southern mainland Australia and Tasmania to Port Stephens, New South Wales.

Heterozostera polychlamys

TYPE LOCALITY: Flinders Bay, Western Australia.
DISTRIBUTION: From Shark Bay, Western Australia, to Investigator Strait, South Australia.
FURTHER READING: Robertson (1984); Kuo (2005); Jacobs & Lee (2009); Waycott et al. (2014).

Heterozostera nigricaulis (Cape Peron, WA).

Heterozostera polychlamys (Nornalup Inlet, WA).

GLOSSARY

Bold type refers to a cross-reference.

abaxial: the position of structures borne on the sides of lateral branches facing away from the main axis.

acropetal: apical growth.

adaxial: the position of structures borne on the sides of lateral branches facing towards the main axis.

adventitious: a branch or **filament** that arises after the primary morphological pattern is established. An accessory structure.

alga (pl. **algae**): general term referring to mostly photosynthetic, unicellular or simply constructed plant-like organisms. A number of divisions are included, many of which are only remotely related to one another. The macroalgae included here belong to the divisions Chlorophyta (green algae), Cyanobacteria (blue-green algae), Rhodophyta (red algae), and Ochrophyta (brown algae).

alternate-distichous: branching pattern in which the lateral branches arise singly from each axial cell (or segment) with a divergence of 180 degrees from preceding branches, such that they lie in one plane (e.g. *Dasyclonium*, p. 226).

amorphous: without a well-defined shape.

anastomose: secondary joining of cells or branches.

annual fronds: fronds lasting one year before being shed or degenerating.

annular: ring-like.

anticlinal: filaments perpendicular to the surface of the thallus.

apex (pl. **apices**): the tip of the thallus or branch; the point most distant from the point of attachment.

apical depression: a pit at the apex.

arcuate: curving towards the base.

articulated: with joints.

asexual reproduction: reproduction not involving sexual fusion or meiosis – generally via **monospores**, **propagules** or **fragmentation**. Progeny produced in this way are identical to the parent plant.

auxiliary cell: a cell in red algae that receives the diploid nucleus following fertilisation.

axial: of the axis.

axis (pl. **axes**): (i) (of habit) refers to the main stem or branch; (ii) (of structure) refers to the central filament (hence uniaxial) or filaments (hence multiaxial).

barbel: a short branch with a posteriorly directed spine, as in the head of a spear.

basal cells: cells found near the point of attachment to the substratum.

basipetally directed: growing towards the base of the plant.

bicuspid: with two points or spines.

bifurcate: dividing equally into two.

calcified tissue: tissue that deposits calcium carbonate, generally external to the cells.

carpogonial: of the **carpogonium**.

carpogonium (pl. **carpogonia**): the female gamete (egg cell) of the Rhodophyta.

carposporangium (pl. **carposporangia**): specialised cell of the **carposporophyte** in which the contents are transformed into **carpospores** and eventually released, leaving behind the wall of the parent cell.

carpospores: the **diploid** spores produced by the **carposporophyte**.

carposporophyte: the **diploid** generation resulting from fertilisation in the Rhodophyta. It is borne on, and initially remains attached to, the female **gametophyte**, but when mature, releases **carpospores** that germinate to produce the **tetrasporophyte**.

carpotetrasporangium: **carposporangium** in which the contents are divided into four **carpospores** as opposed to the more usual one.

carrageenan: an extracellular component of some red algae that has gelling properties and is extracted commercially for use in food preparation and in industry.

cartilaginous: firm but moderately flexible.

cell: the structural unit of organisms.

circinately: with a coiled apex.

clavate: with the shape of a club – narrow at the base and widening gradually to the apex.

coherent: remaining attached.

conceptacles: cavities in the brown algae and the coralline red algae in which reproductive organs are borne.

cortex: the outer layer of the **thallus**, external to the **medulla**.

cortical: refers to filaments or cells that make up the **cortex**.

corticating: similar to **cortical**, but generally referring to filaments or cells that form a secondary **cortex** over the thallus.

cross-walls: cell walls perpendicular to the direction of growth.

cruciately divided: (of tetrasporangia) refers to the pattern of cleavage in which the four spores are produced by successive divisions at right angles (e.g. *Antithamnionella*, p. 175).

cystocarp: reproductive structure arising following fertilisation in the Rhodophyta; includes the **carposporophyte** and the sterile, protective **pericarp** derived from the female **gametophyte**.

cytoplasm: the contents of a cell.

decumbent: growth pattern in which the axes are secondarily attached to the substratum and come to lie parallel to it.

denticulate: with teeth-like structures.

diaphragms: thin sheets of cells that divide internal spaces.

dichotomous: in branching, dividing equally into two as a result of a longitudinal division of the apical cell. Most of what appears to be dichotomous branching in the algae is pseudodichotomous, where the lower lateral branch or cell develops to be equal in size and comparable in position to the primary branch or cell.

diploid: with two sets of chromosomes. The condition of cells after fertilisation and before **meiosis**.

distal: furthest from the point of attachment.

distichous: branching pattern in which the lateral branches are on opposing sides of the axis, the thallus lying in one plane.

dorsiventral: with distinct upper and lower surfaces.

drift: detached plants cast up on the shore.

epilithic: growing on rock.

epiphytic: growing directly on a plant.

estuary: body of water led by a river but also with a marine influence.

fascicle: cluster of filaments or branches.

filamentous: in habit, referring to hair-like plants that are composed of linear series of mostly uncorticated cells, generally not consolidated into a substantial thallus. In morphology, composed of free filaments, not forming a coherent tissue (e.g. filamentous medulla or cortex) (*see also* **pseudoparenchyma**).

filament: a linear series of cells; can be branched or unbranched (e.g. *Cladophora*, p. 354).

flabellate: fan-shaped.

flaccid: limp.

flagellum (pl. **flagella**): a thread-like organelle projecting from the surface of the cell. Used for motility in some unicellular organisms, spores, and gametes.

flexuous: with numerous twists and bends.

fragmentation: vegetative reproduction by disintegration, dispersal of fragments, and subsequent regeneration of plants.

gametangial rays: the radial **gametangia** in *Acetabularia* (p. 405).

gametangium (pl. **gametangia**): cell or specialised structure producing gametes.

gamete: the haploid reproductive cell that unites with another (fertilisation) to form the **diploid zygote**.

gametophyte: the haploid phase of the life history, producing **gametes**.

ganglioid cells: cells with elongate arms or processes.

geniculum (pl. **genicula**): the uncalcified 'joint' region in articulated coralline red algae.

genus (pl. **genera**): taxonomic rank above species.

globose: almost spherical.

gonimoblast: **diploid** tissue of the **carposporophyte** in red algae; produces the **carposporangia**.

haploid: with a single set of chromosomes. The condition of cells after **meiosis** and before fertilisation.

heteromorphic: of two different forms, usually pertaining to a life history in which the various phases are morphologically dissimilar.

holdfast: attachment structure of **algae**.

hyaline: (of cells or filaments) unpigmented.

intercalary: between (usually in reference to cells). Refers to the position of structures or meristems that occur in between normal vegetative cells.

intergeniculum (pl. **intergenicula**): the calcified region in articulated coralline red algae.

intertidal: the area that is covered with water at extreme high tide, and exposed at extreme low tide.

intracellular: within the cell.

iridescent: sparkling.

isodiametric: with sides of equal (or nearly so) dimensions.

laboratory culture: grown under artificial conditions.

lateral: sideways directed, as in branches arising from the main axis.

lenticular: lens-shaped with both sides convex.

life history: the summation of reproductive and morphological phases of a species.

longitudinal: in the direction of growth.

macroalgae: the multicellular **algae**, visible to the naked eye.

medulla: inner region of the **thallus**; can be **filamentous** (e.g. *Halymenia*, p. 122), pseudoparenchymatous (e.g. *Gracilaria*, p. 112) or mostly hollow (e.g. *Botryocladia*, p. 157). Gradations occur between the three conditions.

medullary: pertaining to the **medulla**.

meiosis: occurs in the **sporophyte** prior to the production of spores. A two-stage division of the diploid nucleus where the number of chromosomes is halved.

meristem: a region of actively dividing cells, generally recognisable by their comparatively smaller size. Meristems are generally apical or **intercalary**.

midrib: a central thickened region.

monosiphonous: with a single central filament (siphon).

monospore: an asexual reproductive organ (e.g. *Colaconema daviesii*, p. 9).

mucilage-filled: generally, applied to thalli with a partially hollow medulla that contains a slimy substance (e.g. *Chrysymenia brownii*, p. 160).

mucilaginous: with a slimy or slippery texture; in extreme forms, with a thick coating (e.g. *Trichogloea*, p. 25).

multiaxial: thallus resulting from the apical meristematic activity of a number of filaments, each contributing equally. Generally recognisable by the presence of a cluster of apical cells.

multinucleate: cells with many nuclei.

nemathecium (pl. **nemathecia**): raised fertile patch.

node: position on the main axis where lateral branches are attached.

organelles: subcellular structures.

ostiole: opening to a reproductive cavity.

ovoid: egg shaped, broadest at the base.

parenchyma: tissue composed of relatively equidimensional cells derived from divisions in several planes. Appears in section as closely abutting cells with few intercellular spaces (see also **pseudoparenchyma**).

pedicel: a stalk.

percurrent: main axes that remain discernible from the base to the apex of the plant.

periaxial cells: cells adjacent to and arising from the axial cells.

pericarp: sterile jacket enclosing the **carposporophyte**.

pericentral cells: periaxial cells equal in length and orientation to the parent axial cell, generally from four to many per axial cell and remaining in close association (e.g. *Polysiphonia*, p. 251).

pigmented: coloured.

pinnate: branching pattern in which lateral branches arise on opposite sides of the axis (e.g. the lateral branches of *Acrothamnion*, p. 172).

pit connection: the joint between two adjacent cells in red algae. Can be primary (resulting from cell division) or secondary (resulting from cell fusion).

plurangium (pl. **plurangia**): a many-chambered reproductive structure in the Phaeophyceae that produces spores or gametes.

polychotomously: of branching, with numerous, equal or subequal branches arising from a node (e.g. *Metagoniolithon*, p. 58).

polysiphonous: with many siphons; refers to the condition of many Ceramiales (Rhodophyta) in which the axis is ringed by **pericentral cells**.

primordium: primary cell, or immature state of a branch or structure.

propagules: asexual reproductive organs that are shed as a unit (as opposed to being released from a sporangium).

protuberant: projecting structure, prominently raised from the surface of the thallus.

pseudodichotomous: where a lower lateral branch or cell develops to be equal in size and comparable in position to the primary branch or cell (see also **dichotomous**).

pseudoparenchyma: tissue with the appearance of parenchyma but derived from the secondary anastomosing of filaments (e.g. *Gracilaria*, p. 112).

radial: around a central point or axis.

ramulus (pl. **ramuli**): the smallest or ultimate branches.

receptacles: reproductive branches in the Fucales (Phaeophyceae) that bear conceptacles (e.g. *Scytothalia*, p. 316).

reflexed lines: directed towards the base.

reticulate: net-like.

rhizine: thick-walled filament, found in the Gelidiales (Rhodophyceae) (e.g. *Pterocladia*, p. 42).

rhizoid: accessory filament.

rhizome: an underground horizontal stem in vascular plants. Herein only in reference to the seagrasses (Magnoliophyta).

saccate: hollow, in the form of a sac.

secundly: with lateral branches arising from one side, like a comb.

segregative division: cell division in some green algae in which the cell contents form spherical bodies within the parent cell before the new cells' walls are produced.

septum (pl. **septa**): a dividing wall.

sessile: attached directly, without a stalk.

spermatangium (pl. **spermatangia**): reproductive cell producing the male gamete.

spermatium (pl. **spermatia**): non-motile male gamete in the Rhodophyta.

sporangium (pl. **sporangia**): reproductive structure that produces spores.

spore: reproductive cell released from a **sporangium**.

sporophyte: the diploid phase of the life history; produces **spores**.

stellate: star-shaped.

stichidium (pl. stichidia): specialised branch bearing reproductive structures.

stipe: the stem, between the blade and the **holdfast**.

stolon: a horizontal stem (e.g. *Caulerpa*, p. 369).

sub- (as a prefix): almost (as in subdichotomous) or below (as in subtending).

substratum: the surface or material to which the organism is attached.

subtending cell: the bearing or lower cell.

subtidal: the area below the low-tide mark; never exposed.

symbiont: one of the organisms of a symbiosis.

symbiosis: where two (or more) organisms grow in close association.

sympodial growth: a process in which the apical cell regularly loses its dominance and position to the lateral initial on the subapical cell, which then becomes the new apical cell. This process is ongoing.

temperate: with water temperatures of 10–25°C.

tenacular cells: cells with claw-like processes that form attachments to adjacent cells.

terete: (of morphology) circular in cross-section (e.g. *Sarconema*, p. 100).

terminal cell: ultimate cell.

tetrahedrally divided: (of **tetrasporangia**) the arrangement of spores following a simultaneous division at slightly oblique angles – generally only three spores are visible and they appear to occupy equal thirds of the sporangium. The fourth spore is hidden from view (e.g. *Anotrichium*, p. 178).

tetrasporangium (pl. **tetrasporangia**): the cell on the **diploid**, free-living **sporophyte** that undergoes **meiosis** to produce four **haploid** spores within the parent cell wall. Once released from the associated wall, the tetraspores germinate to produce the **gametophytes** (e.g. *Antithamnionella*, p. 175).

tetrasporophyte: the **diploid** plant arising from the germination of **carpospores** in the Rhodophyta. When fertile, it bears **tetrasporangia**.

thallus (pl. **thalli**): the algal body; also used in relation to other simply constructed, non-vascular plants.

trichoblasts: colourless filaments in the Ceramiales (Rhodophyta).

trichothallic growth: growth in which the **meristematic** region is at the base of an apical tuft of filaments.

turbinate: with the shape of a spinning top.

uniaxial: thallus resulting from the primary **meristematic** activity of a single filament. Generally recognisable by the presence of a single apical cell (e.g. *Anotrichium*, p. 186) and a persistent central axis, although the latter is often obscured. Thalli can be **filamentous**, **parenchymatous**, or **pseudoparenchymatous**.

unangium (pl. **unangia**): a single-chambered **sporangium** in the Phaeophyceae that produces spores.

uninucleate: with a single nucleus per cell.

uniseriate: with cells arranged in a linear series not more than one cell broad.

utricle: inflated cell or terminal portion of filament, generally forming a cortex (e.g. *Codium*, p. 379; *Scinaia*, p. 34).

vegetative propagation: asexual reproduction via **monospores**, **propagules** or **fragmentation**.

vesicle: sac-like structure; can be filled with gas, water or mucilage.

vesicular cells: specialised cells in the Rhodophyta functioning in secretion or storage, recognisable by their refractive contents (e.g. *Botryocladia* p. 157); also known as gland cells.

whorl: a ring of branches, generally of equal length.

zonately divided: (of **tetrasporangia**) refers to the pattern of division in which the four spores are arranged in a linear series (e.g. *Dudresnaya*, p. 81).

zygote: the **diploid** cell resulting from the fusion of **gametes**.

BIBLIOGRAPHY

Abbott, I.A. (1970). *Yamadaella*, a new genus in the Nemaliales (Rhodophyta). *Phycologia* 9: 115-123.

Abbott, I.A. & Huisman, J.M. (2005). Studies in the Liagoraceae (Nemaliales, Rhodophyta) I. The genus *Trichogloea*. *Phycological Research* 53: 149-163.

Adams, N.M. (1994). *Seaweeds of New Zealand*. Canterbury University Press, Christchurch.

Adey, W.H., Townsend, R.A. & Boykins, W.T. (1982). The crustose coralline algae (Rhodophyta: Corallinaceae) of the Hawaiian Islands. *Smithsonian Contributions to the Marine Sciences* 15: [1]-iv, [1]-74.

Allender, B.M. & Kraft, G.T. (1983). The marine algae of Lord Howe Island (New South Wales): the Dictyotales and Cutleriales (Phaeophyta). *Brunonia* 6: 73-130.

Andreakis, N., Costello, P., Zanolla, M., Saunders, G.W. & Mata, L. (2016). Endemic or introduced? Phylogeography of *Asparagopsis* (Florideophyceae) in Australia reveals multiple introductions and a new mitochondrial lineage. *Journal of Phycology* 52: 141-147.

Athanasiadis, A. (1996). Morphology and classification of the Ceramioideae (Rhodophyta) based on phylogenetic principles. *Opera Botanica* 127: 1-221.

Baldock, R.N. (1976). The Griffithsieae group of the Ceramiaceae (Rhodophyta) and its southern Australian representatives. *Australian Journal of Botany* 24: 509-593.

Baldock, R.N. (1998). Tribe Griffithsieae Schmitz 1889: 449. *In*: Womersley, H.B.S., *The Marine Benthic Flora of Southern Australia. Rhodophyta – Part IIIC*, pp. 319-354. State Herbarium of South Australia, Adelaide.

Baldock, R.N. & Womersley, H.B.S. (1968). The genus *Bornetia* (Rhodophyta, Ceramiaceae) and its southern Australian representatives, with description of *Involucrana* gen. nov. *Australian Journal of Botany* 16: 197-216.

Baldock, R.N. & Womersley, H.B.S. (1998). Tribe Bornetieae Baldock & Womersley, tribus nov. *In*: Womersley, H.B.S., *The Marine Benthic Flora of Southern Australia. Rhodophyta – Part IIIC*, pp. 313-319. State Herbarium of South Australia, Adelaide.

Belton, G.S., Huisman, J.M. & Gurgel, C.F.D. (2015). Caulerpaceae. *In*: Huisman, J.M., *Algae of Australia: Marine Benthic Algae of North-western Australia, 1. Green and Brown Algae*, pp. 75-102. ABRS, Canberra; CSIRO Publishing, Melbourne.

Belton, G.S., Draisma, S.G.A., Prud'homme van Reine, W.F., Huisman, J.M. & Gurgel, C.F.D. (2019). A taxonomic reassessment of *Caulerpa* (Chlorophyta, Caulerpaceae) in southern Australia based on *tufA* and *rbcL* sequence data. *Phycologia* 58: 234-253.

Belton, G.S., Prud'homme van Reine, W.F., Huisman J.M., Draisma S.G.A. & Gurgel C.F.D. (2014). Resolving phenotypic plasticity and species designation in the morphologically challenging *Caulerpa racemosa-peltata* complex (Chlorophyta, Caulerpaceae). *Journal of Phycology* 50: 32-54.

Berger, S. & Kaever, M.J. (1992). *Dasycladales: an illustrated monograph of a fascinating algal order*. Georg Thieme Verlag, Stuttgart.

Boedeker, C., Leliaert, F. & Zuccarello, G.C. (2016). Molecular phylogeny of the Cladophoraceae (Cladophorales, Ulvophyceae) with the resurrection of *Acrocladus* Nägeli and *Willeella* Børgesen, and the description of *Lurbica* gen. nov. and *Pseudorhizoclonium* gen. nov. *Journal of Phycology* 52: 905-928.

Bolton, J.J. & Anderson, R.J. (1994). *Ecklonia*. *In*: Akatsuka, I. (ed.), *Biology of Economic Algae*, pp. 385-406. Academic Publishing, The Hague.

Boo, G.H., Le Gall, L., Miller, K.A., Freshwater, D.W., Wernberg, T., Terada, R., Yoon, K.J. & Boo, S.M. (2016a). A novel phylogeny of the Gelidiales (Rhodophyta) based on five genes including the nuclear CesA, with descriptions of *Orthogonacladia* gen. nov. and Orthogonacladiaceae fam. nov. *Molecular Phylogenetics and Evolution* 101: 359-372.

Boo, G.H., Hughey, J.R., Miller, K.A. & Boo, S.M. (2016b). Mitogenomes from type specimens, a genotyping tool for morphologically simple species: ten genomes of agar-producing red algae. *Scientific Reports* 6(35337): 1-11.

Boo, S.M. (2010). Scytosiphonaceae, Petrospongiaceae. *In*: Kim, H.S. & Boo, S.M. (eds), *Algal flora of Korea. vol. 2, no. 1. Heterokontophyta: Phaeophyceae: Ectocarpales. Marine brown algae I*, pp. 155-185. National Institute of Biological Resources, Incheon.

Bucher, K.E. & Norris, J.N. (1992). A new deep-water red alga, *Titanophora submarina* sp. nov. (Gymnophloeaceae, Gigartinales), from the Caribbean Sea. *Phycologia* 31: 180-191.

Cassano, V., G.N. Santos, E.M.S. Pestana, J.M.C. Nunes, M.C. Oliveira & Fujii, M.T. (2019). *Laurencia longiramea* sp. nov. for Brazil and an emendation of the generic delineation of *Corynecladia* (Ceramiales, Rhodophyta). Phycologia 58: 115-127.

Cambridge, M.L. & Kuo, J. (1979). Two new species of seagrasses from Australia, *Posidonia sinuosa* and *P. angustifolia* (Posidoniaceae). *Aquatic Botany* 6: 307-328.

Chiovitti, A., Kraft, G.T., Saunders, G.W., Liao, M.-L. & Bacic, A. (1995). A revision of the systematics of the Nizymeniaceae (Gigartinales, Rhodophyta) based on polysaccharides, anatomy and nucleotide sequences. *Journal of Phycology* 31: 153-166.

Coppejans, E., Leliaert, F., Dargent, O. & De Clerck, O. (2001). Marine green algae (Chlorophyta) from the north coast of Papua New Guinea. *Cryptogamie, Algologie* 22: 375-443.

Coppejans, E., Leliaert, F., Dargent, O., Gunasekara, R., & De Clerck, O. (2009). Sri Lankan seaweeds. Methodologies and field guide to the dominant species. *ABC Taxa* 6: i-viii, 1-265.

Coppejans, E., Leliaert, F., Verbruggen, H., Prathep, A. & De Clerck, O. (2011). *Rhipidosiphon lewmanomontiae* sp. nov. (Bryopsidales, Chlorophyta) a calcified udoteacean alga from the central Indo-Pacific based on morphological and molecular investigations. *Phycologia* 50: 403-412.

Cormaci, M., Furnari, G. & Scammacca, B. (1978). On the fertile tetrasporic phase of *Cottoniella* Boergesen (Ceramiales, Rhodomelaceae, Sarcomenioideae). *Phycologia* 17: 251-256.

Cowan, R.A. & Ducker, S.C. (2007). A history of systematic phycology in Australia. *In*: McCarthy, P.M. & Orchard, A.E. (eds), *Algae of Australia: Introduction*, pp. 1-65. ABRS, Canberra; CSIRO Publishing, Melbourne.

Cremen, M.C.M., Huisman, J.M., Marcelino, V.R. & Verbruggen, H. (2016). Taxonomic revision of *Halimeda* (Bryopsidales, Chlorophyta) in south-western Australia. *Australian Systematic Botany* 29: 41-54.

Cremen, M.C.M., Frederik Leliaert, F., John West, J., Lam, D.W., Shimadae, S., Lopez-Bautista, J.M., Heroen Verbruggen, H. (2019). Reassessment of the classification of Bryopsidales (Chlorophyta) based on chloroplast phylogenomic analyses. *Molecular Phylogenetics and Evolution* 130: 397-405.

Cribb, A.B. (1960). Records of marine algae from south-eastern Queensland. V. *University of Queensland Papers, Department of Botany* 4: 3-31.

Cribb, A.B. (1983). *Marine Algae of the Southern Great Barrier Reef Part I. Rhodophyta.* Australian Coral Reef Society, Brisbane.

Cribb, A.B. (1996). *Seaweeds of Queensland. A Naturalist Guide.* The Queensland Naturalists' Club Inc., Brisbane.

Dampier, W. (1703). *A Voyage to New Holland etc. in the Year 1699 ... Sharks Bay, the Isles and Coast of New Holland.* 3rd edn. James Knapton, London.

D'Archino, R., Nelson, W.A. & Zuccarello, G.C. (2012). *Stauromenia australis*, a new genus and species in the family Kallymeniaceae (Rhodophyta) from southern New Zealand. *Phycologia* 51: 451-460.

Dawson, E.Y. (1956). Some marine algae of the southern Marshall Islands. *Pacific Science* 10: 25-66.

De Clerck, O. (2003). The genus *Dictyota* in the Indian Ocean. *Opera Botanica Belgica* 13: 1-205.

De Clerck, O. & Coppejans, E. (1997). The genus *Dictyota* J.V. Lamour (Dictyotaceae, Phaeophyta) from Indonesia in the Herbarium Wober-van Bosse, including the description of *Dictyota canaliculata* spec. nov. *Blumea* 42: 407-420.

De Clerck, O., Leliaert, F., Verbruggen, H., Lane, C.E., De Paula, J.C., Payo, D.I. & Coppejans, E. (2006). A revised classification of the Dictyoteae (Dictyotales, Phaeophyceae) based on *rbc*L and 26S ribosomal DNA sequence data analyses. *Journal of Phycology* 42: 1271-1288.

De Clerck, O., Verbruggen, H., Huisman, J.M., Faye, E. Leliaert, F., Schils, T. & Coppejans, E. (2008). Systematics and biogeography of the genus *Pseudocodium* (Bryopsidales, Chlorophyta), including the description of *P. natalense* sp. nov. from South Africa. *Phycologia* 47: 225-235.

De Clerck, O., Wynne, M.J. & Coppejans, E. (1999). *Vanvoorstia incipiens* sp. nov. (Delesseriaceae, Rhodophyta) from Tanzania, East Africa. *Phycologia* 38: 394-400.

Desikachary, T.V., Krishnamurthy, V. & Balakrishnan, M.S. (1990). *Rhodophyta. II. Taxonomy part. Part IIA.* Madras Science Foundation, Madras, India.

Díaz-Tapia, P., McIvor, L., Freshwater, D.W., Verbruggen, H., Wynne, M.J. & Maggs, C.A. (2017). The genera *Melanothamnus* Bornet & Falkenberg and *Vertebrata* S.F. Gray constitute well-defined clades of the red algal tribe Polysiphonieae (Rhodomelaceae, Ceramiales). *European Journal of Phycology* 52: 1-20.

Dixon, K.R. (2018). Peyssonneliales. *In*: Huisman, J.M., *Algae of Australia: Marine Benthic Algae of North-western Australia, 2. Red Algae*. pp. 208-244. ABRS, Canberra; CSIRO Publishing, Melbourne.

Dixon, K.R. & Huisman, J.M. (2018). Rhizophyllidaceae. *In*: Huisman, J.M., *Algae of Australia: Marine Benthic Algae of North-western Australia, 2. Red Algae*. pp. 192-199. ABRS, Canberra; CSIRO Publishing, Melbourne.

Dixon, K.R. & Saunders, G.W. (2013). DNA barcoding and phylogenetics of *Ramicrusta* and *Incendia* gen. nov., two early diverging lineages of the Peyssonneliaceae (Rhodophyta). *Phycologia* 52: 82-108.

Dixon, R.R.M. & Huisman, J.M. (2015). Fucales. *In*: Huisman, J.M., *Algae of Australia: Marine Benthic Algae of North-western Australia. 1. Green and Brown Algae*, pp. 245-275. ABRS, Canberra; CSIRO Publishing, Melbourne.

Dixon, R.R.M., Huisman, J.M., Buchanan, J., Gurgel, C.F.D. & Spencer, P. (2012). A morphological and molecular study of austral *Sargassum* (Fucales, Phaeophyceae) supports the recognition of *Phyllotricha* at genus level, with further additions to the genus *Sargassopsis*. *Journal of Phycology* 48: 1119-1129.

Dixon, R.R.M., Mattio, L., Huisman, J.M., Payri, C.E., Bolton, J.J., Gurgel, C.F. (2014). North meets south – Taxonomic and biogeographic implications of a phylogenetic assessment of *Sargassum* subgenera *Arthrophycus* and *Bactrophycus*. *Phycologia* 53: 15-22.

Doty, M.S. (1978). *Izziella abbottae*, a new genus and species among the gelatinous Rhodophyta. *Phycologia* 17: 33-39.

Draisma, S.G.A., Ballesteros, E., Rousseau, F. & Thibaut, T. (2010). DNA sequence data demonstrate the polyphyly of the genus *Cystoseira* and other Sargassaceae genera (Phaeophyceae). *Journal of Phycology* 46: 1329-1345.

Ducker, S.C. (1967). The genus *Chlorodesmis* (Chlorophyta) in the Indo-Pacific Region. *Nova Hedwigia* 13: 145-182, pl. 25-43.

Ducker, S.C. (1969). Additions to the genus *Chlorodesmis* (Chlorophyta). *Phycologia* 8: 17-19.

Ducker, S.C. (1979). The genus *Metagoniolithon* Weber-van Bosse (Corallinaceae, Rhodophyta). *Australian Journal of Botany* 27: 67-101.

Ducker, S.C. (1990). History of Australian marine phycology. *In*: Clayton, M.N. & King, R.J. (eds), *Biology of Marine Plants*, pp. 415-430. Longman Cheshire, Melbourne.

Dumilag, R.V., Liao, L. & Lluisma, A.O. (2014). Phylogeny of *Betaphycus* (Gigartinales, Rhodophyta) as inferred from COI sequences and morphological observations on *B. philippinensis*. *Journal of Applied Phycology* 26. 587-595.

Dumilag, R.V., Nelson, W.A. & Kraft, G.T. (2019). Validation and phylogenetic placement of the Placentophoraceae *fam. nov.* (Gigartinales, Rhodophyta). *Phycologia* (online). DOI: 10.1080/00318884.2018.1564609.

Edgar, G.J. (1997). *Australian Marine Life: the Plants and Animals of Temperate Waters*. Reed Books, Kew.

Edgar, G.J. (2008). *Australian Marine Life: the Plants and Animals of Temperate Waters*. 2nd edn. New Holland Publishers, Sydney.

Edyvane, K.S. & Womersley, H.B.S. (1994). Family Gigartinaceae Kützing 1843: 389. *In*: Womersley, H.B.S., *The Marine Benthic Flora of Southern Australia. Rhodophyta –Part IIIA*, pp. 285-314. ABRS, Canberra.

Farr, T., Nelson, W. & Broom, J. (2003). A challenge to the taxonomy of *Porphyra* in Australia: the New Zealand red alga *Porphyra rakiura* (Bangiales, Rhodophyta) occurs in southern Australia and is distinct from *P. lucasii*. *Australian Systematic Botany* 16: 569-575.

Filloramo, G.V. & Saunders, G.W. (2015). A re-examination of the genus *Leptofauchea* (Faucheaceae, Rhodymeniales) with clarification of species in Australia and the northwest Pacific. *Phycologia* 54: 375-384.

Filloramo, G.V. & Saunders, G.W. (2016). Molecular-assisted alpha taxonomy of the genus *Rhodymenia* (Rhodymeniaceae, Rhodymeniales) from Australia reveals overlooked species diversity. *European Journal of Phycology* 51: 354-367.

Fuhrer, B., Christianson, I.G., Clayton, M.N. & Allender, B.M. (1981). *Seaweeds of Australia*. Reed, Sydney.

Gabriel, D., Driasma, S.G.A., Sauvage, T., Schmidt, W.E., Schils, T., Lim, P.-E., Harris, J. & Fredericq, S. (2016). Multilocus phylogeny reveals *Gibsmithia hawaiiensis* (Dumontiaceae, Rhodophyta) to be a species complex from the Indo-Pacific, with the proposal of *G. eilatensis* sp. nov. *Phytotaxa* 277(1): 1-20.

Gabriel, D., Draisma, S.G.A., Schmidt, W.E., Schils, T., Sauvage, T., Maridakis, C., Gurgel, C.F.D., Harris, D.J. & Fredericq, S. (2017). Beneath the hairy look: the hidden reproductive diversity of the *Gibsmithia hawaiiensis* complex (Dumontiaceae, Rhodophyta). *Journal of Phycology* 53: 1171-1192.

Gabriel, D., Parente, M.I., Neto, A.I., Raposo, M., Schils, T. & Fredericq, S. (2010). Phylogenetic appraisal of the genus *Platoma* (Nemastomatales, Rhodophyta), including life history and morphological observations on *P. cyclocolpum* from the Azores. *Phycologia* 49: 2-21.

Gabriel, D., Schils, T., Parente, M.I., Draisma, S.G.A., Neto, A.I. & Fredericq, S. (2011). Taxonomic studies in the Schizymeniaceae (Nemastomatales, Rhodophyta): on the identity of *Schizymenia* sp. in the Azores and the generic placement of *Nemastoma confusum*. *Phycologia* 50: 109-121.

Gabrielson, P.W. & Kraft, G.T. (1984). The marine algae of Lord Howe Island (N.S.W.): the family Solieriaceae (Gigartinales, Rhodophyta). *Brunonia* 7: 217-251.

Garbary, D.J. & Harper, J.T. (1998). A phylogenetic analysis of the *Laurencia* complex (Rhodomelaceae) of the red algae. *Cryptogamie, Algologie* 19: 185-200.

George, A.S. (1999). *William Dampier in New Holland. Australia's First Natural Historian*. Bloomings Books, Melbourne.

Gordon, E.M. (1972). Comparative morphology and taxonomy of the Wrangelieae, Sphondylothamnieae and Spermothamnieae (Ceramiaceae, Rhodophyta). *Australian Journal of Botany, Supplement* 4: 1-180.

Gordon-Mills, E.M. & Womersley, H.B.S. (1974). The morphology and life history of *Mazoyerella* gen. nov. (*M. arachnoidea* (Harvey) comb. nov.) – Rhodophyta, Ceramiaceae – from southern Australia. *British Phycological Journal* 9: 127-137.

Gordon-Mills, E.M. & Womersley, H.B.S. (1987). The genus *Chondria* C. Agardh (Rhodomelaceae, Rhodophyta) in southern Australia. *Australian Journal of Botany* 35: 477-565.

Guiry, M.D. & Guiry, G.M. (2017). *AlgaeBase*. World-wide electronic publication, National University of Ireland, Galway. http://www.algaebase.org.

Guiry, M.D. & Womersley, H.B.S. (1993). *Capreolia implexa* gen. et sp. nov. in Australia and New Zealand (Gelidiales, Rhodophyta); an intertidal red alga with an unusual life history. *Phycologia* 32: 266-277.

Gurgel, C.F.D. & Fredericq, S. (2004). Systematics of the Gracilariaceae (Gracilariales, Rhodophyta): a critical assessment based on rbcL sequence analysis. *Journal of Phycology* 40: 138-159.

Gurgel, C.F.D., Norris, J.N., Schmidt, W.E., Le, H.N. & Fredericq, S. (2018). Systematics of the Gracilariales (Rhodophyta) including new subfamilies, tribes, subgenera, and two new genera, *Agarophyton* gen. nov. and *Crassa* gen. nov. *Phytotaxa* 374(1): 1–23.

Harper, J.T. & Saunders, G.W. (2002). A re-classification of the Acrochaetiales based on molecular and morphological data, and establishment of the Colaconematales, ord. nov.. *British Phycological Journal* 37: 463-475.

Harvey, A.S., Woelkerling, W.J. & Huisman, J.M. (2018). *Amphiroa*. *In*: Huisman, J.M., *Algae of Australia: Marine Benthic Algae of North-western Australia, 2. Red Algae*. ABRS, Canberra; CSIRO Publishing, Melbourne.

Harvey, A.S., Woelkerling, W. J., Huisman, J.M. & Gurgel, C.F.D. (2013). A monographic account of Australian species of *Amphiroa* (Corallinaceae, Rhodophyta). *Australian Systematic Botany* 26: 81-144.

Harvey, W.H. (1847-1849). *Nereis Australis, or Algae of the Southern Ocean*. Reeve, London.

Harvey, W.H. (1855). Some account of the marine botany of the colony of Western Australia. *Transactions of the Royal Irish Academy* 22 (Science): 525-566.

Harvey, W.H. (1858). *Phycologia Australica*...Vol. 1. Lovell Reeve & Co., London.

Harvey, W.H. (1859a). *Phycologia Australica*...Vol. 2. Lovell Reeve & Co., London.

Harvey, W.H. (1859b). Algae. Part III. Flora Tasmaniae. *In*: Hooker, J.D. (ed.), *The botany of the Antarctic voyage of H.M. discovery ships Erebus and Terror, in the years 1839-1843, under the command of Captain Sir James Clark Ross... Part III. Flora Tasmaniae. Monocotyledones and Acotyledones*, Vol. II, pp. 282-343. Lovell Reeve, London.

Harvey, W.H. (1860). *Phycologia Australica*...Vol. 3. Lovell Reeve & Co., London.

Harvey, W.H. (1862). *Phycologia Australica*...Vol. 4. Lovell Reeve & Co., London.

Harvey, W.H. (1863). *Phycologia Australica*...Vol. 5. Lovell Reeve & Co., London.

Hayden, H.S., Blomster, J., Maggs, C.A., Silva, P.C., Stanhope, M.J. & Waaland, J.R. (2003). Linnaeus was right all along: *Ulva* and *Enteromorpha* are not distinct genera.

European Journal of Phycology 38: 277-294.

Heesch, S., Rindi, F., Guiry, M.D. & Nelson, W.A. (2020). Molecular phylogeny and taxonomic reassessment of the genus *Cladostephus* (Sphacelariales, Phaeophyceae). *European Journal of Phycology* 55: 426-443.

Henne, K.-D. & Schnetter, R. (1999). Revision of the *Pseudobryopsis/Trichosolen* complex (Bryopsidales, Chlorophyta) based on features of gametangial behaviour and chloroplasts. *Phycologia* 38: 114-127.

Hernández-Kantún, J., Sherwood, A.R., Riosmena-Rodriguez, R., Huisman, J.M. & De Clerck, O. (2012). Branched *Halymenia* species (Halymeniaceae, Rhodophyta) in the Indo-Pacific region, including descriptions of *Halymenia hawaiiana* sp. nov. and *H. tondoana* sp. nov. *European Journal of Phycology* 47: 421-432.

Hillis-Colinvaux, L. (1980). Ecology and taxonomy of *Halimeda*: primary producer of coral reefs. *Advances in Marine Biology* 17: 1-327.

Hughey, J.R. & Hommersand, M.H. (2008). Morphological and molecular systematic study of *Chondracanthus* (Gigartinaceae, Rhodophyta) from Pacific North America. *Phycologia* 47: 124-155.

Huisman, J.M. (1986). The red algal genus *Scinaia* (Nemaliales, Galaxauraceae) from Australia. *Phycologia* 25: 271-296.

Huisman, J.M. (1988). *Balliella hirsuta* sp. nov. (Ceramiaceae, Rhodophyta) from Rottnest Island, Western Australia. *Phycologia* 27: 456-462.

Huisman, J.M. (1993). Supplement to the catalogue of marine plants recorded from Rottnest Island. In: *The Marine Flora and Fauna of Rottnest Island, Western Australia*, ed. F.E. Wells, D.I. Walker, H. Kirkman, & R. Lethbridge, 11-18. Western Australian Museum, Perth.

Huisman, J.M. (1994). *Ditria expleta* (Rhodophyta: Rhodomelaceae) a new red algal species from Western Australia. *Japanese Journal of Phycology* 42: 1-9.

Huisman, J.M. (1995). The morphology and taxonomy of *Webervanbossea* De Toni f. (Rhodymeniales, Rhodophyta). *Cryptogamic Botany* 5: 367-374.

Huisman, J.M. (1996). The red algal genus *Coelarthrum* Børgesen (Rhodymeniaceae, Rhodymeniales) in Australian seas, including the description of *Chamaebotrys* gen. nov. *Phycologia* 35: 95-112.

Huisman, J.M. (1997). Marine benthic algae of the Houtman Abrolhos Islands, Western Australia. In: *The Marine Flora and Fauna of the Houtman Abrolhos Islands, Western Australia*, ed. F.E. Wells, 177-237. Western Australian Museum, Perth.

Huisman, J.M. (1999). The vegetative and reproductive morphology of *Nemastoma damaecorne* (Gigartinales, Rhodophyta) from western Australia. *Australian Systematic Botany* 11: 721-728.

Huisman, J.M. (2000). *Marine Plants of Australia*. University of Western Australia Press, Nedlands; ABRS, Canberra.

Huisman, J.M. (2002). The type and Australian species of the red algal genera *Liagora* and *Ganonema* (Liagoraceae, Nemaliales). *Australian Systematic Botany* 15: 773-838.

Huisman, J.M. (2006). *Algae of Australia: Nemaliales*. ABRS, Canberra; CSIRO Publishing, Melbourne.

Huisman, J.M. (2010). Rare Seaweed Rediscovered. *Landscope* 25(4): 39-41.

Huisman, J.M. (2015). *Algae of Australia: Marine Benthic Algae of North-western Australia. 1. Green and Brown Algae*. ABRS, Canberra; CSIRO Publishing, Melbourne.

Huisman, J.M. (2018). *Algae of Australia: Marine Benthic Algae of North-western Australia, 1. Red Algae*. ABRS, Canberra; CSIRO Publishing, Melbourne.

Huisman, J.M. (2019). *Parasitic seaweed*. Landscope 35(1): 22-24.

Huisman, J.M. & Abbott, I.A. (2003). The Liagoraceae (Nemaliales, Rhodophyta) of the Hawaiian Islands I: First record of the genus *Gloiotrichus* for Hawai'i and the Pacific Ocean. *Pacific Science* 57: 267-273.

Huisman, J.M., Abbott, I.A. & Sherwood, A.R. (2004). Large subunit rDNA gene sequences and reproductive morphology reveal *Stenopeltis* as a member of the Liagoraceae, with a description of *Akalaphycus* gen. nov. *European Journal of Phycology* 39: 257-272.

Huisman, J.M., Abbott, I.A., Smith, C.M. (2007). *Hawaiian Reef Plants*. University of Hawai'i Sea Grant College Program, Honolulu.

Huisman, J.M., Boo, G.H. & Boo, S.M. (2018a). Gelidiales. *In*: Huisman, J.M., *Algae of Australia: Marine Benthic Algae of North-western Australia, 2. Red Algae*, pp. 245-264. ABRS, Canberra; CSIRO Publishing, Melbourne.

Huisman, J.M., Boo, G.H. & Boo, S.M. (2018b). The genus *Rosenvingea* (Phaeophyceae: Scytosiphonaceae) in south-west Australia, with the description of *Rosenvingea australis* sp. nov. *Botanica Marina*.

Huisman, J.M., Boo, G.H., Boo, S.M. & Lin, S.-M. (2018c). Galaxauraceae. *In*: Huisman, J.M., *Algae of Australia: Marine Benthic Algae of North-western Australia, 2. Red Algae*, pp. 30-49. ABRS, Canberra; CSIRO Publishing, Melbourne.

Huisman, J.M. & Borowitzka, M.A. (1990). A revision of the Australian species of *Galaxaura* (Rhodophyta, Nemaliales), with a description of *Tricleocarpa* gen. nov. *Phycologia* 29: 150-172.

Huisman, J.M., D'Archino, R., Nelson, W., Boo, S.M. & Petrocelli, A. (2021). Cryptic cryptogam revealed: *Hypnea corona* (Cystocloniaceae, Gigartinales), a new red algal species described from the *Hypnea cornuta* complex. *Pacific Science* 75(2): 263-268.

Huisman, J.M. & De Clerck, O.D. (2018). Halymeniaceae. *In*: Huisman, J.M., *Algae of Australia: Marine Benthic Algae of North-western Australia, 2. Red Algae*, pp. 279-294. ABRS, Canberra; CSIRO Publishing, Melbourne.

Huisman, J.M., De Clerck, O., Prud'homme van Reine, W.F. & Borowitzka, M.A. (2011). *Spongophloea*, a new genus of red algae based on *Thamnoclonium* sect. *Nematophorae* Weber-van Bosse (Halymeniales). *European Journal of Phycology* 46: 1-15.

Huisman, J.M., Dixon, R.R.M., Hart, F.N., Verbruggen, H & Anderson, R. J. (2015). The South African estuarine specialist *Codium tenue* (Bryopsidales, Chlorophyta) discovered in a south-western Australian estuary. *Botanica Marina* 58: 511-521.

Huisman, J.M., Foard, H. & Kraft, G.T. (1993). *Semnocarpa* gen. nov. (Rhodophyta, Rhodymeniales) from southern and western Australia. *European Journal of Phycology* 28: 145-155.

Huisman, J.M. & Gordon-Mills, E.M. (1994). A proposal to resurrect the tribe Monosporeae Schmitz et Hauptfleisch, with a description of *Tanakaella itonoi* sp. nov. (Ceramiaceae,

Rhodophyta) from southern and western Australia. *Phycologia* 33: 81-90.

Huisman, J.M., Harper, J.T. & Saunders, G.W. (2004). Phylogenetic study of the Nemaliales (Rhodophyta) based on large-subunit ribosomal DNA sequences supports segregation of the Scinaiaceae fam. nov. and resurrection of *Dichotomaria* Lamarck. *Phycological Research* 52: 224-234.

Huisman, J.M., Koh, Y.H. & Kim, M.S. (2015). Characterization of *Herposiphonia pectinata* (Decaisne) comb. nov. (Rhodomelaceae, Rhodophyta) from Western Australia, based on morphology and DNA barcoding. *Botanica Marina* 58: 141-150.

Huisman, J.M. & Kraft, G.T. (1984). The genus *Balliella* Itono & Tanaka (Rhodophyta: Ceramiaceae) from eastern Australia. *Journal of Phycology* 20: 73-82.

Huisman, J.M. & Kraft, G.T. (1992). Disposal of auxiliary cell haploid nuclei during post-fertilisation development in *Guiryella repens* gen. et sp. nov. (Ceramiaceae, Rhodophyta). *Phycologia* 31: 127-137.

Huisman, J.M. & Kraft, G.T. (1994). Studies of the Liagoraceae (Rhodophyta) of Western Australia: *Gloiotrichus fractalis* gen. et sp. nov. and *Ganonema helminthaxis* sp. nov. *European Journal of Phycology* 29: 73-85.

Huisman, J.M. & Leliaert, F. (2015). Cladophorales. *In*: Huisman, J.M., *Algae of Australia: Marine Benthic Algae of North-western Australia, 1. Green and Brown Algae*, pp. 32-67. ABRS, Canberra; CSIRO Publishing, Melbourne.

Huisman, J.M., Leliaert, F., Verbruggen, H., & Townsend, R.A. (2009). Marine Benthic Plants of Western Australia's Shelf-Edge Atolls. *Records of the Western Australian Museum Supplement* 77: 50-87.

Huisman, J.M. & Lin, S-.M. (2018a). Liagoraceae. *In*: Huisman, J.M., *Algae of Australia: Marine Benthic Algae of North-western Australia, 2. Red Algae*, pp. 49-77. ABRS, Canberra; CSIRO Publishing, Melbourne.

Huisman, J.M. & Lin, S-.M. (2018b). Yamadaellaceae. *In*: Huisman, J.M., *Algae of Australia: Marine Benthic Algae of North-western Australia, 2. Red Algae*, pp. 77-80. ABRS, Canberra; CSIRO Publishing, Melbourne.

Huisman, J.M. & Lin, S-.M. (2018c). Delesseriaceae. *In*: Huisman, J.M., *Algae of Australia: Marine Benthic Algae of North-western Australia, 2. Red Algae*, pp. 437-439, 458-477. ABRS, Canberra; CSIRO Publishing, Melbourne.

Huisman, J.M. & Millar, A.J.K. (1996). *Asteromenia* (Rhodymeniaceae, Rhodymeniales), a new red algal genus based on *Fauchea peltata*. *Journal of Phycology* 32: 138-145.

Huisman, J.M. & Phillips, J.A. (2015). Dictyotales. *In*: Huisman, J.M., *Algae of Australia: Marine Benthic Algae of North-western Australia, 1. Green and Brown Algae*, pp. 189-216, 233-236. ABRS, Canberra; CSIRO Publishing, Melbourne.

Huisman, J.M., Phillips, J. & Parker, C.M. (2006). *Marine Plants of the Perth Region*. Department of Environment and Conservation, Perth.

Huisman, J.M., Phillips, J.C. & Freshwater, D.W. (2009). Rediscovery of *Gelidiella ramellosa* from near the type locality in Western Australia. *Cryptogamie, Algologie* 30: 1-14.

Huisman, J.M. & Saunders, G.W. (2018a). Kallymeniaceae. *In*: Huisman, J.M., *Algae of Australia: Marine Benthic Algae of North-western Australia, 2. Red Algae*, pp. 184-192. ABRS, Canberra; CSIRO Publishing, Melbourne.

Huisman, J.M. & Saunders, G.W. (2018b). Tsengiaceae. *In*: Huisman, J.M., *Algae of Australia: Marine Benthic Algae of North-western Australia, 2. Red Algae*, pp. 295-297. ABRS, Canberra; CSIRO Publishing, Melbourne.

Huisman, J.M. & Saunders, G.W. (2018c). Sebdeniales. *In*: Huisman, J.M., *Algae of Australia: Marine Benthic Algae of North-western Australia, 2. Red Algae*, pp. 298-303. ABRS, Canberra; CSIRO Publishing, Melbourne.

Huisman, J.M. & Saunders, G.W. (2018d). Hymenocladiaceae. *In*: Huisman, J.M., *Algae of Australia: Marine Benthic Algae of North-western Australia, 2. Red Algae*, pp. 318-324. ABRS, Canberra; CSIRO Publishing, Melbourne.

Huisman, J.M. & Saunders, G.W. (2020a). *Champia pulula* (Champiaceae, Rhodymeniales), a new red algal species from the Perth region, Western Australia. *Nuytsia* 31: 65-68.

Huisman, J.M. & Saunders, G.W. (2020b). Out of the dark: *Leptofauchea lucida* (Faucheaceae, Rhodymeniales), a new red algal species from the Houtman Abrolhos, Western Australia. *Nuytsia* 31: 163-167.

Huisman, J.M. & Saunders, G.W. (2021). Resurrection of *Plocamium pusillum* Sonder (Plocamiaceae, Rhodophyta) from Australia. *Cryptogamie, Algologie* 42 (14): 231-239.

Huisman, J.M. & Saunders, G.W. (2022). Three new species of *Asteromenia* from Australia. *Botanica Marina* https://doi.org/10.1515/bot-2022-0007

Huisman, J.M., Saunders, G.W., Le Gall, L. & Vergés, A. (2016). *Rhytimenia*, a new genus of red algae based on the rare *Kallymenia maculata* (Kallymeniaceae, Rhodophyta). *Phycologia* 55: 299-307.

Huisman, J.M., Saunders, G.W. & Sherwood, A.W. (2006). Recognition of *Titanophycus*, a new genus based on *Liagora valida* Harv. (Liagoraceae, Nemaliales). *In*: Huisman, J.M., *Algae of Australia: Nemaliales*, pp. 116-119. ABRS, Canberra; CSIRO Publishing, Melbourne.

Huisman, J.M. & Schils, T. (2002). A re-assessment of the genus *Izziella* Doty (Liagoraceae, Rhodophyta). *Cryptogamie, Algologie* 23: 237-249.

Huisman, J.M., Sherwood, A.R. & Abbott, I.A. (2003). Morphology, reproduction, and the 18s rRNA gene sequence of *Pihiella liagoraciphila* gen. et sp. nov., (Rhodophyta), the so-called 'monosporangial discs' associated with members of the Liagoraceae (Rhodophyta), and proposal of the Pihiellales ord. nov. *Journal of Phycology* 39: 978-987.

Huisman, J.M., Sherwood, A.R. & Abbott, I.A. (2004). Studies of Hawaiian Galaxauraceae (Nemaliales, Rhodophyta): Large subunit rDNA gene sequences support conspecificity of *G. rugosa* and *G. subverticillata*. *Cryptogamie, Algologie* 25: 337-352.

Huisman, J.M. & Verbruggen, H. (2015a). Halimedaceae. *In*: Huisman, J.M., *Algae of Australia: Marine Benthic Algae of North-western Australia, 1. Green and Brown Algae*, pp. 123-139. ABRS, Canberra; CSIRO Publishing, Melbourne.

Huisman, J.M. & Verbruggen, H. (2015b). Rhipiliaceae. *In*: Huisman, J.M., *Algae of Australia: Marine Benthic Algae of North-western Australia, 1. Green and Brown Algae*, pp. 139-143, 144. ABRS, Canberra; CSIRO Publishing, Melbourne.

Huisman, J.M. & Verbruggen, H. (2020). *Pseudoderbesia eckloniae* sp. nov. (Bryopsidaceae, Ulvophyceae) from Western Australia. *Cryptogamie, Algologie* 41 (3): 19-23.

Huisman, J.M. & Walker, D.I. (1990). A catalogue of the marine plants from Rottnest Island, Western Australia, with notes on their distribution and biogeography. *Kingia* 1: 365-481.

Huisman, J.M. & Woelkerling, W.J. (2018a). Acrochaetiales. *In*: Huisman, J.M., *Algae of Australia: Marine Benthic Algae of North-western Australia, 2. Red Algae*, pp. 18-22. ABRS, Canberra; CSIRO Publishing, Melbourne.

Huisman, J.M. & Woelkerling, W.J. (2018b). Colaconematales. *In*: Huisman, J.M., *Algae of Australia: Marine Benthic Algae of North-western Australia, 2. Red Algae*, pp. 23-28. ABRS, Canberra; CSIRO Publishing, Melbourne.

Huisman, J.M. & Womersley, H.B.S. 1998. Tribe Monosporeae Schmitz & Hauptfleisch 1897: 483, 488. *In*: Womersley, H.B.S., *The Marine Benthic Flora of Southern Australia. Part IIIC*, pp. 300-313. State Herbarium of South Australia, Adelaide.

Huisman, J.M. & Womersley, H.B.S. (2006a). *Helminthocladia*. *In*: Huisman, J.M., *Algae of Australia: Nemaliales*, pp. 48-54. ABRS, Canberra; CSIRO Publishing, Melbourne.

Huisman, J.M. & Womersley, H.B.S. (2006b). *Helminthora*. *In*: Huisman, J.M., *Algae of Australia: Nemaliales*, pp. 54-56. ABRS, Canberra; CSIRO Publishing, Melbourne.

Huisman, J.M. & Womersley, H.B.S. (2006c). *Nothogenia*. *In*: Huisman, J.M., *Algae of Australia: Nemaliales*, pp. 91-93. ABRS, Canberra; CSIRO Publishing, Melbourne.

Indy, J.R., N'Yeurt, A.D.R. & Yasui, H. (2006). Reproductive morphology and taxonomic reappraisal of *Exophyllum wentii* Weber-van Bosse (Rhodomelaceae, Rhodophyta) from Bali Island, Indonesia. *Phycological Research* 54: 308-316.

Jacobs, S.W.L. & Lee, D.H. (2009). New combinations in *Zostera* (Zosteraceae). *Telopea* 12: 419-423.

Johansen, H.W. & Womersley, H.B.S. (1994). *Jania* (Corallinales, Rhodophyta) in southern Australia. *Australian Systematic Botany* 7: 605-625.

Jong, Y.S.D.M. de, Hitipeuw, C. & Prud'Homme van Reine, W.F. (1999). A taxonomic, phylogenetic and biogeographic study of the genus *Acanthophora* (Rhodomelaceae, Rhodophyta). *Blumea* 44: 217-249.

Kamiya, M., Zuccarello, G.C. & West, J.A. (2003). Evolutionary relationships of the genus *Caloglossa* (Delesseriaceae, Rhodophyta) inferred from large-subunit ribosomal RNA gene sequences, morphological evidence and reproductive compatability, with description of a new species from Guatemala. *Phycologia* 42: 478-497.

Kawaguchi, S., Wang, H.-W., Horiguchi, T., Lewis, J.A. & Masuda, M. (2002). Rejection of *Sinkoraena* and transfer of some species of *Carpopeltis* and *Sinkoraena* to *Polyopes* (Rhodophyta, Halymeniaceae). *Phycologia* 41: 619-635.

Keats, D.W., Maneveldt, G.W., Baba, M., Chamberlain, Y.M. & Lewis, J.E. (2009). Three species of *Mastophora* (Rhodophyta: Corallinales, Corallinaceae) in the tropical Indo-Pacific Ocean: *M. rosea* (C.Agardh) Setchell, *M. pacifica* (Heydrich) Foslie, and *M. multistrata*, sp. nov. *Phycologia* 48: 404-422.

Keats, D.W., Steneck, R.S., Townsend, R.A. & Borowitzka, M.A. (1996). *Lithothamnion prolifer* Foslie: a common non-geniculate coralline alga (Rhodophyta; Corallinaceae) from the tropical and subtropical Indo-Pacific. *Botanica Marina* 39: 187-200.

King, R.J. & Puttock, C.F. (1989). Morphology and taxonomy of *Bostrychia* and *Stictosiphonia* (Rhodomelaceae / Rhodophyta). *Australian Systematic Botany* 2: 1-73.

King, R.J. & Puttock, C.F. (1994). Macroalgae associated with mangroves in Australia: Rhodophyta. *Botanica Marina* 37: 181-191.

Kirkendale, L., Saunders, G.W. & Winberg, P. (2013). A molecular survey of *Ulva* (Chlorophyta) in temperate Australia reveals enhanced levels of cosmopolitanism. *Journal of Phycology* 49: 69-81.

Komárek, J. (2013). *Süsswasserflora von Mitteleuropa. Cyanoprokaryota: 3rd Part: Heterocytous Genera*. Vol. 19 Springer Spektrum, Heidelberg.

Komárek, J. & Anagnostidis, K. (2005). *Süsswasserflora von Mitteleuropa. Cyanoprokaryota: 2. Teil/2nd Part: Oscillatoriales*. Vol. 19 Elsevier Spektrum Akademischer Verlag, München.

Komárek, J., Kastovsky, J., Mares, J. & Johansen, J.R. (2014). Taxonomic classification of cyanoprokaryotes (cyanobacterial genera) 2014, using a polyphasic approach. *Preslia* 86: 295-335.

Kraft, G.T. (1972). Preliminary studies of Philippine *Eucheuma* species (Rhodophyta). Part1. Taxonomy and ecology of *Eucheuma arnoldii* Weber-van Bosse. *Pacific Science* 26: 318-334.

Kraft, G.T. (1977a). Studies of marine algae in the lesser-known families of the Gigartinales (Rhodophyta). I. The Acrotylaceae. *Australian Journal of Botany* 25: 97-140.

Kraft, G.T. (1977b). Studies of marine algae in the lesser-known families of the Gigartinales (Rhodophyta). II. The Dicranemaceae. *Australian Journal of Botany* 25: 219-267.

Kraft, G.T. (1978). Studies of marine algae in the lesser-known families of the Gigartinales (Rhodophyta). III. The Mychodeaceae and Mychodeophyllaceae. *Australian Journal of Botany* 26: 515-610.

Kraft, G.T. (1979). Transfer of the Hawaiian red alga *Cladhymenia pacifica* to the genus *Acanthophora* (Rhodomelaceae, Ceramiales). *Japanese Journal of Phycology* 27: 123-135.

Kraft, G.T. (1984a). The red algal genus *Predaea* (Nemastomataceae, Gigartinales) in Australia. *Phycologia* 23: 3-20.

Kraft, G.T. (1984b). Taxonomic and morphological studies of tropical and subtropical species of *Callophycus* (Solieriaceae, Rhodophyta). *Phycologia* 23: 53-71.

Kraft, G.T. (1986a). The green algal genera *Rhipiliopsis* A. & E.S. Gepp and *Rhipiella* gen. nov. (Udoteaceae, Bryopsidales) in Australia and the Philippines. *Phycologia* 25: 47-72.

Kraft, G.T. (1986b). The genus *Gibsmithia* (Dumontiaceae, Rhodophyta) in Australia. *Phycologia* 25: 423-447.

Kraft, G.T. (1988a). *Dotyophycus abbottiae* (Nemaliales), a new red algal species from Western Australia. *Phycologia* 27: 131-141.

Kraft, G.T. (1988b). *Seirospora orientalis* (Callithamnieae, Ceramiales), a new red algal species from the southern Great Barrier Reef. *Japanese Journal of Phycology* 36: 1-11.

Kraft, G.T. (2000). Marine and estuarine benthic green algae (Chlorophyta) of Lord Howe Island, south-western Pacific. *Australian Systematic Botany* 13: 509-648.

Kraft, G.T. (2007). *Algae of Australia. Marine Benthic Algae of Lord Howe Island and the Southern Great Barrier Reef, 1. Green Algae.* ABRS, Canberra; CSIRO Publishing, Melbourne.

Kraft, G.T. (2009). *Algae of Australia. Marine Benthic Algae of Lord Howe Island and the Southern Great Barrier Reef, 2. Brown Algae.* ABRS, Canberra; CSIRO Publishing, Melbourne.

Kraft, G.T., Liao, L.M., Millar, A.J.K., Coppejans, E.G.G., Hommersand, M.H. & Wilson Freshwater, D. (1999). Marine benthic red algae (Rhodophyta) from Bulusan, Sorsogon Province, Southern Luzon, Philippines. *The Philippine Scientist* 36: 1-50.

Kraft, G.T. & Min-Thein, U. (1983). *Claviclonium* and *Antrocentrum*, two new genera of Acrotylaceae (Gigartinales, Rhodophyta) from southern Australia. *Phycologia* 22: 171-183.

Kraft, G.T. & Olsen-Stojkovich, J. (1985). *Avrainvillea calithina* (Udoteaceae, Bryopsidales), a new green alga from Lord Howe Island, NSW, Australia. *Phycologia* 24: 339-345.

Kraft, G.T. & Robins, P.A. (1985). Is the order Cryptonemiales (Rhodophyta) defensible? *Phycologia* 24: 67-77.

Kraft, G.T. & Saunders, G.W. (2017). *Mychodea* and the Mychodeaceae (Gigartinales, Rhodophyta) revisited: molecular analyses shed light on interspecies relationships in Australia's largest endemic genus and family. *Australian Systematic Botany* 30: 230-258.

Kraft, G.T. & Womersley, H.B.S. (1994a). Family Dicranemataceae Kylin 1932: 65. *In*: Womersley, H.B.S., *The Marine Benthic Flora of Southern Australia. Rhodophyta – Part IIIA*, pp. 321-330. ABRS, Canberra.

Kraft, G.T. & Womersley, H.B.S. (1994b). Family Acrotylaceae Schmitz 1892: 18. *In*: Womersley, H.B.S., *The Marine Benthic Flora of Southern Australia. Rhodophyta – Part IIIA*, pp. 363-376. ABRS, Canberra.

Kraft, G.T. & Womersley, H.B.S. (1994c). Family Mychodeaceae Kylin 1932: 62. *In*: Womersley, H.B.S., *The Marine Benthic Flora of Southern Australia. Rhodophyta – Part IIIA*, pp. 450-470. ABRS, Canberra.

Kraft, G.T. & Wynne, M.J. (1996). Delineation of the genera *Struvea* Sonder and *Phyllodictyon* J.E. Gray (Cladophorales, Chlorophyta). *Phycological Research* 44: 129-143.

Kraft, L.G.K., Kraft, G.T. & Waller, R.F. (2010). Investigations into southern Australian *Ulva* (Ulvophyceae, Chlorophyta) taxonomy and molecular phylogeny indicate both cosmopolitanism and endemic cryptic species. *Journal of Phycology* 46: 1257-1277.

Kuo, J. (2005). A revision of the genus *Heterozostera* (Zosteraceae). *Aquatic Botany* 81: 97-140.

Kurihara, A. & Huisman, J.M. (2006). The *Dichotomaria marginata* assemblage in Australia. *In*: Huisman, J.M., *Algae of Australia: Nemaliales*, pp. 120-122, 134-136. ABRS, Canberra; CSIRO Publishing, Melbourne.

Lagourgue, L. & Payri, C.E. (2021). Diversity and taxonomic revision of tribes Rhipileae and Rhipiliopsideae (Halimedaceae, Chlorophyta) based on molecular and morphological data. *Journal of Phycology* 57: 1450-1471.

Larkum, A.W.D. (1995). *Halophila capricorni* (Hydrocharitaceae): a new species of seagrass from the Coral Sea. *Aquatic Botany* 51: 319-328.

Leliaert, F., Huisman, J.M. & Coppejans, E. (2007). Phylogenetic position of *Boodlea vanbosseae* (Siphonocladales, Chlorophyta). *Cryptogamie, Algologie* 28: 337-351.

Leliaert, F., Payo, D.A., Gurgel, C.F.D., Schils, T., Draisma, S.G.A., Saunders, G.W., Kamiya, M., Sherwood, A.R., Lin, S.-M., Huisman, J.M., Le Gall, L., Anderson, R.J., Bolton, J.J., Mattio, L., Zubia, M., Spokes, T., Vieira, C., Payri, C.E., Coppejans, E., D'hondt, S., Verbruggen, H. & De Clerck, O. (2018). Patterns and drivers of species diversity in the Indo-Pacific red seaweed *Portieria*. *Journal of Biogeography* 45: 2299-2313.

Lewis, J.A. (1984). Checklist and Bibliography of Benthic Marine Macroalgae recorded from Northern Australia. I. Rhodophyta. Department of Defence Materials Research Laboratories, Melbourne.

Lewis, J.A. (1985). Checklist and Bibliography of Benthic Marine Macroalgae recorded from Northern Australia. II. Phaeophyta. Department of Defence Materials Research Laboratories, Melbourne.

Lewis, J.A. (1987). Checklist and Bibliography of Benthic Marine Macroalgae recorded from Northern Australia. III. Chlorophyta. Department of Defence Materials Research Laboratories, Melbourne.

Lewis, J.A. & Womersley, H.B.S. (1994). Family Phyllophoraceae Nägeli 1847: 248. *In*: Womersley, H.B.S., *The Marine Benthic Flora of Southern Australia. Rhodophyta – Part IIIA*, pp. 259-270. ABRS, Canberra.

Lin, S.-M., Hommersand, M.H., Fredericq, S. & de Clerck, O. (2009). Characterization of *Martensia* (Delesseriaceae, Rhodophyta) based on a morphological and molecular study of the type species, *M. elegans*, and *M. natalensis* sp. nov. from South Africa. *Journal of Phycology* 45: 678-691.

Lin, S.-M., Hommersand, M.H. & Kraft, G.T. (2001). Characterization of *Hemineura frondosa* and the Hemineureae trib. nov. (Delesseriaceae, Rhodophyta) from southern Australia. *Phycologia* 40: 135-146.

Lin, S.-M., Huisman, J.M. & Ballantine, D.L. (2014). Revisiting the systematics of *Ganonema* (Liagoraceae, Rhodophyta) with emphasis on species from the northwest Pacific Ocean. *Phycologia* 53: 37-51.

Lin, S.-M., Huisman, J.M. & Payri, C. (2013). Characterization of *Liagora ceranoides* (Liagoraceae, Rhodophyta) on the basis of *rbc*L sequence analyses and carposporophyte development, including *Yoshizakia indopacifica* gen. et sp. nov. from the Indo-Pacific region. *Phycologia* 52: 161-170.

Lin, S.-M., Rodríguez-Prieto, C., Huisman, J.M., Guiry, M.D., Payri, C., Nelson, W.A. & Liu, S.-L. (2015). A phylogenetic re-appraisal of the family Liagoraceae sensu lato (Nemaliales, Rhodophyta) based on sequence analyses of two plastid genes and post-fertilisation development. *Journal of Phycology* 51: 546-559.

Lin, S.-M., Yang, S.-Y. & Huisman, J.M. (2011). Systematic revision of the genera *Liagora* and *Izziella* (Liagoraceae, Rhodophyta) from Taiwan based on molecular analyses and carposporophyte development, with the description of two new species *Journal of Phycology* 47: 352-365.

Lin, S.-M., Yang. S.-Y. & Huisman, J.M. (2011). Systematics of *Liagora* with diffuse gonimoblasts based on *rbc*L sequences and carposporophyte development, including the description of the new genera *Neoizziella* and *Macrocarpus* (Liagoraceae, Rhodophyta). *European Journal of Phycology* 46: 249-262.

Lindstrom, S.C., Gabrielson, P.W., Hughey, J.R., Macaya, E.C. & Nelson, W.A. (2015). Sequencing of historic and modern specimens reveals cryptic diversity in *Nothogenia* (Scinaiaceae, Rhodophyte). *Phycologia* 54: 97-108.

Littler, D.S. & Littler, M.M. (1990). Reestablishment of the green alga genus *Rhipidosiphon* Montagne (Udoteaceae, Bryopsidales) with a description of *Rhipidosiphon floridensis* sp. nov. *British Phycological Journal* 25: 33-38.

Littler, D.S. & Littler, M.M. (2003). *South Pacific Reef Plants. A diver's guide to the plant life of the South Pacific Coral Reefs.* OffShore Graphics, Inc., Washington, DC.

Liu, S.-L., Liao, L.M. & Wang, W.-L. (2013). Conspecificity of two morphologically distinct calcified red algae from the northwest Pacific Ocean: *Galaxaura pacifica* and *G. filamentosa* (Galaxauraceae, Rhodophyta). *Botanical Studies* 54:1.

Liu, S.-L., Lin, S.-M. & Chen, P.-C. (2015). Phylogeny, species diversity and biogeographic patterns of the genus *Tricleocarpa* (Galaxauraceae, Rhodophyta) from the Indo-Pacific region, including *T. confertus* sp. nov. from Taiwan. *European Journal of Phycology* 50: 439-456.

Liu, S.-L. & Wang, W.-L. (2009). Molecular systematics of the genus *Actinotrichia* (Galaxauraceae, Rhodophyta) from Taiwan, with a description of *Actinotrichia taiwanica* sp. nov. *European Journal of Phycology* 44: 89-105.

Lucas, A.H.S. (1936). *The Seaweeds of South Australia. Part I. Introduction and the Green and Brown Seaweeds.* Government Printer, Adelaide.

Lucas, A.H.S. & Perrin, E. (1947). *The Seaweeds of South Australia. Part II. The Red Seaweeds.* Government Printer, Adelaide.

Lyra, G.D.M., Costa, E.D.S., Jesus, P.B. de, Matos, J.C.G. de, Caires, T.A., Oliveira, M.C., Oliveira, E.C., Xi, Z., Nunes, J.M.D.C. & Davis, C.C. (2015). Phylogeny of Gracilariaceae (Rhodophyte): evidence from plastid and mitochondrial nucleotide sequences. *Journal of Phycology* 51: 356-366.

Lyra, D. de M., Iha, C., Grassa, C.J., Cai, L., Zhang, H.G., Lane, C., Blouin, N., Oliveira, M.C., Nunes, J. M de C. & Davis, C.C. (2021). Phylogenomics, divergence time estimation and trait evolution provide a new look into the Gracilariales (Rhodophyta). Molecular Phylogenetics and Evolution 165(107294): [1-14].

Macaya, E.C. & Zuccarello, G.C. (2010). DNA barcoding and genetic divergence in the Giant Kelp *Macrocystis* (Laminariales). *Journal of Phycology* 46: 736-742.

MacRaild, G.N. & Womersley, H.B.S. (1974). The morphology and reproduction of *Derbesia clavaeformis* (J.Agardh) DeToni (Chlorophyta). *Phycologia* 13: 83-93.

Masuda, M. & Guiry, M.D. (1995). The reproductive morphology of *Platoma cyclocolpum* (Nemastomataceae, Gigartinales) from Gran Canaria, Canary Islands. *Cryptogamie, Algologie* 15: 191-212.

McDonald, J., Huisman, J.M., Hart, F.N., Dixon, R.R.M. & Lewis, J.A. (2015). The first detection of the invasive macroalga *Codium fragile* ssp. *fragile* in Western Australia. *Bioinvasion Records* 4: 75-80.

McGregor, G.B. (2007). *Freshwater Cyanoprokaryota of North-Eastern Australia 1: Oscillatoriales*. ABRS, Canberra.

Maggs, C.A., Ward, B.A., McIvor, L.M., Evans, C.M., Rueness, J. & Stanhope, M.J. (2002). Molecular analyses elucidate the taxonomy of fully corticated, nonspiny species of *Ceramium* (Ceramiaceae, Rhodophyta) in the British Isles. *Phycologia* 41: 409-420.

Metti, Y., Huisman, J.M. & Millar, A.J.K. (2018). *Laurencia*. *In*: Huisman, J.M., *Algae of Australia: Marine Benthic Algae of North-western Australia, 2. Red Algae*, pp. 521-530. ABRS, Canberra; CSIRO Publishing, Melbourne.

Metti, Y., Millar, A.J.K. & Steinberg, P. (2015). A new molecular phylogeny of the *Laurencia* complex (Rhodophyta, Rhodomelaceae) and a review of key morphological characters result in a new genus, *Coronaphycus*, and a description of *C. novus*. *Journal of Phycology* 51: 929-942.

Millar, A.J.K. (1990). Marine red algae of the Coffs Harbour region, northern New South Wales. *Australian Systematic Botany* 3: 293-593.

Millar, A.K.J. (2003). The world's first recorded extinction of a seaweed. *Proceedings of the International Seaweed Symposium* 17: 313-318.

Millar, A.J.K. & Freshwater, D.W. (2005). Morphology and molecular phylogeny of the marine algal order Gelidiales (Rhodophyta) from New South Wales, including Lord Howe and Norfolk Islands. *Australian Systematic Botany* 18: 215-263.

Millar, A.J.K. & Guiry, M.D. (1989). Morphology and life history of *Predaea kraftiana* sp. nov. (Gymnophloeaceae, Rhodophyta) from Australia. *Phycologia* 28: 409-421.

Millar, A.J.K. & Kraft, G.T. (1984). The red algal genus *Acrosymphyton* (Dumontiaceae, Cryptonemiales) in Australia. *Phycologia* 23: 135-145.

Millar, A.J.K. & Kraft, G.T. (1993). Catalogue of marine and freshwater red algae (Rhodophyta) of New South Wales, including Lord Howe Island, south-western Pacific. *Australian Systematic Botany* 6: 1-90.

Millar, A.J.K. & Kraft, G.T. (1994a). Catalogue of marine brown algae (Phaeophyta) of New South Wales, including Lord Howe Island, south-western Pacific. *Australian Systematic Botany* 7: 1-46.

Millar, A.J.K. & Kraft, G.T. (1994b). Catalogue of marine benthic green algae (Chlorophyta) of New South Wales, including Lord Howe Island, south-western Pacific. *Australian Systematic Botany* 7: 419-453.

Millar, A.J.K. & Kraft, G.T. (2001). Monograph of the green macroalgal genus *Rhipilia* (Udoteaceae, Halimedales), with a description of *R. crassa* sp. nov. from Australia and the Philippines. *Phycologia* 40: 21-34.

Millar, A.J.K. & Wynne, M.J. (1992). *Patulophycus eclipes* gen. et sp. nov. (Delesseriaceae, Rhodophyta) from the Southwestern Pacific. *Systematic Botany* 17: 409-416.

Millar, A.J.K. & Xia, B.M. (1997). Studies on terete species of Australian *Gracilaria*. *In*: Abbott, I.A. (ed.) *Taxonomy of Economic Seaweeds.* Vol.6, pp.103-109. California Sea Grant College System, La Jolla, California.

Min-Thein, U. & Womersley, H.B.S. (1976). Studies on southern Australian taxa of Solieriaceae, Rhabdoniaceae and Rhodophyllidaceae

(Rhodophyta). *Australian Journal of Botany* 24: 1-166.

Mshiqeni, K.E. (1976). New records of *Hypneocolax stellaris* f. *orientalis* Weber-van Bosse, a parasitic red alga. *Nova Hedwigia* 27: 829-834.

Mshigeni, K.E. & Papenfuss, G.F. (1980). New records of the occurrence of the red algal genus *Titanophora* (Gigartinales: Gymnophlaeaceae) in the western Indian Ocean, with observations on the anatomy of the species found. *Botanica Marina* 23: 779-789.

Nam, K.W. (1999). Morphology of *Chondrophycus undulata* and *C. parvipapillata* and its implications for the taxonomy of the *Laurencia* (Ceramiales, Rhodophyta) complex. *European Journal of Phycology* 34: 455-468.

Nam, K.W. & Saito, Y. (1991). *Laurencia similis* (Ceramiales, Rhodophyta), a new species from Queensland, Australia. *British Phycological Journal* 26: 375-382.

Nelson, W.A. (2013). *New Zealand Seaweeds. An Illustrated Guide*. Te Papa Press, Wellington.

Nelson, W.A., Kim, S.Y., D'Archino, R. & Boo, S.M. (2013). The first record of *Grateloupia subpectinata* from the New Zealand region and comparison with *G. prolifera*, a species endemic to the Chatham Islands. *Botanica Marina* 56: 507-513.

Nelson, W.A., Payri. C.E., Sutherland, J.E. & Dalen, J. (2013). The genus *Melanthalia* (Gracilariales, Rhodophyta): new insights from New Caledonia and New Zealand. *Phycologia* 52: 426-436.

Nelson, W.A. & Ryan, K.G. (1986). *Palmophyllum umbracola* sp. nov. (Chlorophyta) from offshore islands of northern New Zealand. *Phycologia* 25: 168-177.

Nielsen, R., Petersen, G., Seberg, O., Daugbjerg, N., O'Kelly, C.J. & Wysor, B. (2013). Revision of the genus *Ulvella* (Ulvellaceae, Ulvophyceae) based on morphology and *tuf*A gene sequences of species in culture, with *Acrochaete* and *Pringsheimiella* placed in synonymy. *Phycologia* 52: 37-56.

Ni-Ni-Win, Sun, Z.M., Hanyuda, T., Kurihara, A., Millar, A.J.K., Gurgel, C.F.D. & Kawai, H. (2014). Four newly recorded species of the calcified marine brown macroalgal genus *Padina* (Dictyotales, Phaeophyceae) for Australia. *Australian Systematic Botany* 26: 448-464.

Nizamuddin, M. & Womersley, H.B.S. (1967). The morphology and taxonomy of *Myriodesma* (Fucales). *Nova Hedwigia* 12: 373-383.

Noble, J.M. & Kraft, G.T. (1983). Three new species of parasitic red algae (Rhodophyta) from Australia: *Holmsella australis* sp. nov., *Meridiocolax bracteata* sp. nov. and *Trichidium pedicellatum* gen. et sp. nov. *British Phycological Journal* 18: 391-413.

Noble, J.M. & Kraft, G.T. (2007). Halimedaceae Halimeda. *In*: Kraft, G.T., *Algae of Australia. Marine benthic algae of Lord Howe Island and the southern Great Barrier Reef, 1. Green algae*, pp. 189-221. ABRS, Canberra; CSIRO Publishing, Melbourne.

Norris, R.E. (1987). *Lenormandiopsis* (Rhodomelaceae), newly recorded from Africa, with a description of *L. nozawae* sp. nov. and comparison with other species. *Japanese Journal of Phycology* 35: 81-90.

Norris, R.E. (1988). Structure and reproduction of *Amansia* and *Melanamansia* gen nov. (Rhodophyta, Rhodomelaceae) on the southeastern African coast. *Journal of Phycology* 24: 209-223.

Norris, R.E. (1991). Some unusual marine red algae (Rhodophyta) from South Africa. *Phycologia* 30: 582-596.

Norris, R.E. (1991). The structure, reproduction and taxonomy of *Vidalia* and *Osmundaria* (Rhodophyta, Rhodomelaceae). *Journal of the Linnean Society of London, Botany* 106: 1-40.

Norris, R.E. (1993). Taxonomic studies on Ceramiaceae (Ceramiales, Rhodophyta) with predominantly basipetal growth of corticating filaments. *Botanica Marina* 36: 389-398.

Norris, R.E. (1994). Notes on some Hawaiian Ceramiaceae (Rhodophyceae), including two new species. *Japanese Journal of Phycology* 42: 149-155.

Norris, R.E. (1995). *Melanamansia glomerata*, comb. nov., and *Amansia rhodantha*, two hitherto confused species of Indo-Pacific Rhodophyceae. *Taxon* 44: 65-68.

Norris, R.E. & Aken, M.E. (1985). Marine benthic algae new to South Africa. *South African Journal of Botany* 51: 55-65.

Nozawa, Y. (1970). Systematic anatomy of the red algal genus *Rhodopeltis*. *Pacific Science* 24: 99-133.

Okamura, K. (1912). *Icones of Japanese Algae*. K. Okamura, Tokyo.

Papenfuss, G.F. & Edelstein, T. (1974). The morphology and taxonomy of the red alga *Sarconema* (Gigartinales: Solieriaceae). *Phycologia* 13: 31-43.

Parnell, J. & Huisman, J.M. (2006). Typification of species based on specimens in the herbarium of W.H. Harvey (TCD). *In*: Huisman, J.M., *Algae of Australia: Nemaliales*, pp. 112-115. ABRS, Canberra; CSIRO Publishing, Melbourne.

Parsons, M.J. (1975). Morphology and taxonomy of the Dasyaceae and the Lophothalieae (Rhodomelaceae) of the Rhodophyta. *Australian Journal of Botany* 23: 549-713.

Parsons, M.J. (1980). The morphology and taxonomy of *Brongniartella* Bory *sensu* Kylin (Rhodomelaceae, Rhodophyta). *Phycologia* 19: 273-295.

Parsons, M.J. & Womersley, H.B.S. (1998). Family Dasyaceae Kützing 1843: 413, 414. *In*: Womersley, H.B.S., *The Marine Benthic Flora of Southern Australia. Rhodophyta – Part IIIC*, pp. 422-510. State Herbarium of South Australia, Adelaide.

Payo, D.A., Leliaert, F., Verbruggen, H., D'hondt, S., Calumpong, H.P. & De Clerck, O. (2013). Extensive cryptic species diversity and fine-scale endemism in the marine red alga *Portieria* in the Philippines. *Proceedings of the Royal Society of London B Biological Sciences* 280: 20122660.

Penrose, D.L. (1996a). Genus *Hydrolithon* (Foslie) Foslie 1909: 55. *In*: Womersley, H.B.S., *The Marine Benthic Flora of Southern Australia. Rhodophyta – Part IIIB*, pp. 255-266. ABRS, Canberra.

Penrose, D.L. (1996b). Genus *Spongites* Kützing. *In*: Womersley, H.B.S., *The Marine Benthic Flora of Southern Australia. Rhodophyta – Part IIIB*, pp. 273-280. ABRS, Canberra.

Phillips, J.A. (2000). Systematics of the Australian species of *Dictyopteris* (Dictyotales, Phaeophyceae). *Australian Systematic Botany* 13: 283-324.

Phillips, J.A. & Huisman, J.M. (1998). *Dictyopteris serrata* (Dictyotales, Phaeophyceae). A poorly known algal species newly recorded from Australia. *Botanica Marina* 41: 43-49.

Phillips, J.A., King, R.J., Tanaka, J. & Mostaert, A. (1993). *Stoechospermum*

(Dictyotales, Phaeophyceae): a poorly known algal genus newly recorded in Australia. *Phycologia* 32: 395-398.

Phillips, J.A. & Price, I.R. (1997). A catalogue of Phaeophyta (brown algae) from Queensland, Australia. *Australian Systematic Botany* 10: 683-721.

Phillips, L.E. (2002a). Taxonomy and molecular phylogeny of the red algal geuns *Lenormandia* (Rhodomelaceae, Ceramiales). *Journal of Phycology* 38: 184-208.

Phillips, L.E. (2002b). Taxonomy of *Adamsiella* L.E. Phillips et W.A. Nelson, gen. nov. and *Epiglossum* Kützing. *Journal of Phycology* 38: 209-229.

Phillips, L.E. (2006). A re-assessment of the species previously included in *Lenormandiopsis* including the description of *Aneurianna* gen. nov. (Rhodomelaceae, Ceramiales). *Cryptogamie, Algologie* 27: 213-232.

Phillips, L.E. & De Clerck, O. (2005). The terete and sub-terete members of the red algal tribe Amansieae (Ceramiales, Rhodomelaceae). *Cryptogamie, Algologie* 26: 5-33.

Popolizio, T.R., Schneider, C.W. & Lane, C.E. (2015). A molecular evaluation of the Liagoraceae sensu lato (Nemaliales, Rhodophyta) in Bermuda including *Liagora nesophila* sp. nov. and *Yamadaella grassyi* sp. nov. *Journal of Phycology* 51: 637-658.

Price, I.R., Huisman, J.M. & Borowitzka, M.A. (1998). Two new species of *Caulerpa* (Caulerpales, Chlorophyta) from the west coast of Australia. *Phycologia* 37: 10-15.

Price, I.R. & Kraft, G.T. (1991). Reproductive development and classification of the red algal genus *Ceratodictyon* (Rhodymeniales, Rhodophyta). *Phycologia* 30: 106-116.

Price, I.R. & Scott, F.J. (1992). *The Turf Algal Flora of the Great Barrier Reef, Part I, Rhodophyta*. James Cook University, Townsville.

Racault, M.-F.L.P., Fletcher, R.L., De Reviers, B., Cho, G.Y., Boo, S.M., Parente, M.I. & Rousseau, F. (2009). Molecular phylogeny of the brown algal genus *Petrospongium* Nägeli ex Kütz. (Phaeophyceae) with evidence for Petrospongiaceae fam. nov. *Cryptogamie, Algologie* 30: 111-123.

Ramirez, M.E., Nuñez, J.D., Ocampo, E.H., Matula, C.V., Suzuki, M., Hashimoto, T. & Cledón, M. (2012). *Schizymenia dubyi* (Rhodophyta, Schizymeniaceae), a new introduced species in Argentina. *New Zealand Journal of Botany* 50: 51-58.

Reedman, D.J. & Womersley, H.B.S. (1976). Southern Australian species of *Champia* and *Chylocladia* (Rhodymeniales: Rhodophyta). *Transactions of the Royal Society of South Australia* 100: 75-104.

Richards, Z.T. & Huisman, J.M. (2014). Coral-mimicking alga *Eucheuma arnoldii* found at Ashmore Reef. *Coral Reefs* 33: 441.

Robertson, E.L. (1984). Seagrasses. *In*: Womersley, H.B.S., *The Marine Benthic Flora of Southern Australia. Part 1*, pp. 57-122. Flora and Fauna Handbooks Committee, Adelaide.

Robins, P.A. & Kraft, G.T. (1985). Morphology of the type and Australian species of *Dudresnaya* (Dumontiaceae, Rhodophyta). *Phycologia* 24: 1-34.

Rodríguez-Prieto, C., Huisman, J.M. & Lin, S.-M. (2022). Molecular phylogeny of foliose *Halymenia* and *Austroepiphloea* (Halymeniaceae, Rhodophyta) from the Indo-Pacific, with the description of *Halymenia taiwanensis* sp. nov. *Phycologia* 61: 384-395.

Rothman, M.D., Mattio, L., Wernberg, T., Anderson, R.J., Uwai, S., Mohring, M.B. & Bolton, J.J. (2015). A molecular investigation of the genus *Ecklonia* (Phaeophyceae, Laminariales) with special focus on the southern hemisphere. *Journal of Phycology* 51: 236-246.

Saito, Y. & Womersley, H.B.S. (1974). The southern Australian species of *Laurencia* (Ceramiales: Rhodophyta). *Australian Journal of Botany* 22: 815-874.

Santelices, B. & Hommersand, M. (1997). *Pterocladiella*, a new genus in the Gelidiaceae (Gelidiales, Rhodophyta). *Phycologia* 36: 114-119.

Santiañez, W.J.E., Macaya, E.C., Lee, K.M., Cho, G.Y., Boo, S.M. & Kogame, K. (2018). Taxonomic reassessment of the Indo-Pacific Scytosiphonaceae (Phaeophyceae): *Hydroclathrus rapanuii* sp. nov. and *Chnoospora minima* from Easter Island, with proposal of *Dactylosiphon* gen. nov. and *Pseudochnoospora* gen. nov. *Botanica Marina* 61: 47-64.

Santiañez, W.J.E. & Wynne, M.J. (2020). Establishment of *Mimica* gen. nov. to accommodate the anaxiferous species of the economically important red seaweed *Eucheuma* (Solieriaceae, Rhodophyta). *Phytotaxa* 439(2): 167-170.

Saunders, G.W., Birch, T.C. & K.R. Dixon (2015). A DNA barcode survey of *Schizymenia* (Nemastomatales, Rhodophyta) in Australia and British Columbia reveals overlooked diversity including *S. tenuis* sp. nov. and *Predaea borealis* sp. nov. *Botany* 93: 859-871.

Saunders, G.W., Huisman, J.M., Vergés, A., Kraft, G.T. & Le Gall, L. (2017). Phylogenetic analyses support recognition of ten new genera, ten new species and 16 new combinations in the family Kallymeniaceae (Gigartinales, Rhodophyta). *Cryptogamie, Algologie* 38: 79-132.

Saunders, G.W. & Kraft, G.T. (1994). Small-subunit rRNA gene sequences from representatives of selected families of the Gigartinales and Rhodymeniales (Rhodophyta). 1. Evidence for the Plocamiales ord. nov. *Canadian Journal of Botany* 72: 1250-1263.

Saunders, G.W. & Kraft, G.T. (1995). The phylogenetic affinities of *Notheia anomala* (Fucales, Phaeophyceae) as determined from partial small-subunit rRNA gene sequences. *Phycologia* 34: 383-389.

Saunders, G.W. & Kraft, G.T. (1996). Small-subunit rRNA gene sequences from representatives of selected families of the Gigartinales and Rhodymeniales (Rhodophyta). 2. Recognition of the Halymeniales ord. nov. *Canadian Journal of Botany* 74: 694-707.

Saunders, G.W. & Kraft, G.T. (2002). Two new Australian species of *Predaea* (Nemastomataceae, Rhodophyta) with taxonomic recommendations for an emended Nemastomatales and expanded Halymeniales. *Journal of Phycology* 38: 1245-1260.

Saunders, G.W., Lane, C.E., Schneider, C.W. & Kraft, G.T. (2006). Unraveling the *Asteromenia peltata* species complex with clarification of the genera *Halichrysis* and *Drouetia* (Rhodymeniaceae, Rhodophyta). *Canadian Journal of Botany* 84: 1581-1607.

Saunders, G.W. & Lehmkuhl, K.V. (2005). Molecular divergence and morphological diversity among four cryptic species of *Plocamium* (Plocamiales,

Florideophyceae) in northern Europe. *European Journal of Phycology* 40: 293-312.

Saunders, G.W. & McDonald, B. (2010). DNA barcoding reveals multiple overlooked Australian species of the red algal order Rhodymeniales (Florideophyceae), with resurrection of *Halopeltis* J. Agardh and description of *Pseudohalopeltis* gen. nov. *Botany* 88: 639-667.

Schils, T. & Coppejans, E. (2002). Gelatinous red algae of the Arabian Sea, including *Platoma heteromorphum* sp. nov. (Gigartinales, Rhodophyta). *Phycologia* 41: 254-267.

Schils, T., Huisman, J.M. & Coppejans, E. (2003). *Chamaebotrys erectus* sp. nov. (Rhodymeniales, Rhodophyta) from the Socotra Archipelago, Yemen. *Botanica Marina* 46: 2-8.

Schmidt, W.E., Gurgel, C.F.D. & Fredericq, S. (2016). Taxonomic transfer of the red algal genus *Gloiosaccion* to *Chrysymenia* (Rhodymeniaceae, Rhodymeniales), including the description of a new species, *Chrysymenia pseudoventricosa*, for the Gulf of Mexico. *Phytotaxa* 243(1): 54-70.

Schneider, C.W. (2000). Notes on the marine algae of the Bermudas. 5. Some Delesseriaceae (Ceramiales, Rhodophyta), including the first record of *Hypoglossum barbatum* Okamura from the Atlantic Ocean. *Botanica Marina* 43: 455-466.

Schneider, C.W., Freshwater, D.W. & Saunders, G.W. (2012). First report of *Halopeltis* (Rhodophyta, Rhodymeniaceae) from the non-tropical Northern Hemisphere: *H. adnata* (Okamura) comb. nov. from Korea, and *H. pellucida* sp. nov. and *H. willisii* sp. nov. from the North Atlantic. *Algae* 27: 95-108.

Schneider, C.W., Popolizio, T.R., Kraft, L.G.K. & Saunders, G.W. (2019). New species of *Gulenia* and *Howella* *gen. nov.* (Halymeniaceae, Rhodophyta) from the mesophotic zone off Bermuda. *Phycologia* 58: 690-697.

Scott, F.J. (2017). *Marine Plants of Tasmania*. Tasmanian Museum and Art Gallery, Hobart.

Scott, F.J., Wetherbee, R. & Kraft, G.T. (1982). The morphology and development of some prominently stalked southern Australian Halymeniaceae (Cryptonemiales, Rhodophyta). I. *Cryptonemia kallymenioides* (Harvey) Kraft comb. nov. and *C. undulata* Sonder. *Journal of Phycology* 18: 245-257.

Scott, F.J., Wetherbee, R. & Kraft, G.T. (1984). The morphology and development of some prominently stalked southern Australian Halymeniaceae (Cryptonemiales, Rhodophyta). II. The sponge-associated genera *Thamnoclonium* Kuetzing and *Codiophyllum* Gray. *Journal of Phycology* 20: 286-295.

Shepley, E.A. & Womersley, H.B.S. (1983). The Dumontiaceae (Cryptonemiales, Rhodophyta) of southern Australia. *Transactions of the Royal Society of South Australia* 107: 201-217.

Sherwood, A.R., Kurihara, A. & Conklin, K.Y. (2011). Molecular diversity of Amansieae (Ceramiales, Rhodophyta) from the Hawaiian Islands: a multi-marker assessment reveals high diversity within *Amansia glomerata*. *Phycological Research* 59: 16

Silva, P.C., Basson, P.W & Moe, R.L. (1996). Catalogue of the Benthic Marine Algae of the Indian Ocean. *University of California Publications in Botany* Volume 79.

Silva, P.C. & Chacana, M.E. (2015). Codiaceae. *In*: Huisman, J.M., *Algae of Australia: marine benthic algae of north-western Australia. 1. Green and brown algae*, pp. 103-113. ABRS, Canberra; CSIRO Publishing, Melbourne.

Sonder, O.G.W. (1845). Nova algarum genera et species, quas in itinere ad oras occidentales Novae Hollandiae, collegit L. Priess, Ph. Dr. *Botanische Zeitung* 3: 49-57.

Sonder, O.G.W. (1846-1848). Algae. *In*: Lehmann, C., *Plantae Preissianae ... quas in Australasia occidentali et meridionali-occidentali annis 1838-1841 collegit Ludovicus Preiss*. Vol. 2, pp. 148-160 (1846), 161-195 (1848). Hamburg.

Sun, Z., Hanyuda, T., Lim, P.-E., Tanaka, J., Gurgel, C.F.D. & Kawai, H. (2012). Taxonomic revision of the genus *Lobophora* (Dictyotales, Phaeophyceae) based on morphological evidence and analyses *rbcL* and *cox3* gene sequences. *Phycologia* 51: 500-512.

Tan, P.-L., Lim, P.-E., Lin, S.-M., Phang, S.-M., Draisma, G.A. & Liao, L.M. (2015). Foliose *Halymenia* species (Halymeniaceae, Rhodophyta) from Southeast Asia, including a new species, *Halymenia malaysiana* sp. nov. *Botanica Marina* 58: 203-217.

Tanaka, T. & Itono, H. (1969). Studies on the genus *Neurymenia* (Rhodomelaceae) from southern Japan and vicinities. *Memoirs of the Faculty of Fisheries, Kagoshima University* 18: 7-27.

Thurstan, R.H., Brittain, Z., Jones, D.S., Cameron, E., Dearnaley, J. & Bellgrove, A. (2018). Aboriginal uses of seaweeds in temperate Australia: an archival assessment. *Journal of Applied Phycology* 30: 1821-1832.

Townsend, R.A. & Huisman, J.M. (2018). 'Coralline Algae'. *In*: Huisman, J.M., *Algae of Australia: Marine Benthic Algae of North-western Australia, 2. Red Algae*, pp. 86-97, 105-138, 143-146. ABRS, Canberra; CSIRO Publishing, Melbourne.

Townsend, R.A., Woelkerling, W.J., Saunders, G.W. & Huisman, J.M. (2018). *Rhizolamellia*. *In*: Huisman, J.M., *Algae of Australia: Marine Benthic Algae of North-western Australia, 2. Red Algae*, pp. 139-142. ABRS, Canberra; CSIRO Publishing, Melbourne.

Tronchin, E.M. & Freshwater, D.W. (2007). Four Gelidiales (Rhodophyta) new to southern Africa, *Aphanta pachyrrhiza* gen. et sp. nov., *Gelidium profundum* sp. nov., *Pterocladiella caerulescens* and *P. psammophila* sp. nov. *Phycologia* 46: 325-348.

Turland, N.J., Wiersema, J.H., Barrie, F.R., Greuter, W., Hawksworth, D.L., Herendeen, P.S., Knapp, S., Kusber, W.-H., Li, D.-Z., Marhold, K., May, T.W., McNeill, J., Monro, A.M., Prado, J., Price, M.J. & Smith, G.F. (2018). *International Code of Nomenclature for algae, fungi, and plants (Shenzhen Code)*. Regnum Vegetabile, 159.

Umezaki, I. & Watanabe, M. (1994). Enumeration of the Cyanophyta (blue-green algae) of Japan. 2. Nostocales and Stigonematales. *Japanese Journal of Phycology* 42: 301-324.

Uwai, S., Kogame, K. & Masuda, M. (2002). Conspecificity of *Elachista nigra* and *Elachista orbicularis* (Elachistaceae, Phaeophyceae). *Phycological Research* 50: 217-226.

Verlaque, M., Durand, C., Huisman, J.M., Boudouresque, C.F. & Le Parco, Y. (2003). On the identity and origin of the invasive Mediterranean *Caulerpa racemosa* (Caulerpales, Chlorophyta). *European Journal of Phycology* 38: 325-339.

Verlaque, M., Langar, H., Ben Hmida, A., Pergent, C. & Pergent, G. (2017). Introduction of a new potential invader into the Mediterranean Sea: the Indo-Pacific *Avrainvillea amadelpha* (Montagne) A.Gepp & E.Gepp (Dichotomosiphonaceae, Ulvophyceae). *Cryptogamie, Algologie* 38: 267-281.

Vieira, C., Camacho, O., Wynne, M.J., Mattio, L., Anderson, R.J., Bolton, J.J., Sansón, M., D'hondt, S., Leliaert, F., Fredericq, S. & Payri, C. (2016). Shedding new light on old algae: Matching names and sequences in the brown algal genus *Lobophora* (Dictyotales, Phaeophyceae). *Taxon* 65: 689-707.

Waycott, M., McMahon, K., Mellors, J., Calladine, A. & Kleine, D. (2004). *A Guide to Tropical Seagrasses of the Indo-West Pacific*. James Cook University, Townsville.

Waycott, M., McMahon, K. & Lavery, P. (2014). *A Guide to Southern Temperate Seagrasses*. CSIRO Publishing, Melbourne.

Weber, X.A., Edgar, G.J., Banks, S.C., Waters, J.M. & Fraser, C.I. (2017). A morphological and phylogenetic investigation into divergence among sympatric Australian southern bull kelps (*Durvillaea potatorum* and *D. amatheiae* sp. nov.). *Molecular Phylogenetics and Evolution* 107: 630-643.

Whitton, B.A. (2011). Cyanobacteria (Cyanophyta). *In*: John, D.M., Whitton, B.A. & Brook, A.J. (eds), *The freshwater algal flora of the British Isles*. An identification guide to freshwater and terrestrial algae. Second edition, pp. 31-158. Cambridge University Press, Cambridge.

Wilks, K.M. & Woelkerling, W.J. (1994). An account of southern Australian species of *Phymatolithon* (Corallinaceae, Rhodophyta) with comments on *Leptophytum*. *Australian Systematic Botany* 7: 183-223.

Wiriyadamrikul, J., Geraldino, P.J.L., Huisman, J.M., Lewmanomont, K. & Boo, S.M. (2013a). Molecular diversity of the calcified red algal genus *Tricleocarpa* (Galaxauraceae, Nemalionales) with the description of *T. jejuensis* and *T. natalensis*. *Phycologia* 52: 338-351.

Wiriyadamrikul, J., Lewmanomont, K. & Boo, S.M. (2013b). Molecular diversity and morphology of the genus *Actinotrichia* (Galaxauraceae, Rhodophyta) from the western Pacific, with a new record of *A. robusta* in the Andaman Sea. *Algae* 28: 53-62.

Wiriyadamrikul, J., Wynne, M.J. & Boo, S.M. (2014). Phylogenetic relationships of *Dichotomaria* (Nemalionales, Rhodophyta) with the proposal of *Dichotomaria intermedia* (R.C.Y. Chou) comb. nov. *Botanica Marina* 57: 27-40.

Withell, A.F., Millar, A.J.K. & Kraft, G.T. (1994). Taxonomic studies of the genus *Gracilaria* (Gracilariales, Rhodophyta) from Australia. *Australian Systematic Botany* 7: 281-352.

Woelkerling, W.J. (1980). Studies on *Metamastophora* (Corallinaceae, Rhodophyta). I. *M. flabellata* (Sonder) Setchell: morphology and anatomy. *British Phycological Journal* 15: 201-225.

Woelkerling, W. J. (1996a). Subfamily Melobesioideae. *In*: Womersley, H.B.S., *The Marine Benthic Flora of Southern Australia. Rhodophyta – Part IIIB*, pp. 164-210. ABRS, Canberra.

Woelkerling, W.J. (1996b). Subfamily Lithophylloideae Setchell 1943 (as Lithophylleae). *In*: Womersley, H.B.S., *The Marine Benthic Flora of Southern Australia. Rhodophyta – Part* IIIB, pp. 214-237. ABRS, Canberra.

Woelkerling, W.J. (1996c). Mastophoroideae (excluding *Hydrolithon, Neogoniolithon, Pneophyllum, Spongites*). *In*: Womersley, H.B.S., *The Marine Benthic Flora of Southern Australia. Rhodophyta – Part* IIIB, pp. 237-255. ABRS, Canberra.

Woelkerling, W., Harvey, A. & de Reviers, B. (2012). The neotypification and taxonomic status of *Amphiroa crassa* Lamouroux (Corallinales, Rhodophyta). *Cryptogamie, Algologie* 33: 339-358.

Woelkerling, W.J. & Womersley, H.B.S. (1994). Order Acrochaetiales Feldmann 1953: 12. *In*: Womersley, H.B.S., *The Marine Benthic Flora of Southern Australia. Rhodophyta – Part IIIA*, pp. 42-76. ABRS, Canberra.

Wollaston, E.M. (1968). Morphology and taxonomy of Southern Australian genera of Crouanieae Schmitz (Ceramiaceae, Rhodophyta). *Australian Journal of Botany* 16: 217-417.

Wollaston, E.M. (1990). Recognition of the genera *Spongoclonium* Sonder and *Lasiothalia* Harvey (Ceramiaceae, Rhodophyta) in southern Australia. *Botanica Marina* 33: 19-30.

Wollaston, E.M. (1992). Morphology and taxonomy of *Thamnocarpus* (Ceramiaceae, Rhodophyta) in southern Australia and east Africa. *Phycologia* 31: 138-146.

Wollaston, E.M. & Womersley, H.B.S. (1998a). Tribe Crouanieae Schmitz 1889:451. *In*: Womersley, H.B.S., *The Marine Benthic Flora of Southern Australia. Rhodophyta – Part IIIC*, pp. 42-67. State Herbarium of South Australia, Adelaide.

Wollaston, E.M. & Womersley, H.B.S. (1998b). Tribe Antithamnieae Hommersand 1963:330. *In*: Womersley, H.B.S., *The Marine Benthic Flora of Southern Australia. Rhodophyta – Part* IIIC, pp. 98-129. State Herbarium of South Australia, Adelaide.

Womersley, H.B.S. (1946). Studies on the marine algae of southern Australia. Introduction and No 1 The genera *Isactis* and *Rivularia* (Myxophyceae). *Transactions of the Royal Society of South Australia* 70: 127-136.

Womersley, H.B.S. (1965). The Helminthocladiaceae (Rhodophyta) of southern Australia. *Australian Journal of Botany* 13: 451-487.

Womersley, H.B.S. (1971). *Palmoclathrus*, a new deep water genus of Chlorophyta. *Phycologia* 10: 229-233.

Womersley, H.B.S. (1978). Southern Australian species of *Ceramium* Roth (Rhodophyta). *Australian Journal of Botany* 29: 205-257.

Womersley, H.B.S. (1979). Southern Australian species of *Polysiphonia* Greville (Rhodophyta). *Australian Journal of Botany* 27: 459-528.

Womersley, H.B.S. (1984). *The Marine Benthic Flora of Southern Australia, Part I*. Flora and Fauna Handbooks Committee, Adelaide.

Womersley, H.B.S. (1987). *The Marine Benthic Flora of Southern Australia, Part II*. Flora and Fauna Handbooks Committee, Adelaide.

Womersley, H.B.S. (1994). *The Marine Benthic Flora of Southern Australia. Rhodophyta – Part IIIA*. ABRS, Canberra.

Womersley, H.B.S. (1996). *The Marine Benthic Flora of Southern Australia. Rhodophyta – Part IIIB*. ABRS, Canberra.

Womersley, H.B.S. (1998). *The Marine Benthic Flora of Southern Australia. Rhodophyta – Part IIIC.* State Herbarium of South Australia, Adelaide.

Womersley, H.B.S. (2003). *The Marine Benthic Flora of Southern Australia. Rhodophyta – Part IIID.* ABRS, Canberra; State Herbarium of South Australia, Adelaide.

Womersley, H.B.S. & Bailey, A. (1987). Family Chordariaceae Greville. *In*: Womersley, H.B.S., *The Marine Benthic Flora of Southern Australia, Part II.* Phaeophyta and Chrysophyta (*Vaucheria*), pp. 103-127. Flora and Fauna Handbooks Committee, Adelaide.

Womersley, H.B.S. & Cartledge, S.A. (1975). The southern Australian species of *Spyridia* (Ceramiaceae: Rhodophyta). *Transactions of the Royal Society of South Australia* 99: 221-233.

Womersley, H.B.S. & Conway, E. (1975). *Porphyra* and *Porphyropsis* (Rhodophyta) in southern Australia. *Transactions of the Royal Society of South Australia* 99: 59-70.

Womersley, H.B.S. & Guiry, M.D. (1994). Order Gelidiales Kylin 1923: 132. *In*: Womersley, H.B.S., *The Marine Benthic Flora of Southern Australia. Rhodophyta – Part IIIA*, pp. 118-142. ABRS, Canberra.

Womersley, H.B.S. & Johansen, H.W. (1996a). Subfamily Amphiroideae Johansen 1969: 47. *In*: Womersley, H.B.S., *The Marine Benthic Flora of Southern Australia. Rhodophyta – Part IIIB*, pp.283-288. ABRS, Canberra.

Womersley, H.B.S. & Johansen, H.W. (1996b). Subfamily Corallinoideae (Areschoug) Foslie 1908: 19. *In*: Womersley, H.B.S., *The Marine Benthic Flora of Southern Australia. Rhodophyta – Part IIIB*, pp. 288-317. ABRS, Canberra.

Womersley, H.B.S. & Johansen, H.W. (1996c). Subfamily Metagoniolithoideae Johansen 1969. *In*: Womersley, H.B.S., *The Marine Benthic Flora of Southern Australia. Rhodophyta – Part IIIB*, pp. 317-323. ABRS, Canberra.

Womersley, H.B.S. & Kraft, G.T. (1994). Family Nemastomataceae Schmitz 1892: 2, nom. cons. *In*: Womersley, H.B.S., *The Marine Benthic Flora of Southern Australia. Rhodophyta – Part IIIA*, pp. 270-285. ABRS, Canberra.

Womersley, H.B.S. & Lewis, J.A. (1994). Family Halymeniaceae Bory 1828: 158. *In*: Womersley, H.B.S., *The Marine Benthic Flora of Southern Australia. Rhodophyta – Part IIIA*, pp. 167-218. ABRS, Canberra.

Womersley, H.B.S. & Norris, R.E. (1971). The morphology and taxonomy of Australian Kallymeniaceae (Rhodophyta). *Australian Journal of Botany* (Suppl. 2): 1-62.

Womersley, H.B.S. & Parsons, M.J. (2003). Tribe Lophothalieae Schmitz & Falkenberg 1897: 445. *In*: Womersley, H.B.S., *The Marine Benthic Flora of Southern Australia. Rhodophyta – Part IIID*, pp. 235-282. ABRS, Canberra; State Herbarium of South Australia, Adelaide.

Womersley, H.B.S. & Phillips, L.E. (2003a). Genus *Lenormandia* Sonder 1845: 54, *nom. cons. In*: Womersley, H.B.S., *The Marine Benthic Flora of Southern Australia. Rhodophyta – Part IIID*, pp. 380-390. ABRS, Canberra; State Herbarium of South Australia, Adelaide.

Womersley, H.B.S. & Phillips, L.E. (2003b). Genus *Epiglossum* Kützing 1849: 878. *In*: Womersley, H.B.S., *The Marine Benthic Flora of Southern Australia. Rhodophyta – Part IIID*, pp. 401-404. ABRS, Canberra; State Herbarium of South Australia, Adelaide.

Womersley, H.B.S. & Shepley, E.A. (1959). Studies on the *Sarcomenia* group of the Rhodophyta. *Australian Journal of Botany* 7: 168-223.

Womersley, H.B.S., Wilson, S.M. & Kraft, G.T. (2003a). Genus *Protokuetzingia* Falkenberg in Schmitz & Falkenberg 1897: 469. *In*: Womersley, H.B.S., *The Marine Benthic Flora of Southern Australia. Rhodophyta – Part IIID*, pp. 374-376. ABRS, Canberra; State Herbarium of South Australia, Adelaide.

Womersley, H.B.S., Wilson, S.M. & Kraft, G.T. (2003b). Genus *Kuetzingia* Sonder 1845: 54. *In*: Womersley, H.B.S., *The Marine Benthic Flora of Southern Australia. Rhodophyta – Part IIID*, pp. 376-378. ABRS, Canberra; State Herbarium of South Australia, Adelaide.

Womersley, H.B.S. & Wollaston, E.M. (1998a). Tribe Heterothamnieae Wollaston 1968: 407. *In*: Womersley, H.B.S., *The Marine Benthic Flora of Southern Australia. Rhodophyta – Part IIIC*, pp. 156-208. State Herbarium of South Australia, Adelaide.

Womersley, H.B.S. & Wollaston, E.M. (1998b). Tribe Callithamnieae Schmitz 1889:450. *In*: Womersley, H.B.S., *The Marine Benthic Flora of Southern Australia. Rhodophyta – Part IIIC*, pp. 231-269. State Herbarium of South Australia, Adelaide.

Womersley, H.B.S. & Wollaston, E.M. (1998c). Tribe Compsothamnieae Schmitz 1889:450. *In*: Womersley, H.B.S., *The Marine Benthic Flora of Southern Australia. Rhodophyta – Part IIIC*, pp. 270-286. State Herbarium of South Australia, Adelaide.

Womersley, H.B.S. & Wollaston, E.M. (1998d). Tribe Spongoclonieae Schmitz 1889:450. *In*: Womersley, H.B.S., *The Marine Benthic Flora of Southern Australia. Rhodophyta – Part IIIC*, pp. 286-300. State Herbarium of South Australia, Adelaide.

Wynne, M.J. (1988). A reassessment of the *Hypoglossum* group (Delesseriaceae, Rhodophyta), with a critique of its genera. *Helgoländer Meeresuntersuchungen* 42: 511-534.

Wynne, M.J. (1995). Benthic marine algae from the Seychelles collected during the R/V *Te Vega* Indian Ocean Expedition. *Contributions of the University of Michigan Herbarium* 20: 261-346.

Wynne, M.J. (1999). New records of benthic marine algae from the Sultanate of Oman. *Contributions from the University of Michigan Herbarium* 22: 189-208.

Wynne, M.J. (2003). *Leveillea major* sp. nov. (Rhodomelaceae, Rhodophyta) from the Sultanate of Oman. *Botanica Marina* 46: 357-365.

Wynne, M.J. (2014). *The red algal families Delesseriaceae and Sarcomeniaceae.* Koeltz Scientific Books, Königstein.

Wynne, M.J. (2017). The reinstatement of *Lychaete* J.Agardh (Ulvophyceae, Cladophoraceae). *Notulae Algarum* 31: 1-4.

Wynne, M.J. & Huisman, J.M. (1998). First report of *Yamadaella caenomyce* (Liagoraceae, Rhodophyta) from the Atlantic Ocean, with descriptive notes and comments on nomenclature. *Caribbean Journal of Science* 34: 280-285.

Yee, N.R., Millar, A.J.K. & Huisman, J.M. (2015). Sporochnales. *In*: Huisman, J.M., *Algae of Australia: Marine Benthic Algae of North-western Australia, 1. Green and Brown Algae*, pp. 237-242. ABRS, Canberra; CSIRO Publishing, Melbourne.

Yee, N.R., Millar, A.J.K. & Kraft, G.T. (2009). Sporochnales, *In*: Kraft, G.T., *Algae of Australia: Marine benthic algae of Lord Howe Island and the*

Southern Great Barrier Reef, 2. Brown Algae, pp. 97-107. ABRS, Canberra; CSIRO Publishing, Melbourne.

Yoshida, T. & Mikami, H. (1994). Observations on *Vanvoorstia spectabilis* Harvey and *V. coccinea* Harvey (Delesseriaceae, Rhodophyta) from southern Japan. *Japanese Journal of Phycology* 42: 11-20.

Yoshida, T. & Yoshida, M. (1983). Observations on *Ditria zonaricola* (Okamura) comb. nov. based on *Herpopteros zonaricola* Okamura (Rhodophyta, Rhodomelaceae). *Journal of the Faculty of Science, Hokkaido University, Series V (Botany)* 13: 39-48.

Yoshizaki, M. (1987). The structure and reproduction of *Patenocarpus paraphysiferus* gen. *et* sp. nov. (Dermonemataceae, Nemaliales, Rhodophyta). *Phycologia* 26: 47-52.

Zanolla, M., Carmona, R., De La Rosa, J., Salvador, N., Sherwood, A.R., Andreakis, N. & Altamirano, M. (2014). Morphological differentiation of cryptic lineages within the invasive genus *Asparagopsis* (Bonnemaisoniales, Rhodophyta). *Phycologia* 53: 233-242.

Zuccarello, G.C., Critchley, A.T., Smith, J., Sieber, V., Lhonneur, G.B. & West, J.A. (2006). Systematics and genetic variation in commercial *Kappaphycus* and *Eucheuma* (Solieriaceae, Rhodophyta). *Journal of Applied Phycology* 18: 643-651.

Zuccarello, G.C., West, J.A., & Kikuchi, N. (2008). Phylogenetic relationships within the Stylonematales (Stylonematophyceae, Rhodohyta): biogeographic patterns do not apply to *Stylonema alsidii*. *Journal of Phycology* 44: 384-393.

TAXONOMIC INDEX